I0470809

FEMA 306

EVALUATION OF EARTHQUAKE DAMAGED CONCRETE AND MASONRY WALL BUILDINGS

Basic Procedures Manual

Prepared by:

Applied Technology Council (ATC-43 Project)

555 Twin Dolphin Drive, Suite 550
Redwood City, California 94065

Prepared for:

The Partnership for Response and Recovery

Washington, D.C.

Funded by:

Federal Emergency Management Agency

1998

Applied Technology Council

The Applied Technology Council (ATC) is a nonprofit, tax-exempt corporation established in 1971 through the efforts of the Structural Engineers Association of California. ATC is guided by a Board of Directors consisting of representatives appointed by the American Society of Civil Engineers, the Structural Engineers Association of California, the Western States Council of Structural Engineers Associations, and four at-large representatives concerned with the practice of structural engineering. Each director serves a three-year term.

The purpose of ATC is to assist the design practitioner in structural engineering (and related design specialty fields such as soils, wind, and earthquake) in the task of keeping abreast of and effectively using technological developments. ATC also identifies and encourages needed research and develops consensus opinions on structural engineering issues in a nonproprietary format. ATC thereby fulfills a unique role in funded information transfer.

Project management and administration are carried out by a full-time Executive Director and support staff. Project work is conducted by a wide range of highly qualified consulting professionals, thus incorporating the experience of many individuals from academia, research, and professional practice who would not be available from any single organization. Funding for ATC projects is obtained from government agencies and from the private sector in the form of tax-deductible contributions.

1998-1999 Board of Directors

Notice

This report was prepared under Contract EMW-95-C-4685 between the Federal Emergency Management Agency and the Partnership for Response and Recovery.

For further information concerning this document or the activities of the ATC, contact the Executive Director, Applied Technolgy Council, 555 Twin Dolphin Drive, Suite 550, Redwood City, California 94065; phone 650-595-1542; fax 650-593-2320; e-mail atc@atcouncil.org.

Preface

Following the two damaging California earthquakes in 1989 (Loma Prieta) and 1994 (Northridge), many concrete wall and masonry wall buildings were repaired using federal disaster assistance funding. The repairs were based on inconsistent criteria, giving rise to controversy regarding criteria for the repair of cracked concrete and masonry wall buildings. To help resolve this controversy, the Federal Emergency Management Agency (FEMA) initiated a project on evaluation and repair of earthquake damaged concrete and masonry wall buildings in 1996. The project was conducted through the Partnership for Response and Recovery (PaRR), a joint venture of Dewberry & Davis of Fairfax, Virginia, and Woodward-Clyde Federal Services of Gaithersburg, Maryland. The Applied Technology Council (ATC), under subcontract to PaRR, was responsible for developing technical criteria and procedures (the ATC-43 project).

The ATC-43 project addresses the investigation and evaluation of earthquake damage and discusses policy issues related to the repair and upgrade of earthquake-damaged buildings. The project deals with buildings whose primary lateral-force-resisting systems consist of concrete or masonry bearing walls with flexible or rigid diaphragms, or whose vertical-load-bearing systems consist of concrete or steel frames with concrete or masonry infill panels. The intended audience is design engineers, building owners, building regulatory officials, and government agencies.

The project results are reported in three documents. The FEMA 306 report, *Evaluation of Earthquake Damaged Concrete and Masonry Wall Buildings, Basic Procedures Manual,* provides guidance on evaluating damage and analyzing future performance. Included in the document are component damage classification guides, and test and inspection guides. FEMA 307, *Evaluation of Earthquake Damaged Concrete and Masonry Wall Buildings, Technical Resources*, contains supplemental information including results from a theoretical analysis of the effects of prior damage on single-degree-of-freedom mathematical models, additional background information on the component guides, and an example of the application of the basic procedures. FEMA 308, *The Repair of Earthquake Damaged Concrete and Masonry Wall Buildings*, discusses the policy issues pertaining to the repair of earthquake damaged buildings and illustrates how the procedures developed for the project can be used to provide a technically sound basis for policy decisions. It also provides guidance for the repair of damaged components.

The project also involved a workshop to provide an opportunity for the user community to review and comment on the proposed evaluation and repair criteria. The workshop, open to the profession at large, was held in Los Angeles on June 13, 1997 and was attended by 75 participants.

The project was conducted under the direction of ATC Senior Consultant Craig Comartin, who served as Co-Principal Investigator and Project Director. Technical and management direction were provided by a Technical Management Committee consisting of Christopher Rojahn (Chair), Craig Comartin (Co-Chair), Daniel Abrams, Mark Doroudian, James Hill, Jack Moehle, Andrew Merovich (ATC Board Representative), and Tim McCormick. The Technical Management Committee created two Issue Working Groups to pursue directed research to document the state of the knowledge in selected key areas: (1) an Analysis Working Group, consisting of Mark Aschheim (Group Leader) and Mete Sozen (Senior Consultant) and (2) a Materials Working Group, consisting of Joe Maffei (Group Leader and Reinforced Concrete Consultant), Greg Kingsley (Reinforced Masonry Consultant), Bret Lizundia (Unreinforced Masonry Consultant), John Mander (Infilled Frame Consultant), Brian Kehoe and other consultants from Wiss, Janney, Elstner and Associates (Tests, Investigations, and Repairs Consultant). A Project Review Panel provided technical overview and guidance. The Panel members were Gregg Borchelt, Gene Corley, Edwin Huston, Richard Klingner, Vilas Mujumdar, Hassan Sassi, Carl Schulze, Daniel Shapiro, James Wight, and Eugene Zeller. Nancy Sauer and Peter Mork provided technical editing and report production services, respectively. Affiliations are provided in the list of project participants.

The Applied Technology Council and the Partnership for Response and Recovery gratefully acknowledge the cooperation and insight provided by the FEMA Technical Monitor, Robert D. Hanson.

Tim McCormick
PaRR Task Manager

Christopher Rojahn
ATC-43 Principal Investigator
ATC Executive Director

Table of Contents

List of Figures

List of Tables

List of Test and Inspection Guides

(See Section 3.8)

List of Component Damage Classification Guides

Concrete (See Chapter 5)

Reinforced Masonry (See Chapter 6)

Unreinforced Masonry (See Chapter 7)

Infilled Frames (See Chapter 8)

Prologue

This document is one of three to result from the ATC-43 project funded by the Federal Emergency Management Agency (FEMA). The goal of the project is to develop technically sound procedures to evaluate the effects of earthquake damage on buildings with primary lateral-force-resisting systems consisting of concrete or masonry bearing walls or infilled frames. The procedures are based on the knowledge derived from research and experience in engineering practice regarding the performance of these types of buildings and their components. The procedures require thoughtful examination and review prior to implementation. The ATC-43 project team strongly urges individual users to read all of the documents carefully to form an overall understanding of the damage evaluation procedures and repair techniques.

Before this project, formalized procedures for the investigation and evaluation of earthquake-damaged buildings were limited to those intended for immediate use in the field to identify potentially hazardous conditions. ATC-20, *Procedures for Postearthquake Safety Evaluation of Buildings*, and its addendum, ATC-20-2 (ATC, 1989 and 1995) are the definitive documents for this purpose. Both have proven to be extremely useful in practical applications. ATC-20 recognizes and states that in many cases, detailed structural engineering evaluations are required to investigate the implications of earthquake damage and the need for repairs. This project provides a framework and guidance for those engineering evaluations.

What have we learned?

The project team for ATC-43 began its work with a thorough review of available analysis techniques, field observations, test data, and emerging evaluation and design methodologies. The first objective was to understand the effects of damage on future building performance. The main points are summarized below.

- **Component behavior controls global performance.**

    Recently developed guidelines for structural engineering seismic analysis and design techniques focus on building displacement, rather than forces as the primary parameter for the characterization of seismic performance. This approach models the building as an assembly of its individual components. Force-deformation properties (e.g., elastic stiffness, yield point, ductility) control the behavior of wall panels, beams, columns, and other components. The component behavior, in turn, governs the overall displacement of the building and its seismic performance. Thus, the evaluation of the effects of damage on building performance must concentrate on how component properties change as a result of damage.

- **Indicators of damage (e.g., cracking, spalling) are meaningful only in light of the mode of component behavior.**

    Damage affects the behavior of individual components differently. Some exhibit ductile modes of post-elastic behavior, maintaining strength even with large displacements. Others are brittle and lose strength abruptly after small inelastic displacements. The post-elastic behavior of a structural component is a function of material properties, geometric proportions, details of construction, and the combination of demand actions (axial, flexural, shearing, torsional) imposed upon it. As earthquake shaking imposes these actions on components, the components tend to exhibit predominant modes of behavior as damage occurs. For example, if earthquake shaking and its associated inertial forces and frame distortions cause a reinforced concrete wall panel to rotate at each end, statics defines the relationship between the associated bending moments and shear force. The behavior of the panel depends on its strength in flexure relative to that in shear. Cracks and other signs of damage must be interpreted in the context of the mode of component behavior. A one-eighth-inch crack in a wall panel on the verge of brittle shear failure is a very serious condition. The same size crack in a flexurally-controlled panel may be insignificant with regard to future seismic performance. This is, perhaps, the most important finding of the ATC-43 project: the significance of cracks and other signs of damage, with respect to the future performance of a building, depends on the mode of behavior of the components in which the damage is observed.

- **Damage may reveal component behavior that differs from that predicted by evaluation and design methodologies.**

When designing a building or evaluating an undamaged building, engineers rely on theory and their own experience to visualize how earthquakes will affect the structure. The same is true when they evaluate the effects of actual damage after an earthquake, with one important difference. If engineers carefully observe the nature and extent of the signs of the damage, they can greatly enhance their insight into the way the building actually responded to earthquake shaking. Sometimes the actual behavior differs from that predicted using design equations or procedures. This is not really surprising, since design procedures must account conservatively for a wide range of uncertainty in material properties, behavior parameters, and ground shaking characteristics. Ironically, actual damage during an earthquake has the potential for improving the engineer's knowledge of the behavior of the building. When considering the effects of damage on future performance, this knowledge is important.

- **Damage may not significantly affect displacement demand in future larger earthquakes.**

One of the findings of the ATC-43 project is that prior earthquake damage does not affect maximum displacement response in future, larger earthquakes in many instances. At first, this may seem illogical. Observing a building with cracks in its walls after an earthquake and visualizing its future performance in an even larger event, it is natural to assume that it is worse off than if the damage had not occurred. It seems likely that the maximum displacement in the future, larger earthquake would be greater than if it had not been damaged. Extensive nonlinear time-history analyses performed for the project indicated otherwise for many structures. This was particularly true in cases in which significant strength degradation did not occur during the prior, smaller earthquake. Careful examination of the results revealed that maximum displacements in time histories of relatively large earthquakes tended to occur after the loss of stiffness and strength would have taken place even in an undamaged structure. In other words, the damage that occurs in a prior,

smaller event would have occurred early in the subsequent, larger event anyway.

What does it mean?

The ATC-43 project team has formulated performance-based procedures for evaluating the effects of damage. These can be used to quantify losses and to develop repair strategies. The application of these procedures has broad implications.

- **Performance-based damage evaluation uses the actual behavior of a building, as evidenced by the observed damage, to identify specific deficiencies.**

The procedures focus on the connection between damage and component behavior and the implications for estimating actual behavior in future earthquakes. This approach has several important benefits. First, it provides a meaningful engineering basis for measuring the effects of damage. It also identifies performance characteristics of the building in its pre-event and damaged states. The observed damage itself is used to calibrate the analysis and to improve the building model. For buildings found to have unacceptable damage, the procedures identify specific deficiencies at a component level, thereby facilitating the development of restoration or upgrade repairs.

- **Performance-based damage evaluation provides an opportunity for better allocation of resources.**

The procedures themselves are technical engineering tools. They do not establish policy or prescribe rules for the investigation and repair of damage. They may enable improvements in both private and public policy, however. In past earthquakes, decisions on what to do about damaged buildings have been hampered by a lack of technical procedures to evaluate the effects of damage and repairs. It has also been difficult to investigate the risks associated with various repair alternatives. The framework provided by performance-based damage evaluation procedures can help to remove some of these roadblocks. In the long run, the procedures may tend to reduce the prevailing focus on the loss caused by damage from its pre-event conditions and to increase the focus on what the damage reveals about future building performance. It makes little sense to implement unnecessary repairs to buildings that would perform relatively well even in a damaged condition. Nor is it wise to neglect buildings in which the component behavior reveals serious hazards regardless of the extent of damage.

- **Engineering judgment and experience are essential to the successful application of the procedures.**

ATC-20 and its addendum, ATC-20-2, were developed to be used by individuals who might be somewhat less knowledgeable about earthquake building performance than practicing structural engineers. In contrast, the detailed investigation of damage using the performance-based procedures of this document and the companion FEMA 307 report (ATC, 1998a) and FEMA 308 report (ATC, 1998b) must be implemented by an experienced engineer. Although the documents include information in concise formats to facilitate field operations, they must not be interpreted as a "match the pictures" exercise for unqualified observers. Use of these guideline materials requires a thorough understanding of the underlying theory and empirical justifications contained in the documents. Similarly, the use of the simplified direct method to estimate losses has limitations. The decision to use this method and the interpretation of the results must be made by an experienced engineer.

- **The new procedures are different from past damage evaluation techniques and will continue to evolve in the future.**

The technical basis of the evaluation procedures is essentially that of the emerging performance-based seismic and structural design procedures. These will take some time to be assimilated in the engineering community. The same is true for building officials. Seminars, workshops, and training sessions are required not only to introduce and explain the procedures but also to gather feedback and to improve the overall process. Additionally, future materials-testing and analytical research will enhance the basic framework developed for this project. Current project documents are initial editions to be revised and improved over the years.

In addition to the project team, a Project Review Panel has reviewed the damage evaluation and repair procedures and each of the three project documents. This group of experienced practitioners, researchers, regulators, and materials industry representatives reached a unanimous consensus that the products are technically sound and that they represent the state of knowledge on the evaluation and repair of earthquake-damaged concrete and masonry wall buildings. At the same time, all who contributed to this project acknowledge that the recommendations depart from traditional practices. Owners, design professionals, building officials, researchers, and all others with an interest in the performance of buildings during earthquakes are encouraged to review these documents and to contribute to their continued improvement and enhancement. Use of the documents should provide realistic assessments of the effects of damage and valuable insight into the behavior of structures during earthquakes. In the long run, they hopefully will contribute to sensible private and public policy regarding earthquake-damaged buildings.

1. Introduction and Overview

1.1 Purpose

The purpose of this document is to provide practical criteria and guidance for evaluating earthquake *damage* to buildings with primary lateral-force-resisting systems consisting of concrete or masonry walls or *infilled frames*. The procedures in this manual are intended to characterize the observed damage caused by the earthquake in terms of the loss in building performance capability. This information may be used to facilitate the settlement of insurance claims, the development of strategies for *repair*, or other purposes. The intended users of this document are primarily practicing engineers with experience in concrete and masonry design in seismic regions. Information in this document also may be useful to building owners, building officials insurance adjusters, and government agencies; however these users should consult with a qualified engineer for interpretation or specific application of the document.

1.2 Scope

Concrete and masonry wall buildings include those with vertical-load *bearing wall* panels, with and without openings. This document also applies to buildings with vertical-load-bearing frames of concrete or steel that incorporate masonry or concrete infill panels to resist horizontal forces. For both types of buildings, the procedures and criteria in this document address:

a. The investigation and documentation of damage caused by earthquakes

b. The classification of the damage for building *components* according to mode of structural behavior and *severity of damage*

c. The evaluation of the effects of the damage on the performance of the building during future earthquakes

d. The development of hypothetical measures that would restore the performance of the building to that of its condition immediately before the damaging earthquake

Evaluating of the effects of earthquake damage on future seismic performance entails the *relative performance analysis* of the building in its damaged and pre-event states for one or more *seismic performance objectives*. If the expected performance of the damaged building is significantly worse than that anticipated for the building in its pre-event condition, conceptual *performance restoration measures* are developed on a component level to generate global performance nearly equivalent to the pre-event condition. Performance restoration measures rely on the technical analysis of potential component actions. The document also includes a simplified *direct method* for generating an approximate scope for performance restoration measures for some cases. Although performance restoration measures specified by either method are essentially hypothetical physical repairs, they are not recommended for actual implementation solely on the basis of these damage evaluation procedures. The selection of appropriate repairs for an earthquake-damaged building typically requires consideration of a wider range of technical and policy issues. This process is summarized in a companion document, FEMA 308: *The Repair of Earthquake Damaged Concrete and Masonry Wall Buildings* (ATC, 1998b).

The procedures for damage evaluation in this document are technical; however, their use requires policy considerations including the selection of performance objectives as benchmarks for measuring changes in seismic performance. This document does not specify or limit the use of the damage evaluations, nor does it impose damage repair scope or procedures. Users should not infer otherwise.

Earthquakes can cause damage to the structural and nonstructural components of buildings. This document addresses structural damage. The direct evaluation of nonstructural damage is not included. The effects of structural damage on potential future nonstructural damage can be addressed indirectly by the selection of appropriate seismic performance objectives for the evaluation procedure.

The term *damage*, when used in this document, refers to damage to the building caused by the earthquake. It is important to note that prior effects of environmental deterioration, service conditions, and previous earthquakes are considered to be *pre-existing conditions* and not part of the damage to be evaluated. This distinction is covered further in the presentation of the evaluation procedures.

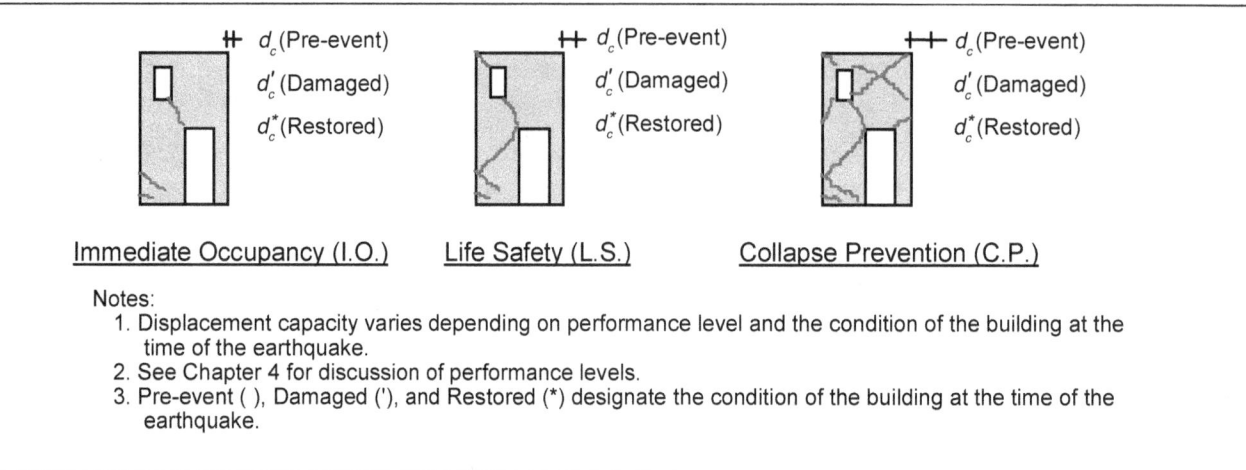

Notes:
1. Displacement capacity varies depending on performance level and the condition of the building at the time of the earthquake.
2. See Chapter 4 for discussion of performance levels.
3. Pre-event ( ), Damaged ('), and Restored (*) designate the condition of the building at the time of the earthquake.

Figure 1-1   *Global Displacement Capacities for Various Performance Levels. Capacities will vary, depending on damage level and restoration measure.*

1.3   Basis

The evaluation procedure assumes that when an earthquake causes damage to a building, a competent engineer can assess the effects, at least partially, through visual inspection augmented by investigative tests, structural analysis, and knowledge of the building construction. By determining how the structural damage has changed structural properties, it is feasible to develop potential actions (performance restoration measures) that, if implemented, would restore the damaged building to a condition such that its future earthquake performance would be essentially equivalent to that of the building in its pre-event condition. The costs associated with these conceptual performance restoration measures quantify the loss associated with the earthquake damage.

The damage evaluation procedure measures the effects of damage by comparing the relative capability of pre-event, damaged, and restored models of a building to meet seismic performance objectives for future earthquakes. The analysis technique is to compare a *global displacement capacity* limit, d_c, to a global displacement demand, d_d, for the building model (see Figures 1-1 and 1-2). Both of these global displacement parameters are controlled by the force-deformation properties of all the individual structural components of the building model. The procedure includes techniques for modifying these component force-deformation properties to account for the effects of both the observed damage and potential restoration measures.

The damage evaluation criteria build, to the extent possible, on existing performance-based procedures in the FEMA 273 and FEMA 274 reports, *NEHRP Guidelines for the Seismic Rehabilitation of Buildings* (ATC, 1997a) and companion *Commentary* (ATC, 1997b), and the ATC-40 Report, *Seismic Evaluation and Retrofit of Concrete Buildings* (ATC, 1996). This document adapts the existing state of knowledge rather than developing completely new techniques. This approach contributes to consistency of language, nomenclature, and technical concepts among emerging procedures intended for use by structural engineering practitioners. The intent is to improve the application of the existing knowledge and techniques by using observations of earthquake damage to calibrate analytical models of component behavior.

Two principal research efforts augment the basic procedures:

- An Analysis Working Group has investigated the theoretical effects of prior damage on the displacement response of single-degree-of-freedom models subjected to earthquakes in an effort to verify and/or modify current methods of predicting displacement demand. The implications of the results from this investigation for damage evaluation are reflected in Section 4.4.4 of this volume. A summary report on the results is included in FEMA 307: *Evaluation of Earthquake Damaged Concrete and Masonry Wall Buildings, Technical Resources* (ATC, 1998a).

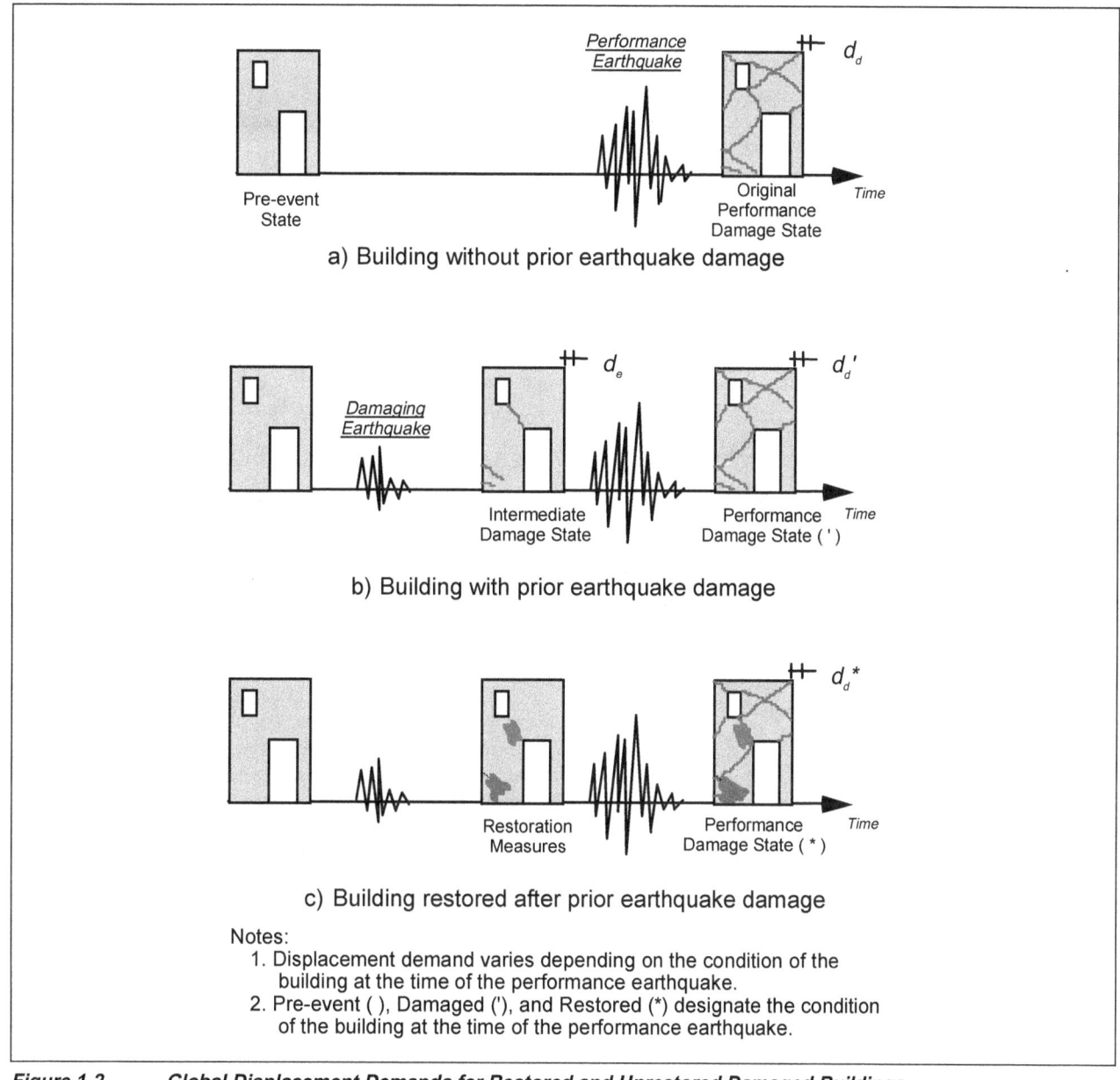

a) Building without prior earthquake damage

b) Building with prior earthquake damage

c) Building restored after prior earthquake damage

Notes:
1. Displacement demand varies depending on the condition of the building at the time of the performance earthquake.
2. Pre-event ( ), Damaged ('), and Restored (*) designate the condition of the building at the time of the performance earthquake.

Figure 1-2   ***Global Displacement Demands for Restored and Unrestored Damaged Buildings.***

- A Materials Working Group has assembled tests and investigative techniques to document the effects of earthquake damage. This effort produced the Test and Inspection Guides included in Chapter 3 of this volume. This group also used existing research results to develop recommended modifications to component force-deformation relationships for nonlinear structural analysis to include the effects of damage. The results are the Component Damage Classification Guides included in Chapters 5 through 8. Additional background information including that

forming the basis of the Component Guides is in FEMA 307 (ATC, 1998a). Finally this group assembled information on repair techniques commonly applied to earthquake damage in concrete and masonry wall buildings. These are documented in a companion document, FEMA 308: *The Repair of Earthquake Damaged Concrete and Masonry Wall Buildings* (ATC, 1998b).

In the past, there has been a tendency to gauge the effect of earthquake damage by estimating the loss of lateral-force-resisting capacity of the structure (Hanson, 1996). It has been suggested by some that this loss can be related to the observed width and extent of concrete and masonry cracks in the damaged structure. There has been widespread disagreement on the significance of cracking on capacity and skepticism on the suitability of force capacity as a parameter for measuring damage. The procedure in this document is based on global displacement and component deformation capacities rather than force capacities. This approach facilitates a more meaningful engineering assessment of the effects of damage on future performance.

1.4 Overview of the Damage Investigation and Evaluation Procedures

This section briefly summarizes the damage investigation and evaluation procedures, referring as necessary to specific chapters. One objective is to provide the practicing engineer with a road map for the use of the document in real-life applications. Another equally important objective is to provide a basic exposure to the process for owners, building officials, disaster assistance personnel, and others with an interest in the results who may not be familiar with the technical details.

1.4.1 Introduction and Overview

Chapter 1 summarizes the purpose, basis, and scope of the document. The technical basis of the damage investigation and evaluation procedures are reviewed. A step-by-step outline presents these basic procedures. Brief synopses are included for subsequent chapters.

1.4.2 Characteristics of Concrete and Masonry Wall Buildings

Chapter 2 presents a summary of the characteristic features of concrete and masonry wall buildings. The chapter introduces the concept of structural systems, *elements* and components that is used throughout the evaluation process. The discussion includes the distinction between bearing walls and infilled frames. The effect of the dimensional and material characteristics of the wall components and the importance of this concept for the investigation of the damage caused by an earthquake are discussed. This chapter also illustrates the formulation of an *inelastic lateral mechanism* for a building based on the properties of its individual components. Additionally, the chapter

discusses how observed damage can be used to enhance and augment the model used for the investigation process.

1.4.3 Investigation of Earthquake Damage

The initial effort in the evaluation of damage to a specific building concentrates on investigating and documenting the damage that has occurred to a building during the earthquake (see Figure 1-3). Investigation procedures are given in Chapter 3. The objective is to assemble the basic information in a format that facilitates its use in evaluating the effects of the damage on future seismic performance. The primary steps in the investigation are summarized below.

1.4.3.1 Assemble Information

The first step in the investigation is a compilation of basic information on the damaging earthquake and the building.

A. Damaging Earthquake

Performance-based evaluations rely on a comparison between the capacity of a building to sustain lateral movement and the demand for lateral movement imposed by the *performance ground motion*. Information about the performance characteristics of a building can be derived from estimating the displacement demand that the damaging earthquake placed on it. For example, the decision regarding repair or upgrading of a building with moderate damage is affected by the magnitude of shaking that caused the damage. Section 3.1 provides a summary of suggestions for characterizing the *damaging ground motion* at the site for subsequent analysis.

B. Building Data

A discussion of the common configuration characteristics and components of concrete and masonry wall buildings is given in Chapter 2. The focus of the damage investigation is on the structural components that make up the vertical- and lateral-force-resisting system for the specific building under investigation. The construction drawings for the building, soils reports, prior building inspections, and other relevant reports and documents are the primary sources of the pertinent information (see Section 3.2). Basic information about the building includes its age, size, and use. If it was inspected after the damaging earthquake for posting purposes, these data can be

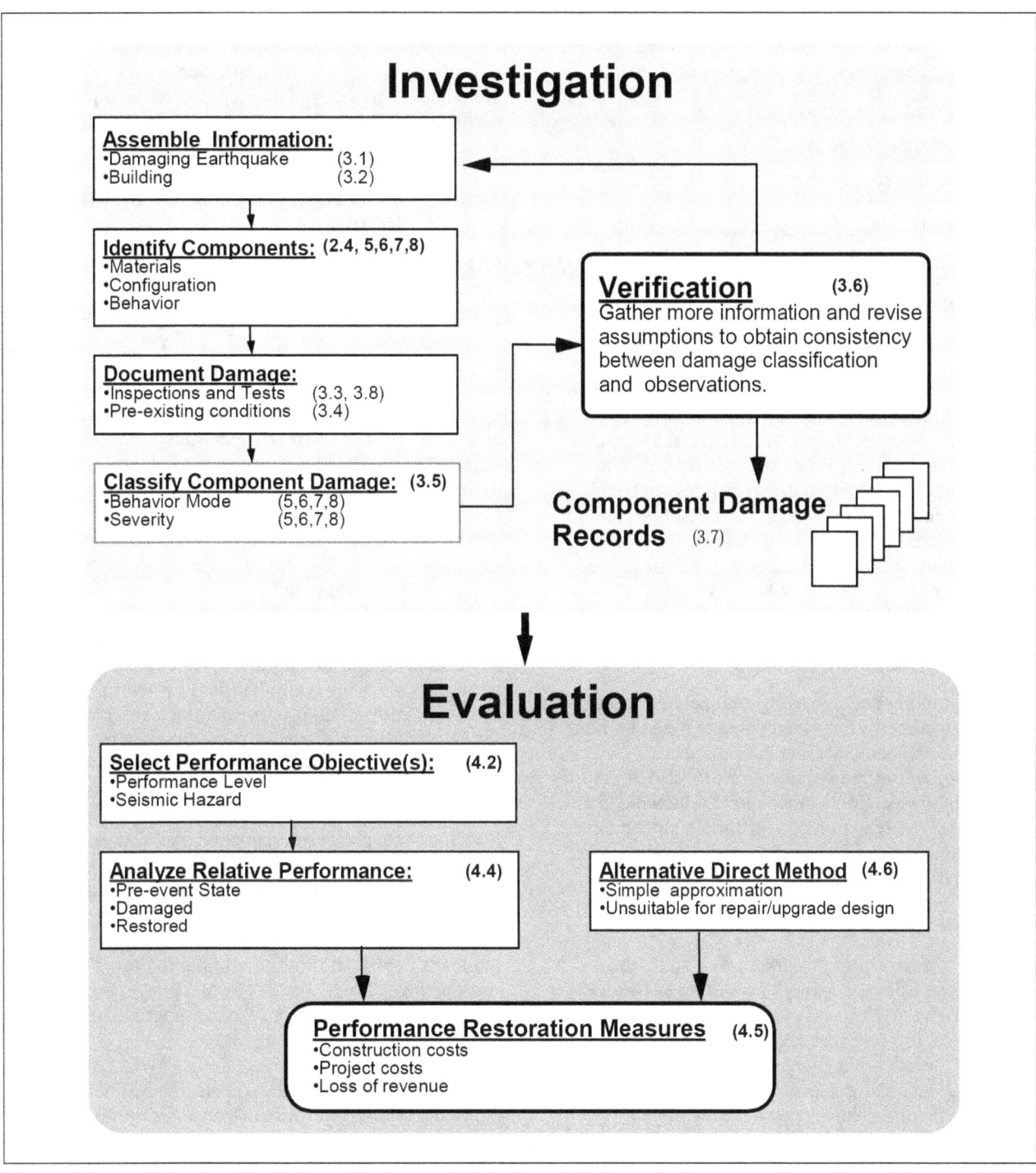

Figure 1-3 *Flowchart for the Investigation and Evaluation of Earthquake Damage to Concrete and Masonry Wall Buildings. (Section numbers are indicated.)*

useful. If records of the operation and maintenance are available, they can be useful in distinguishing between pre-existing conditions and damage caused by the earthquake.

C. Performance Objectives

The evaluation procedures are based on the performance objective for the building (see Section 4.2). Although it is possible to investigate and document damage without choosing a performance objective, it is worthwhile to consider this issue early in the evaluation process.

1.4.3.2   Identify Components

The engineer identifies basic structural components by anticipating the governing mechanism of inelastic behavior for each element in the structural system. This process normally requires some basic calculations to compare the relative strength and stiffness of the individual components of the structure . For each type of wall material (reinforced concrete, reinforced masonry, and unreinforced masonry) and for infilled frames, there are a number of basic component types. These are compiled in Chapters 5, 6, 7, and 8.

1.4.3.3   Document Damage

After assembling and reviewing available data, the engineer documents the actual damage based on field inspections and tests. Section 3.8 provides a compilation of outline specifications for different types of tests and investigative procedures. It includes guidance on the selection of appropriate procedures, equipment and personnel requirements, report format, and interpretation of results.

1.4.3.4   Classify Component Damage

For each component of the structural system, the engineer classifies the damage according to *behavior mode* and severity. The various behavior modes for each material and framing type are tabulated in Component Damage Classification Guides in Chapters 5, 6, 7, and 8. The engineer also categorizes the severity of damage for each type of damage encountered within any component.

1.4.3.5   Verification

The investigation of damage is a cyclic process. Information from the field can help the engineer determine component type based on actual behavior. Calculations and analyses can also help with the interpretation of field data. In some cases, the engineer may decide to conduct further tests to resolve conflicting data. Properly implemented, the process concludes with a reasonable representation of the actual damage and a basic understanding of the response of the structure to the earthquake shaking.

1.4.4   Evaluation of Earthquake Damage

Chapter 4 provides guidance on how to evaluate the significance of the observed damage. A seismic performance objective (see Section 4.2) consists of a specific *performance level* (e.g., *collapse prevention*, *life safety*, or *immediate occupancy*) for a specific seismic hazard (probability of shaking of a given intensity, or a deterministic event). The damage evaluation procedure uses a specified performance objective as a benchmark to gauge the effects of damage. The selection of applicable performance objectives for a building is a policy decision that depends on its age, size, use, and other considerations. For some cases, consideration of multiple performance objectives is appropriate.

Once the effects of the damaging ground motion on all of the components are tabulated, the engineer quantifies these effects for the entire building by determining the scope of actions that, if implemented, would restore the future seismic performance of the building to that of its pre-event state. These are performance restoration measures and they are the subject of Chapter 4. These measures are formulated by detailed analysis of the building in its pre-event, damaged, and restored conditions (i.e., relative performance analysis). In some cases a simplified approach (i.e., direct method) may be applicable to generate an estimate of loss. The selection of the appropriate method for a building depends on a number of considerations, including the severity of the earthquake, the extent and type of damage, and the likely course of action for repair or upgrade of the building.

The performance restoration measures determined by either the relative performance analysis or the direct method represent the conceptual physical changes to the damaged structure that would be required to restore the performance to the level that existed before the damaging earthquake. The loss in future seismic performance caused by the damaging earthquake is measured by the hypothetical costs to implement these measures. The total loss includes indirect costs, such as design and management fees and loss of use of the facility, in addition to direct construction costs that would be associated with the performance restoration measures if they were to be implemented.

Section 4.4 addresses the technical aspects of seismic performance analysis of concrete and masonry wall buildings. This quantitative procedure uses nonlinear analysis techniques to estimate the performance of the building in future earthquakes in its pre-event, damaged and restored states. The force-deformation characteristics of components are modified to account for damage according to recommendations in the Component Damage Classification Guides in Chapters 5 through 8. In order to determine the scope of the performance restoration measures, the engineer analyzes selective component restoration measures as well as the possible addition of supplemental components with the objective of restoring the seismic performance to that of the pre-event building.

1.4.5   Component Information

1.4.5.1   Component Damage Classification Guides

Chapters 5, 6, 7, and 8 provide a compilation of Component Guides for use in the damage evaluation process. These assist the engineer in identifying the structural components, determining behavior modes, and gauging damage severity. The guides also provide information on how damage affects the force-deformation characteristics of the components. This information is for use in the performance analysis. Recommendations for measures to restore structural properties are also tabulated. The component guides are classified according to structural system. The four classifications are:

- Concrete (Chapter 5)

- Reinforced masonry (Chapter 6)

- Unreinforced masonry (Chapter 7)

- Infilled frames (Chapter 8)

1.4.6   Terms and Symbols

A conscientious effort has been made to utilize concepts and language that are familiar to practicing engineers. This document, however, introduces terms whose definitions are not necessarily in common use. Such items, italicized at their first occurrence, are defined in the Glossary.

To the extent possible this document uses common symbols and notation that are familiar to practicing engineers. New symbols are required in some instances. These are listed at the end of this document. Symbols related primarily to specific materials are listed at the end of Chapter 5 for concrete, Chapter 6 for reinforced masonry, Chapter 7 for unreinforced masonry, and Chapter 8 for infilled frames.

1.4.6.1   Test and Inspection Guides

Section 3.8 presents information on common tests and inspection methods for investigation of earthquake damage to concrete, masonry wall, and infill frame buildings. It includes summaries of the required equipment and personnel, and the objectives and limitations of the procedures are reviewed. Reference and resource materials are listed.

1.4.7   Related Documents

FEMA 307: *Evaluation of Earthquake Damaged Concrete and Masonry Wall Buildings, Technical Resources* (ATC, 1998a)

FEMA 307 provides additional detailed information on the basis and use of the damage-evaluation procedures of FEMA 306. Background information on the development of the Component Guides is included for each material type and for infilled frames. It is essential that the engineer understand this information both for the general application of the procedures and for special cases when the typical component data must be modified to suit actual conditions. A summary of the analytical studies on the effects of damage on the global response of buildings is provided. This information is the basis for the recommendations on determining seismic displacement demand contained in FEMA 306. Finally, damage evaluation of a specific building is presented as a practical illustration of the application of the procedures.

FEMA 308: *The Repair of Earthquake Damaged Concrete and Masonry Wall Buildings* (ATC, 1998b)

This document supplements the evaluation procedures with a summary of policy considerations on the repair of earthquake-damaged concrete and masonry wall buildings. A model framework for repair policy is developed from past experience with damaging earthquakes. The use of the information from the evaluation process within this framework is illustrated for both the private and public sectors. The alternatives for repairing and upgrading earthquake-damaged buildings are reviewed along with potentially applicable standards and methodologies. Outline specifications for typical repair techniques are provided. Information on the objectives and limitations of the procedures is summarized. Reference standards and quality assurance measures are tabulated. These Repair Guides are also intended for use in the damage evaluation process to assist in the development of performance restoration measures.

ATC-20: *Procedures for the Post Earthquake Safety Evaluation of Buildings* (ATC, 1989)

ATC-20 is the standard for the safety investigation of buildings immediately following an earthquake. The intent of the document is to determine by visual observation of damage whether buildings are safe to occupy shortly after the earthquake. There are three levels of possible evaluation implied in ATC 20. The first level, Rapid Evaluation, is an inspection of the damage, which is intended to be implemented by building officials, engineers, architects, inspectors, or other individuals with a general familiarity with building construction. Questionable structures may be then subject to Detailed Evaluation by a structural engineer. If a structure cannot be appraised effectively by visual techniques alone, an Engineering Evaluation is required. At the time that ATC-20 was published, guidelines for Engineering Evaluations were not available. The procedures in FEMA 306 may be effectively utilized by qualified structural engineers to fill this gap. Consequently, FEMA 306 supplements the provisions of ATC-20.

1.5   Limitations

The procedures and criteria for the evaluation of damage in this document have been developed based on the current state of the knowledge on nonlinear inelastic behavior of structures and structural components. The state of knowledge varies by material, component type, and mode of behavior as discussed in Chapters 5, 6, 7, and 8 and FEMA 307. This knowledge will expand over time. The evaluation procedures and the information on component behavior must be adapted appropriately to reflect new information.

The interpretation of damage and the performance of buildings subject to earthquakes benefits from considerable experience and expert judgment. These procedures and criteria provide a framework for an engineer to apply experience and to formulate judgments on the effects of earthquake damage on future performance. The limitations of the procedures notwithstanding, the relative validity of results for a given situation are predominantly dependent on the capabilities of the engineer or engineers. The procedures should not be applied by non-engineering personnel (e.g., inspectors, insurance adjusters, claims managers).

In the past, other methodologies have been used to evaluate buildings damaged in earthquakes and to design repairs. If the procedures and criteria of this document are applied retroactively to such buildings, the results may be different. Any difference is not necessarily a reflection on the competency of the individual or firm responsible for the original work. Prior repairs should be judged on the basis of the procedures and criteria that were available at the time of the work.

2. Characteristics of Concrete And Masonry Wall Buildings

This chapter describes the basic design and construction features of concrete and masonry wall buildings. Descriptions of typically encountered structural components for various material types serve as a guide for the user when investigating actual buildings.

The evaluation of damage to a building requires an understanding on the part of the engineer of the way in which it supports gravity loads, resists earthquake forces, and accommodates related displacements. It is helpful to imagine the global building structure as an assembly of elements (see Figure 2-1). An element is a vertical or a horizontal portion of a building that acts to resist lateral and/or vertical loads. Common vertical elements in concrete and masonry wall buildings include structural walls and combined frame-wall (infilled) elements. Common floor or roof horizontal elements are reinforced concrete or wood diaphragms. For evaluation and analysis purposes, each element acts in its own plane to transmit seismic actions through the building in a three-dimensional global assembly of two-dimensional elements. Although out-of-plane seismic actions can act on elements at the same time, these actions are conventionally considered separately.

Elements are themselves assemblies of individual components such as beams, slabs, columns, joints, and others. The global performance of the structural system is an aggregation of the performance of its components.

For seismic performance analysis, structural properties (force-deformation relationships) and acceptability criteria (deformation limits) are specified for components. The global behavior of the building depends on these component properties. Evaluation procedures tabulate damage type and severity for components. The identification of components (see Section 2.4) of the lateral-force-resisting elements normally requires some basic engineering analysis and consideration of the type of damage that may have occurred.

2.1 Typical Vertical Elements

Concrete and masonry wall buildings rely primarily on the walls as vertical elements for lateral seismic resistance. The construction of these elements varies by material and the basic system for vertical load transfer. Behavior and damage characteristics of the walls during earthquakes depend on the physical dimensions and configuration of the wall elements including openings and penetrations.

2.1.1 Bearing Walls and Infilled Frames

In concrete and masonry wall buildings, there are two basic systems through which vertical loads are transmitted from the roofs and floors to the foundations:

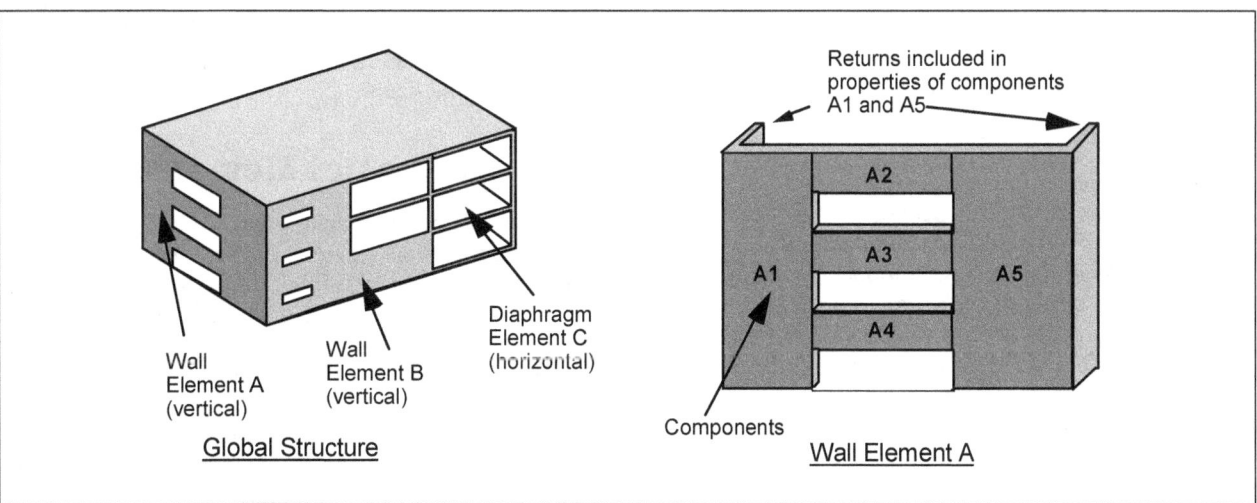

Figure 2-1   *Global Structure, Lateral-Force-Resisting Elements, and Components.*

bearing walls and infilled frames (see Figure 2-2). Bearing walls may support a portion of adjacent vertical load, as well as their own weight. In some areas of a bearing wall building, supplemental frames, columns, and/or flat slabs might support a portion of the vertical load. The walls themselves can be made of reinforced or unreinforced concrete or masonry.

Infilled frames differ from bearing walls in that they always include a vertical load carrying frame of concrete or steel beams and columns. Wall panels are placed within the frame. The infill can be reinforced or unreinforced concrete or masonry. To be effective at resisting in-plane lateral loads, the infill must be in contact with the surrounding frame. In basic configuration (e.g., distribution of elements within a building, extent of openings in walls), bearing wall and infilled frame buildings often appear similar. Reinforced concrete or masonry bearing walls can have boundary elements that are wider than the wall itself that resemble beams or columns of a frame. Their details of construction and behavior of bearing walls and infilled frames under lateral loads, however, can be quite different. The basic components of bearing wall and infilled frame buildings also differ from one another, as detailed further in Chapters 5, 6, 7, and 8.

2.1.2 Wall Elevations

The elevations of Figure 2-3 illustrate three general categories of concrete and masonry wall element configurations. Each of these configurations may be built of bearing wall or infilled frame construction. Cantilevered walls are those that act predominantly as vertical beams restrained at their foundation level. This is not to imply fixity at the base. In fact, many wall elements are sensitive to foundation movements caused by uplift, soil displacements, or deformations of foundation components, as discussed in Section 2.1.3.

Coupled walls or wall elements are those with a generally regular pattern of openings that form a configuration of vertical (*piers*) and horizontal (*spandrels* or coupling beams) components similar to a frame element. The inelastic action of a coupled wall element consequently depends on the relative strength and stiffness of the pier and spandrel components.

Perforated walls or wall elements may also exhibit an irregular pattern of openings in contrast to coupled walls. If the total area of opening relative to wall area is small, their behavior tends toward that of cantilevered walls. This behavior is illustrated by the strongly

coupled perforated wall in Figure 2-4. When there is a relatively large proportion of wall openings, behavior tends toward that of coupled walls with irregular (semi-vertical and semi-horizontal) components. This behavior is illustrated by the weakly coupled perforated wall in Figure 2-4. The modeling of perforated walls requires judgment and experience. Strut and tie models can be used to analyze walls with an irregular pattern of penetrations (Pauley and Priestley, 1992). Observations of damage after an earthquake can provide valuable evidence to assist the engineer in formulating a model to reflect actual behavior.

When walls intersect to form L-shaped, T-shaped, C-shaped, or similar sections, typically the entire section is considered as an integral unit and a single component. The contribution of flanges and wall returns should be considered in evaluating the strength of the component, based on the guidelines given in Chapters 5 through 8.

2.1.3 Foundation Effects

Foundation flexibility and deformation affect the earthquake response of many concrete and masonry wall buildings. Foundation effects tend to reduce the force demand on the primary lateral-force-resisting elements such as *shear walls*. At the same time, however, the rotational flexibility of the base of the shear walls often results in larger lateral displacements of the entire structure. The larger drifts can lead to damage in the beams, columns, or slabs. There is evidence of this type of damage from past earthquakes. Fixed-base analysis techniques do not adequately model these effects. FEMA 273/274 (ATC, 1997a,b) and ATC-40 (ATC, 1996) contain recommendations for modeling foundation elements and components similarly to other structural components.

2.2 Horizontal Elements

Horizontal elements (diaphragms) typically interconnect vertical elements at floor and roof levels in concrete and masonry wall buildings. Reinforced concrete slabs and the associated framing comprise relatively rigid diaphragms. These rigid diaphragms are characteristic of many concrete and masonry wall buildings. For analysis purposes, the flexibility of these diaphragms is often neglected, and the vertical elements are assumed to be rigidly linked at floor and roof levels. While this assumption is tolerable for most buildings, concrete diaphragms are not always rigid and can be

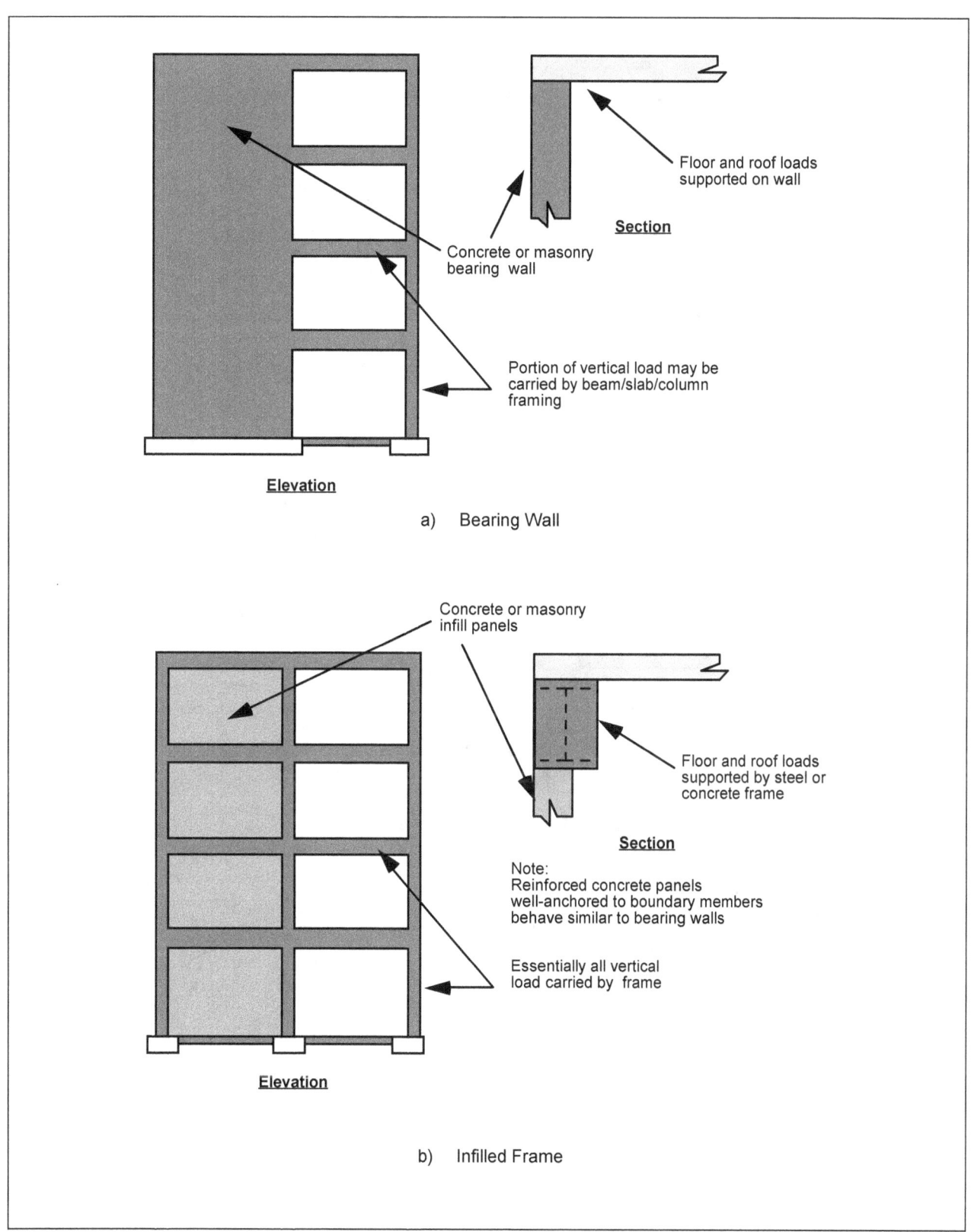

Figure 2-2 *Characteristics of Bearing Walls and Infilled Frames*

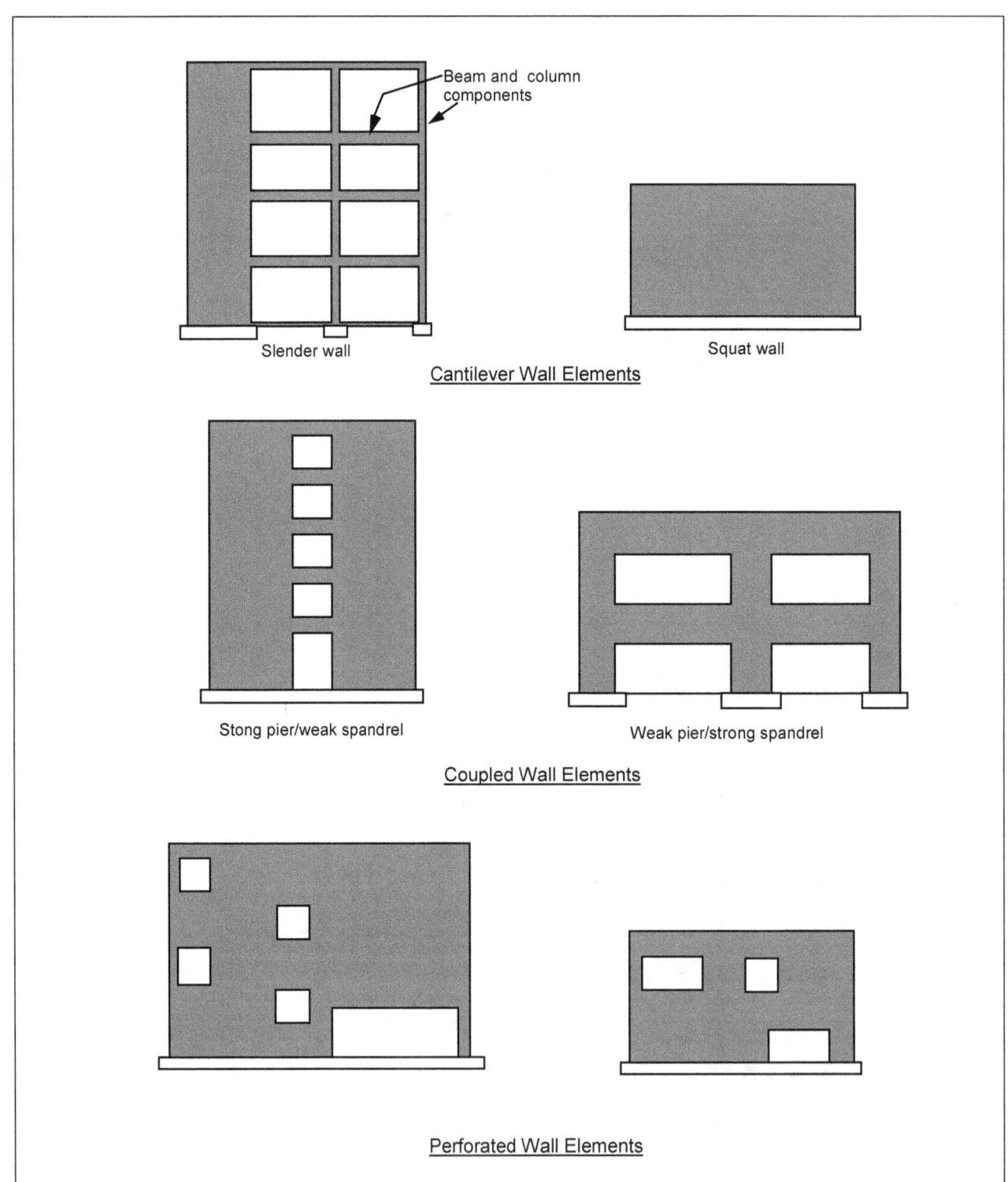

Figure 2-3   *Three General Categories of Concrete and Masonry Wall Configurations*

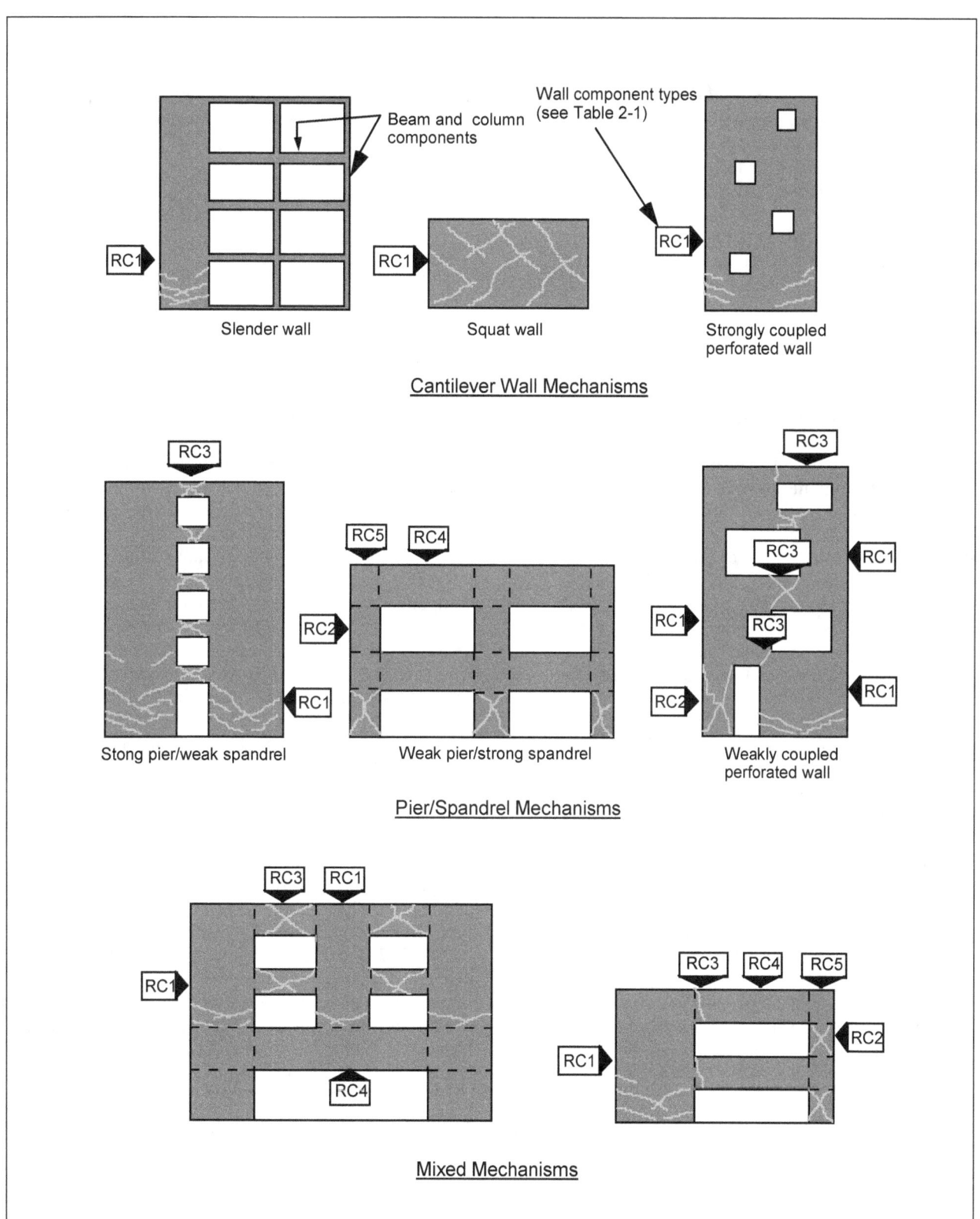

Figure 2-4 *Example Wall Mechanisms and Components*

damaged in earthquakes. Such damage has been observed, and repair may be required in some cases.

Many unreinforced masonry and precast (tilt-up) reinforced concrete bearing wall buildings have flexible diaphragms of wood sheathing. Walls resist the in-plane lateral loads that are distributed based on the tributary area. Connections between flexible diaphragms and walls are frequently the weak links in the lateral load path of the building, for forces both parallel and perpendicular to the wall. These connections are not addressed specifically in this document, but damage evaluations should consider the potential at these locations. Guidance may be found in FEMA 273/274.

2.3 Three-Dimensional Considerations

The interpretation of earthquake damage in concrete and masonry wall buildings can be complicated by the three-dimensional response of the buildings.

- Global horizontal torsion of the building can affect the distribution of damage to vertical elements. Analysis techniques contained in FEMA 273/274 and ATC-40 that can account for this effect are helpful for damage evaluation. However, the magnitude of the actual torsional response may differ from the estimates (actual plus accidental torsion) conventionally used for design. Careful interpretation of the distribution of damage in the field is required to interpret the torsional behavior.

- Damage to individual elements and components can be due to actions from either, or both, orthogonal directions. For example, a shear wall element acting parallel to one orthogonal direction may include a perpendicular return at either or both ends. Damage to the perpendicular return can be due to forces in either direction and must be carefully interpreted.

- Wall elements and components are subject to both in-plane and out-of-plane earthquake forces. Cracking or other damage due to out-of-plane forces can be misinterpreted as an in-plane effect. If cracks are evident on only one side of a wall element, they may be due to out-of-plane forces.

As a separate issue, parapets and other building appendages can pose serious risks, particularly in unreinforced masonry buildings.

2.4 Identification of Components

The procedures for damage evaluation focus on the components of the building that resist earthquake shaking. The identification of these components is central to the overall evaluation process. The ultimate identification of components for an earthquake-damaged building entails a combination of theoretical analysis and observation of the damage itself.

At the beginning of the evaluation process, the engineer identifies basic components by anticipating the governing inelastic lateral mechanism for each element in the lateral-force-resisting system. This analysis consists of determining the relevant stiffness and ultimate strength (flexure, shear, axial) of each component to anticipate the behavior and geometry of the mechanism that would form as the element is displaced laterally by a monotonically increasing lateral load pattern. Reinforced concrete wall component types are summarized in Table 2-1 and Figure 2-4. The component strength and load patterns are initially assumed using conventional sources including FEMA 273/274, ATC-40, and consensus design standards. FEMA 273/274 and ATC-40 also provide guidance on foundation components.

For each basic material, there are a number of component types. Chapters 5, 6, 7, and 8 provide a compilation of component data by material and framing type. The data in these chapters are supplemented in FEMA 307 by expanded information on component behavior that is based on available test data and theoretical techniques that go beyond conventional design standards. This resource material is useful when the effects of damage are introduced into the evaluation process, as discussed in Section 3.5.

Table 2-1   *Component Types for Reinforced Concrete Walls*

Component Type		Description
RC1	Cantilever wall or stronger wall pier	This type of component is stronger than beam or spandrel components that may frame into it, so that nonlinear behavior (and damage) is generally concentrated at the base, with a flexural plastic hinge or shear failure. This category includes isolated (cantilever) walls. If the component has a major setback or reduction of reinforcement above the base, this location should be also checked for nonlinear behavior.
RC2	Weaker wall pier	This type of component is weaker than the spandrels to which it connects. Damage is characterized by flexural hinging at the top and bottom of the pier, or by shear failure.
RC3	Weaker spandrel or coupling beam	This type of component is weaker than the wall piers to which it connects. Damage is characterized by hinging at each end, shear failure, or sliding shear failure.
RC4	Stronger spandrel	This type of component should not suffer damage because it is stronger than attached piers. If such a component is damaged, it should be re-classified as RC3.
RC5	Pier-spandrel panel zone	This component is a pier-spandrel connection zone. High shear forces in this zone can cause cracking. Severe damage is uncommon in reinforced concrete and masonry.

3. Investigation of Earthquake Damage

This chapter describes the investigation and documentation of earthquake damage to concrete and masonry wall or infill frame buildings. The objectives of the investigation are listed below.

- To gather information on the characteristics of the damaging ground motion at the building site

- To verify the general physical characteristics of the building, including its geometry and mass

- To identify structural components and elements of the lateral-force-resisting system

- To determine structural properties of the components in sufficient detail for structural analysis purposes

- To observe and record damage to the components

- To distinguish, to the extent possible, between damage caused by the earthquake and damage that may have existed before

The process includes the assembly and review of available existing information relating to the characteristics of the earthquake, assembly and review of information on the structural condition of the building both immediately before and after the earthquake, inspections and tests to characterize the nature and extent of damage, and the documentation and interpretation of the results of the investigation.

3.1 Characteristics of the Damaging Earthquake

During the evaluation of damage to concrete or masonry wall buildings, information on the characteristics of the damaging earthquake can lead to valuable insight on the performance characteristics of the structure. For example, if the ground motion caused by the earthquake can be estimated quantitatively, the analysis techniques summarized in Chapter 4 can provide an estimate of the resulting maximum displacement of the structure. This displacement, in conjunction with the theoretical capacity curve, indicates an expected level of component damage. If the observed component damage is similar to that predicted, the validity of the theoretical model is verified in an approximate manner. If the damage differs, informed adjustments can be made to the model.

A general process for gathering information and evaluating the effects of a damaging earthquake is outlined below:

1. Collect information on the damaging earthquake. If strong motion data is available, it is preferable to use data

   a. from a record taken at or very near to the site, or

   b. from contour maps of ground motions parameters, such as those shown in Figures 3-2, 3-3, and 3-4, created from a spatial interpolation of all nearby strong-motion data.

   If strong-motion data is not available, contours of intensity (e.g., Modified Mercalli Intensity) could be used to estimate spectral accelerations.

   Attenuation relationships can also be used to estimate ground-motion parameters. However, the scatter inherent in such relationships can lead to a large uncertainty in the prediction of ground motion for an individual site.

   In all cases, site soil conditions should be considered in the estimate of ground motion.

2. Formulate an approximate response spectrum for the site (see Figures 3-1 through 3-4). The example in the figures uses the acceleration at a period of 0.3 second to define the acceleration response regime. The 1997 *NEHRP Recommended Provisions for New Buildings* (BSSC, 1997) uses 0.2 second. Either approach may be used depending on the available data.

3. Generate a capacity curve for the structure at the time of the damaging earthquake (see Chapter 4)

4. Use *nonlinear static procedures* to estimate the maximum global displacement, d_e, that the damaging earthquake should have generated for the structure.

5. Estimate the expected component damage for the maximum global displacement of d_e and compare to the observed damage.

3.2 Review of Existing Building Data

The data collection process begins with the acquisition of documents describing the pertinent conditions of the building. Review of construction drawings simplifies

---

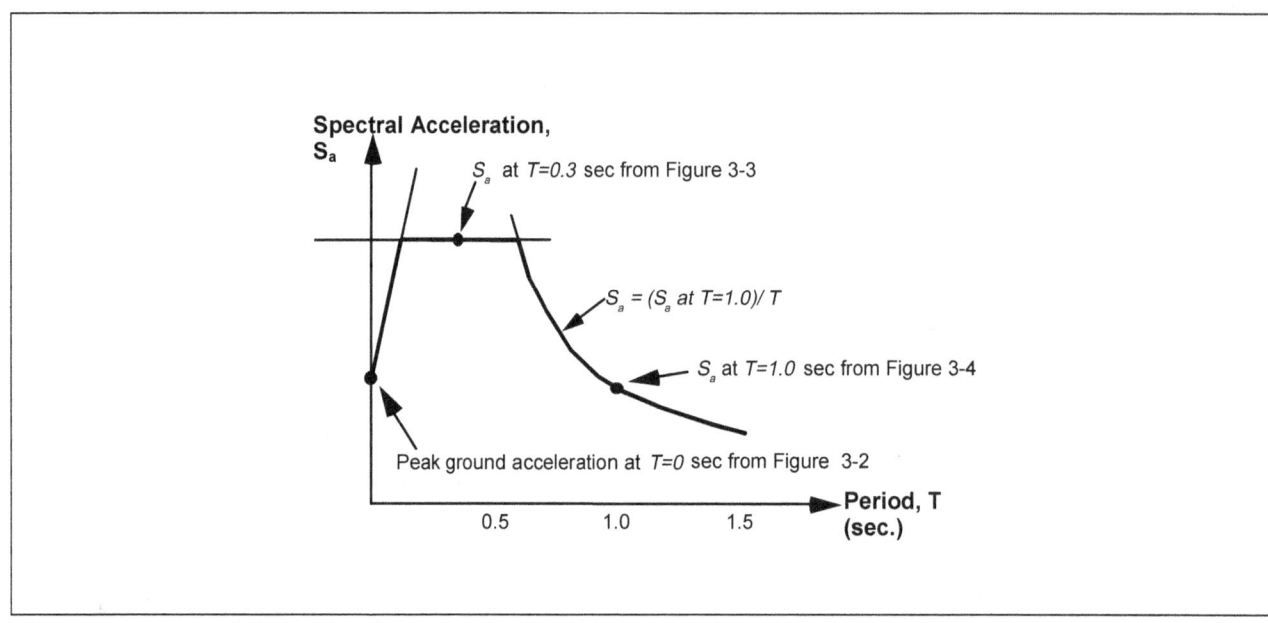

Figure 3-1   Parameters Needed and Form of Approximate Site Response Spectrum

field work and leads to a more complete understanding of the building. Original architectural and structural construction drawings are central to an effective and efficient evaluation of damage. Potential sources of these and other documents include the current and previous building owners, building departments, and the original architects or engineers. Drawings may also be available from architects or engineers who have performed prior evaluations for the building. In addition to construction drawings, it is helpful to assemble the following documents if possible:

• Site seismicity/geotechnical reports

• Structural calculations

• Construction specifications

• Contractors' shop drawings and other construction records

• Foundation reports

• Prior building assessments

Review of the existing building information serves several purposes. If reviewed before field investigations, the information facilitates the analytical identification of structural components, as discussed in Section 2.4.

This preliminary analysis also helps to guide the field investigation to components that are likely to be damaged. Existing information can also help to distinguish between damage caused by the earthquake and pre-existing damage. Finally, the scope of the field inspection and testing program depends on the accuracy and availability of existing structural information. For example, if structural drawings reliably detail the size and placement of reinforcing, expensive and intrusive tests to verify conditions in critical locations may be unnecessary.

3.3   Assessing the Consequences of the Damaging Earthquake

Methods for inspecting and testing concrete and masonry wall buildings for earthquake damage fall into two general categories, nondestructive and intrusive. Nondestructive techniques do not require any removal of the integral portions of the components. In some cases, however, it may be necessary to remove finishes in order to conduct the procedure. In contrast, intrusive techniques involve extraction of structural materials for the purpose of testing or for access to allow inspection of portions of a component. Table 3-1 summarizes the types of inspections and tests that apply to concrete and masonry wall buildings.

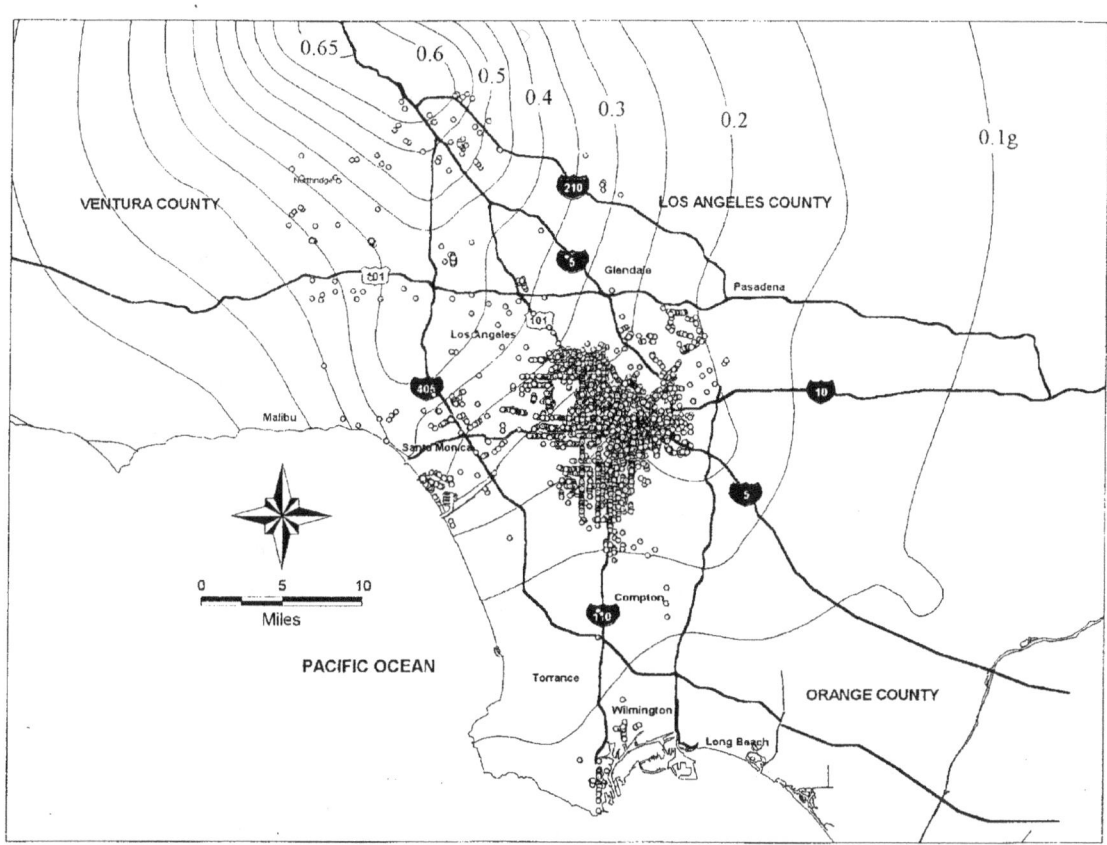

Figure 3-2 *Peak Ground Acceleration Contours for 1994 Northridge, California, Earthquake (from NIST, 1997, "dots" indicate locations of a particular building type)*

Section 3.8 provides guides for each procedure. Each guide includes a basic background for the practicing engineer on selecting and implementing appropriate procedures based on the actual conditions encountered in the field. Each guide consists of the following information:

Test Name and ID	For reference and identification
Test Type	Nondestructive (NDE) or Intrusive (IT)
Materials	Applicability to reinforced concrete, reinforced masonry, and/or unreinforced masonry
Description	Basic overview of the objectives and scope of the procedure
Equipment	A summary of the tools, instrumentation, or devices required
Execution	General sequence of operations
Reporting Requirements	Format for reporting of results
Personnel Qualifications	Skill level and specialized training that may be required
Limitations	Restrictions on the type of information that can be gained and advice on the interpretation of results
References	Applicable standards, detailed specifications, or sources of additional information

Table 3-1   ***Summary of Inspection and Test Procedures***

Structural Or Material Property	Material			Test ID (Section 3.8)	Test Type
	Reinf. Conc.	Reinf. Mas.	URM		
Crack Location and Size	✓	✓	✓	NDE 1	Visual observation
Spall Location and Size	✓	✓	✓	NDE 1	Visual observation
	✓	✓	✓	NDE 2	Sounding
Location of Interior Cracks or Delaminations	✓	✓	✓	NDE 6	Impact echo
	✓			NDE 7	Spectral Analysis of Surface Waves
	✓	✓	✓	IT 1	Selective removal
Reinforcing Bar Buckling or Fracturing	✓	✓		NDE 1	Visual observation
	✓	✓		IT 1	Selective removal
Relative Age of Cracks	✓	✓	✓	IT 2	Petrography
Relative Compressive Strength	✓	✓	✓	NDE 3	Rebound hammer
Compressive Strength	✓	✓	✓	IT 3	Material extraction and testing
Reinforcing Bar Location and Size	✓	✓		NDE 4	Rebar detector
	✓	✓		NDE 8	Radiography
	✓	✓		NDE 9	Penetrating radar
	✓	✓		IT 1	Selective removal
Strength of Reinforcing Bar	✓	✓		IT 3	Material extraction and testing
Wall Thickness	✓	✓	✓	NDE 1	Visual observation
	✓	✓	✓	NDE 6	Impact echo
	✓	✓	✓	IT 1	Selective removal
Presence of Grout in Masonry Cells		✓	✓	NDE 2	Sounding
		✓	✓	NDE 6	Impact echo
		✓	✓	NDE 7	Spectral Analysis of Surface Waves
		✓	✓	IT 1	Selective removal
Strength of Masonry		✓	✓	IT 3	Material extraction and testing
			✓	IT 4, 5	In situ testing
Mortar Properties		✓	✓	IT 2	Petrography
			✓	IT 4, 5	In situ testing

NDE: Nondestructive

IT: Intrusive

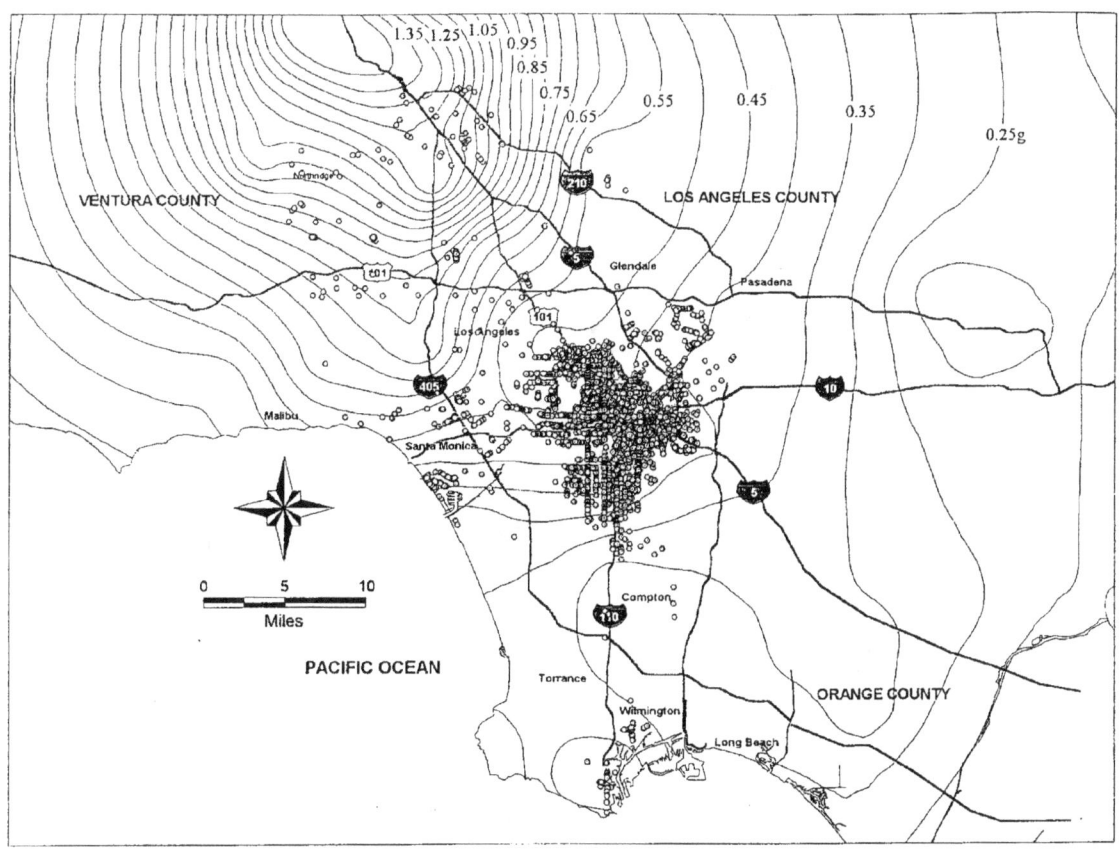

Figure 3-3   *Spectral Acceleration Contours for T=0.3 sec., 1994 Northridge, California, Earthquake (from NIST, 1997, "dots" indicate locations of a particular building type)*

The procedures included in Section 3.8 are those that are generally accessible to the practicing engineering community and that have been used successfully on projects that required evaluation of existing concrete and masonry structures. They are not, however, an exhaustive list. Other more sophisticated or specialized techniques may be useful in specific instances.

The overall scope of the type and number of tests and inspections depends on a number of factors including:

• The completeness of existing documentation. If accurate and complete documentation of the structural conditions is available, the scope of the investigation may be relatively small.

• The nature and extent of the damage. Pervasive or diverse damage trigger more extensive

investigations. Buildings with damage that may have occurred prior to the earthquake may require a greater degree of attention to distinguish between pre-existing conditions and earthquake damage.

• The quality of construction. If the field conditions differ routinely from construction documents, more investigative work will be required. If in-place material quality is inconsistent, more tests of individual components will be necessary.

• The correlation between analytical information and field observation. If calculations to identify critical components and expected damage give results that are corroborated by the actual damage, then fewer tests and inspections are warranted.

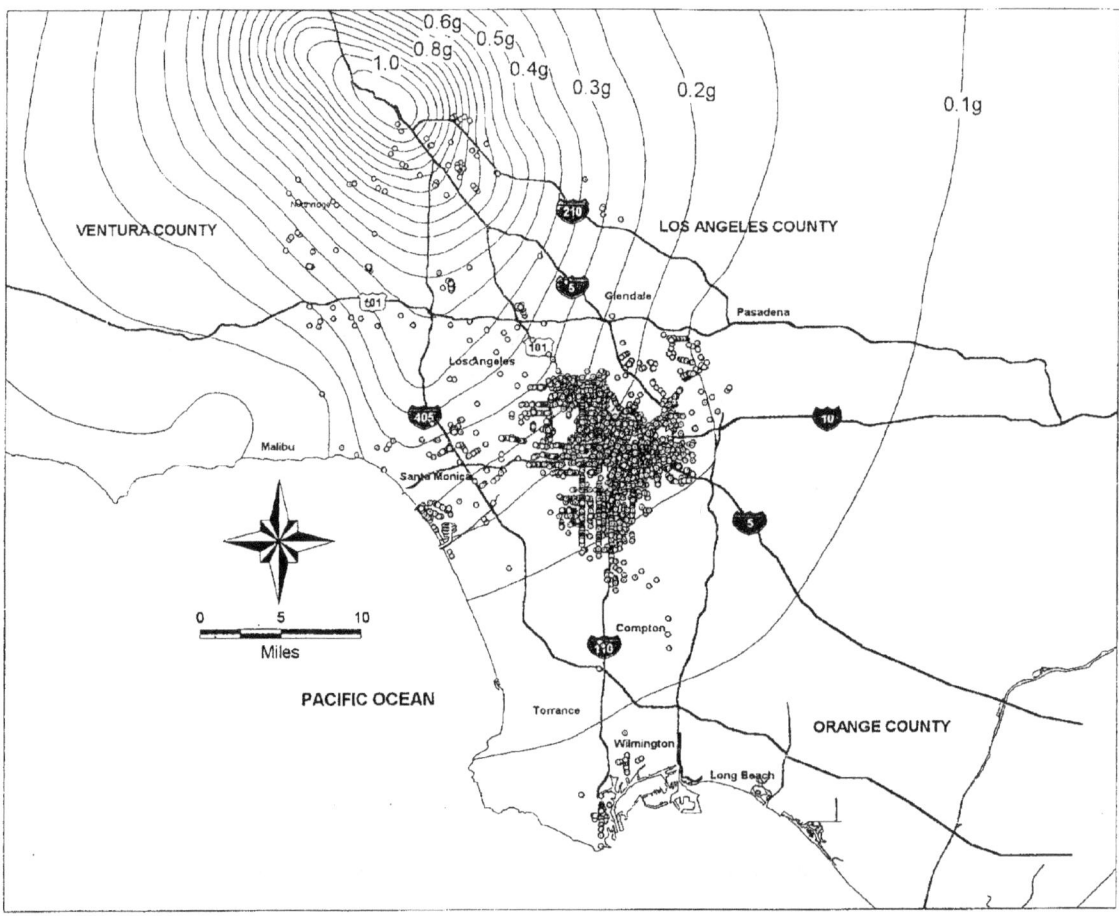

Figure 3-4 *Spectral Acceleration Contours for T=1.0 sec., 1994 Northridge, California, Earthquake (from NIST, 1997, "dots" indicate locations of a particular building type)*

- The degree of accessibility to critical areas for visual examination.

In general, the scope of the investigation can vary considerably among individual buildings. A plan for the investigation should begin with relatively simple and inexpensive procedures. The goal should be to visually inspect all the elements and components of the lateral-load-resisting system. In some cases, finishes may prevent the examination of certain elements and components. If analysis suggests that damage is likely to have occurred in hidden areas, finishes should be removed for inspection at critical locations. As the investigation proceeds, the scope can be expanded, if necessary, based on the results of visual inspections and

comparison with analytical predictions of behavior. When testing is needed to obtain material properties for a relative performance analysis, the number of tests required to quantify the in-place properties of the materials may be based on the guidelines provided in FEMA 273/274 and ATC-40.

3.4   Pre-existing Conditions

Interpretation of the findings of damage observations requires care and diligence. When evaluating damage to a concrete or masonry wall, an engineer should consider all possible causes in an effort to distinguish between that attributable to the damaging earthquake and that which occurred earlier (pre-existing conditions). ACI

224.1R (ACI Committee 224, 1994) discusses possible causes of cracking in reinforced concrete. Some of the causes described are also applicable to reinforced and unreinforced masonry construction. Since the evaluation of earthquake damaged buildings is typically conducted within weeks or months of the event, cracking and spalling caused by earthquakes is normally relatively recent damage. cracks associated with drying shrinkage or a previous earthquake, on the other hand, would be relatively old. General guidance for assessing the relative age of cracks based on visual observations is as follows.

Recent cracks typically have the following characteristics:

• Small, loose edge spalls

• Light, uniform color of concrete or mortar within crack

• Sharp, uneroded edges

• Little or no evidence of carbonation

Older cracks typically have the following characteristics:

• Paint or soot inside crack

• Water, corrosion, or other stains seeping from crack

• Previous, undisturbed patches over crack

• Rounded, eroded edges

• Deep carbonation

Evaluating the significance of damage requires an understanding of the structural behavior of the wall during the earthquake. The evaluating engineer must consider the implications of the observations with respect to the overall behavior of the building and the results of analytical calculations. The behavior must be correlated with the damage. If the observed damage is not reasonably consistent with the overall seismic behavior of the structure, the crack may have been caused by an action other than the earthquake.

3.5 Component Damage Classification

For each component of the structural system, the engineer classifies the damage according to behavior mode. Behavior mode indicates the predominant type of damage that a component sustains, or has the potential to sustain, in response to earthquake forces and displacements. The behavior mode depends on the relative strength of the component part for various actions (e.g., shear or moment). For each component, the engineer also classifies the severity of damage as follows:

Insignificant: Damage does not significantly affect structural properties in spite of a minor loss of stiffness. Restoration measures are cosmetic unless the performance objective requires strict limits on nonstructural component damage in future events.

Slight: Damage has a small effect on structural properties. Relatively minor structural restoration measures are required for restoration for most components and behavior modes.

Moderate: Damage has an intermediate effect on structural properties. The scope of restoration measures depends on the component type and behavior mode. Measures may be relatively major in some cases.

Heavy: Damage has a major effect on structural properties. The scope of restoration measures is generally extensive. Replacement or enhancement of some components may be required.

Extreme: Damage has reduced structural performance to unreliable levels. The scope of restoration measures generally requires replacement or enhancement of components.

Chapters 5, 6, 7, and 8 address the classification of damage for components of reinforced concrete, reinforced masonry, unreinforced masonry, and infilled frames, respectively. Guidance is tabulated according to component type and behavior mode, to assist the engineer in identifying and for assessing the severity of the damage based on the observed conditions and calculations of component properties.

The information and guidance for each typical component type are summarized in tabular form in the Component Damage Classification Guides (Component Guides) at the end of each chapter. The intention is to provide practical assistance in a concise format for use by an engineer in applying the evaluation procedures. Component Guides are not intended to be used by inexperienced or unqualified observers of damage. The identification of components and the determination of modes of behavior requires a thorough understanding of the technical basis of the damage evaluation procedures.

The format of the Component Guides is similar for all components.

Behavior Mode

A brief summary of how to distinguish the particular behavior mode both by observation of the damage and by analysis is provided. These relate to damage inspection procedures (Sections 3.3 and 3.8) and the component evaluation techniques (Chapters 5, 6, 7, and 8).

Description of Damage

The central column in the tabular layout of the Component Guides contains descriptive information on the typical damage for the particular component. These data consist of sketches and verbal criteria relating the observed damage to the various damage severity classifications.

Severity

The left hand column of the Component Guides designates the severity of damage for the five categories described above in this Section. This column also contains the recommended component modification factors (λ - factors) for damaged components. These are used to change the basic properties of the components to reflect the effects of damage in a relative performance analysis (Section 4.4.3).

Performance Restoration Measures

The right hand column in the Component Guides tabulates performance restoration measures intended to restore, as much as possible, the structural properties of the component. In cases where complete restoration is not possible, component modification factors for the restored component (λ*) are tabulated. The use of the performance restoration measures for damage evaluation is discussed in Section 4.5. The specific repair techniques are summarized in FEMA 308: *The Repair of Earthquake Damaged Concrete and Masonry Wall Buildings.*

It is important to recognize that the Component Guides in Chapters 5, 6, 7, and 8 are representative of typically encountered conditions. Judgment is required to adapt and apply this information to specific conditions. The Component Guides were developed from a review of available empirical and theoretical data. Included with the Component Guides for each material is guidance on their use and the evaluation of component behavior. FEMA 307 provides additional technical background information and identifies resources for component identification and damage classification.

3.6   Verification

In practice, the investigation of damage and identification of components may be an iterative process. As presented in Chapter 2, the initial identification of components is based on relative strength and stiffness, and the anticipated inelastic lateral mechanism. Information from the field helps the engineer verify the component type based on actual behavior. For example, Figure 3-5 illustrates two possible inelastic lateral mechanisms for the same element. Theoretical calculations may predict one mechanism and therefore certain types of component damage. Observations of damage in the field, however, may lead to a different conclusion regarding the basic mechanism and component identification. There are several sources of discrepancies between analysis and observation, described below.

1. The distribution of the lateral forces from the damaging earthquake might have differed from that used in the analysis to generate the inelastic lateral mechanism. In such a case, the component behavior modes observed in the field might differ from those predicted analytically because of the relative magnitudes of component actions. For example, the use of a conventional upper triangular distribution of lateral load for a cantilevered shear wall might predict a flexural behavior mode in which the ultimate moment capacity at the base of the wall is attained before reaching the shear capacity. If a shearing behavior mode is encountered in the field, it may indicate a more rectangular or trapezoidal lateral

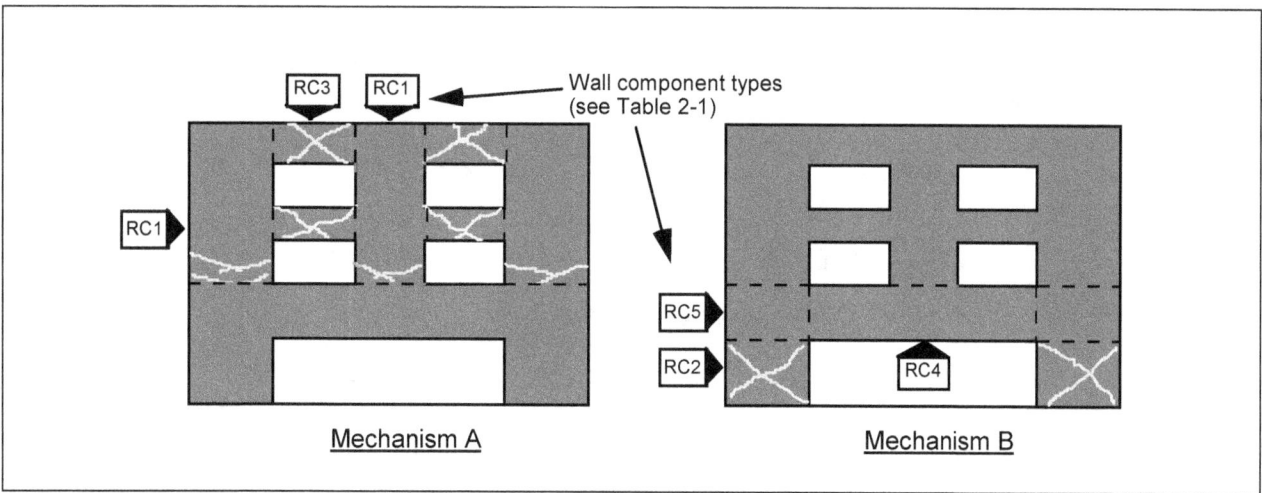

Figure 3-5 **Different Inelastic Lateral Mechanisms and Components for Same Wall Element**

load distribution, which would tend to lower the shear span (M/V) for the component. Also, a uniform distribution of seismic forces is more likely to cause a story mechanism than an inverted triangular distribution

2. The strength of components for various actions may differ from that predicted analytically. This could lead to different component types and/or behavior modes being a better representation of actual behavior. Many of the conventional theoretical formulations for component strength are intended for use as design equations. As such, they reflect an appropriate degree of conservatism and are suitable for a wide range of applications. The damage evaluation process differs fundamentally from design. The objective is to use theory and observation to assess the actual strength and behavior of the structural components. Figure 3-6 illustrates the difference between design strength and expected strength for a flexural component. The component data in FEMA 307 provide resources for alternative formulations based on available empirical and theoretical research on actual behavior and material properties. If an alternative strength estimate correlates more closely with observed behavior and specific conditions, it is appropriate to use that estimate for evaluation purposes. This is not to imply that the estimate is then applicable for general design purposes.

3. The severity and significance of damage depends heavily on ductility and behavior mode (see Figure 3-7). Some components exhibit mixed behavior modes, as shown in the moderate-ductility example in Figure 3-7(b). The component initially exhibits flexural behavior, but there is a transition to shear-controlled behavior at higher deformations. This type of behavior is not unusual, and it can be difficult to identify. Chapters 5 through 8 and FEMA 307 provide additional information and guidance on this point.

4. The overall intensity of the damaging ground motion might differ from that assumed in the analysis. The maximum global displacement that actually occurred during the earthquake, d_e, could be larger or smaller than that predicted. This would tend to produce a correspondingly greater or lesser overall severity of component damage. Component type and behavior mode would not be affected in the absence of other differences.

Resolution of these discrepancies entails adjustments to the analysis and the structural model so that the resulting component types, behavior modes, and severity of damage match the observed conditions. In some cases, the engineer may decide to conduct further tests or investigations to resolve conflicting data. Properly implemented, the process concludes with a reasonable and consistent representation of the governing behavior modes and the actual damage, as is necessary for an accurate understanding of the response of the structure to the damaging ground motion.

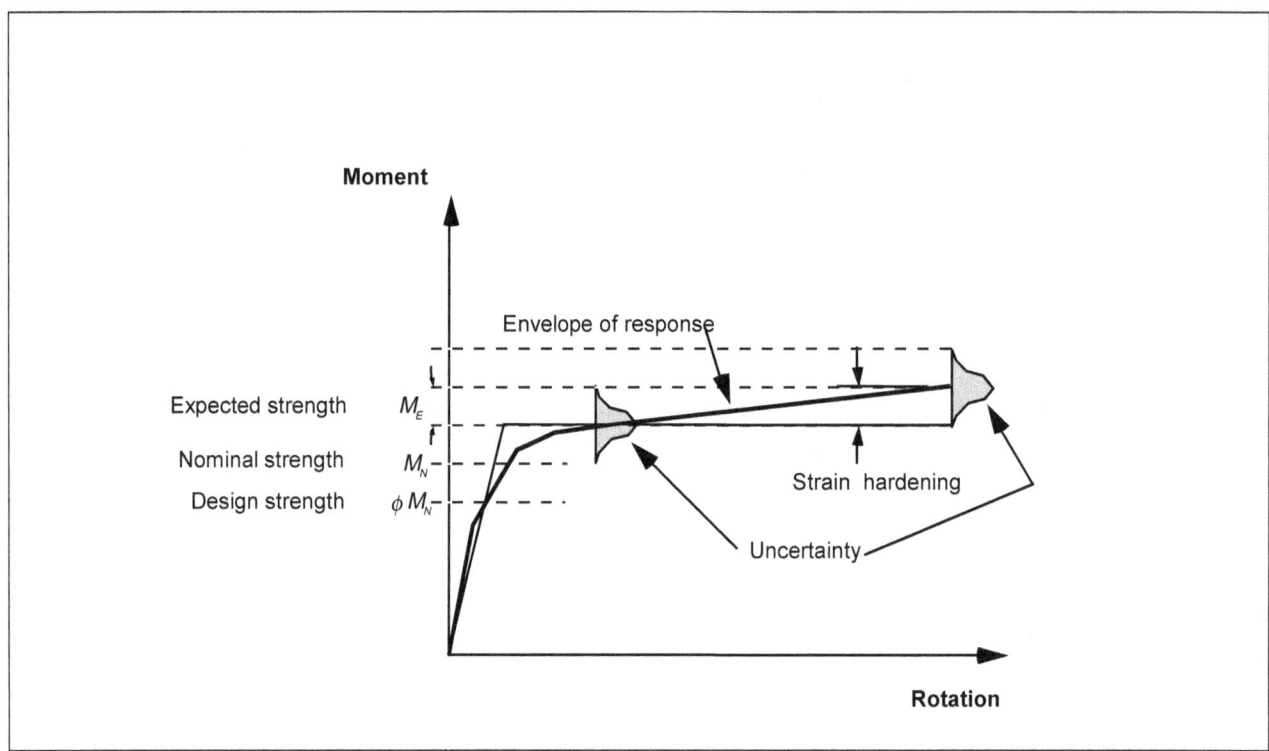

Figure 3-6   *Relationship between design strength and expected strength*

3.7   Documentation

Documentation of the results of the investigation should be complete and unambiguous. Plan drawings should show the location of elements and components and the locations and dates of tests. Elevations of critical elements and components should also be included where appropriate. Test results should be tabulated in accordance with the recommendations in the guides.

Crack maps, sketches, and photographs, keyed to the plan drawings, should record all visual observations. Results of the investigation should be organized to focus on structural components and behavior modes. This organization facilitates the generation of Component Damage Record forms, shown in Figure 3-8.

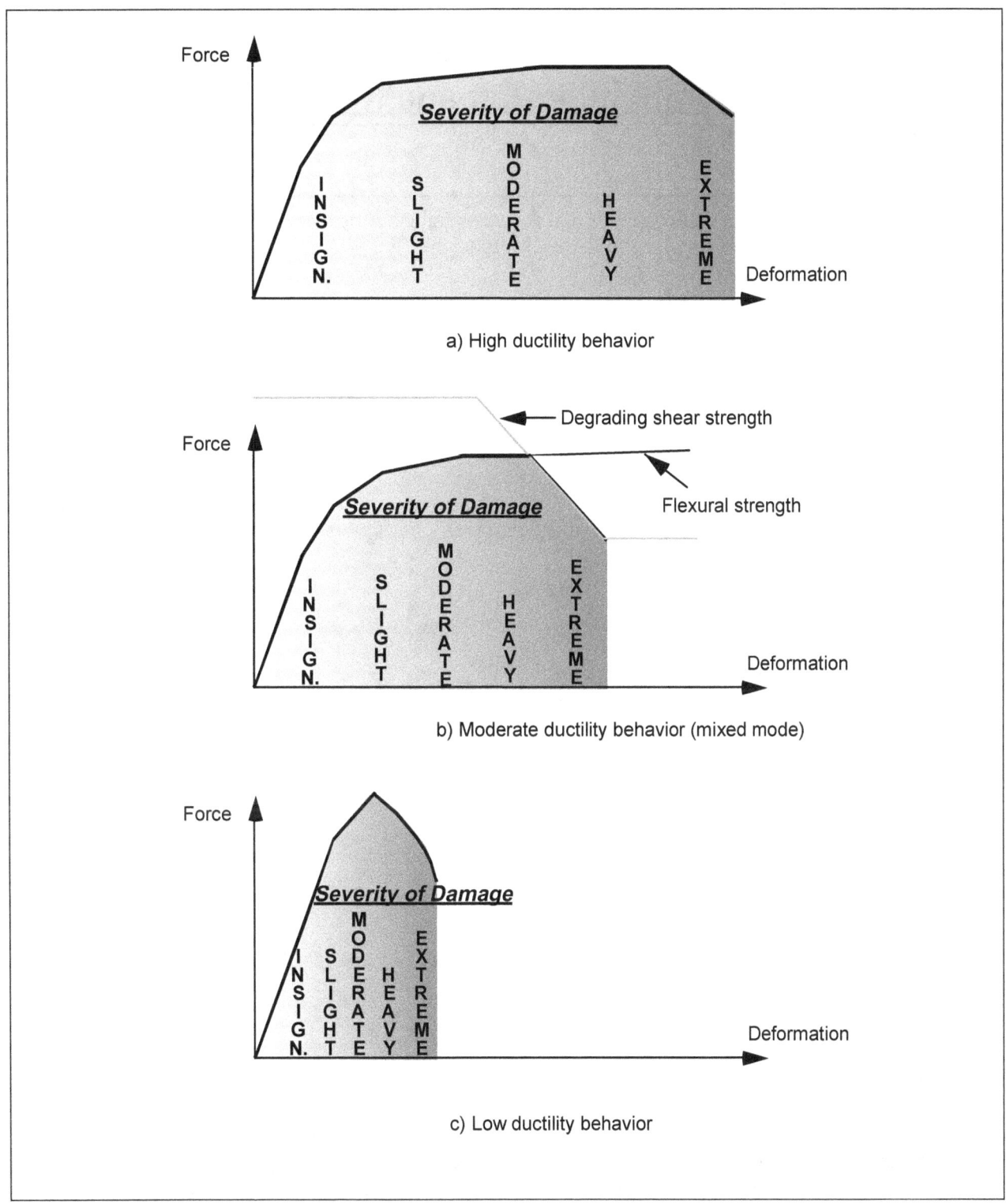

Figure 3-7   *Component force-deformation behavior, ductility, and severity of damage*

Component Damage Record		
Component ID:	Component Type:	Location:
Sketch and description of damage types and severities (attach supplemental data if necessary):		
Test results summary (attach detail):		
Building:	Engineer:	Inspection date:
FEMA 306: Evaluation and Repair of Earthquake Damaged Concrete and Masonry Wall Buildings		

Figure 3-8   Example Component Damage Record

3.8   Test and Inspection Guides

This section provides guidelines for the use of typical tests and inspections to assess the consequences of earthquake damage to concrete and masonry wall buildings as discussed in Section 3.3.

TEST AND INVESTIGATION GUIDE	Test Type:	Nondestructive
NDE 1 VISUAL INSPECTION	Materials:	Concrete, Reinforced Masonry, Unreinforced Masonry

Description

Visual inspection is perhaps the most useful test available in the assessment of earthquake damage to concrete and masonry walls. Generally, earthquake damage to concrete and masonry walls is visible on the exposed surface. Observable types of damage include cracks, spalls and delaminations, permanent lateral displacement, and buckling or fracture of reinforcement. Visual inspection can also be useful for estimating the drift experienced by the building.

Visual inspection should always accompany other testing methods that are used. Findings from the visual inspections should be used as a basis for determining locations for conducting further testing. The observed damage should be documented on sketches. The patterns of damage can then be interpreted to assess the behavior of the wall during the earthquake.

Equipment

The materials and equipment typically required for a visual inspection are a tape measure, a flashlight, a crack comparator, a pencil, and a sketchpad.

A tape measure is used to measure the dimensions of the wall and, if necessary, to measure the lengths of the cracks. Tape measures that are readily available from a hardware store, with lengths of 20 to 50 feet, are sufficiently accurate for damage evaluation.

Flashlights are used to aid in lighting the areas to be inspected. In postearthquake evaluations, electric power may not be completely available, so supplemental lighting should be supplied.

In a visual inspection, the engineer uses a crack comparator or a tape measure to measure the width of cracks at representative locations. Two types of crack comparators are generally available: thin clear plastic cards, which have specified widths denoted on the card and small, hand-held magnifying lenses with a scale marked on the surface. Plastic card comparators have gradated lines to a minimum width of about 0.002 inches. Magnifying lens comparators are accurate to about 0.001 inch (ACI Committee 201, 1994a).

The engineer uses a sketchpad to prepare a representation of the wall elevation, indicating the locations of the cracks, spalling, or other damage. All significant features of the wall should be recorded, including the dimensions of openings, the finishes on the wall, and the presence of nonstructural elements that may affect the repairs. The sketch should be supplemented with photographs or video tape.

Detailed examination of the surface of a crack can be accomplished with a portable microscope, which allows for magnified viewing of the surface of the cracks. Portable microscopes are available with magnifications of 18- to 36-fold. An external light source is needed for viewing. A camera adapter may be available for photographic documentation.

TEST AND INVESTIGATION GUIDE
continued

$$\boxed{\text{NDE 1}}$$

Execution

The initial steps in the visual observation of earthquake damage are to identify the location of the wall in the building and to determine the dimensions of the wall (height, length, and thickness). A tape measure is used for quantifying the overall dimensions of the wall. A sketch of the wall elevation should then be prepared. The sketch should include sufficient detail to depict the dimensions of the wall, it should be roughly to scale, and it should be marked with the wall location (See example on page 33).

Observable damage such as cracks, spalling, and exposed reinforcing bars should be indicated on the sketch. Sketches should be made in sufficient detail to indicate the approximate orientation and width of cracks. Crack width is measured using the crack comparator or tape measure at representative locations along significant cracks. Avoid holes and edge spalls when measuring crack widths. Crack widths typically do not change abruptly over the length of a crack. If the wall is accessible from both sides, the opposite side of the wall should be checked to evaluate whether the cracks extend through the thickness of the wall and to verify that the crack widths are consistent.

Photographs can be used to supplement the sketches. If the cracks are small, they may not show up in the photographs, except in extreme close-up shots. Paint, markers, or chalk can be used to highlight the location of cracks in photographs. However, photographs with highlighted crack should always be presented with a written disclaimer that the cracks have been highlighted and that the size of the cracks cannot be inferred from the photograph.

During a visual inspection, the engineer should carefully examine the wall for the type of damage and possible causes. ACI 201 is a guide that describes conditions that might be observed when surveying concrete walls. Indications that the cracks or spalls may be recent or that the damage may have occurred prior to the earthquake should be noted. The guidelines in Section 3.4 can be helpful for assessing the relative age of the cracks.

Visual observation of the nonstructural elements in the building can also be very useful in assessing the overall severity of the earthquake, the interstory displacements experienced by the building, and the story accelerations. Full-height nonstructural items such as partitions and facades should be inspected for evidence of interstory movement such as recent scrapes, cracked windows, or crushed wallboard.

Personnel Qualifications

Visual inspection of concrete and masonry walls should be performed by an engineer or trained technician. Engineers and technicians should have previous experience in identifying damage to concrete and masonry structures and should be familiar with the use of a tape measure and crack comparator. Engineers and technicians should also have sufficient training to be able to distinguish between recent damage and damage that may have been pre-existing. For this type of assessment, the person conducting the inspection should understand how the structure is designed and how earthquake, gravity, and other forces may have acted on the wall.

TEST AND INVESTIGATION GUIDE

continued

$$\boxed{\text{NDE 1}}$$

Limitations

The width of a crack can vary substantially along its length. Both the plastic card and the magnifying crack comparators can produce a reasonable estimate of the width of a crack. The magnifying comparators are generally more accurate when measuring small (<0.001 inches) crack widths. The plastic cards can sometimes overestimate the crack width due to the lighting conditions. With either type of comparator, the crack width is only measured at representative locations to determine repair thresholds. The measurements should be used primarily to compare damage levels among walls. The crack comparators may not be necessary when the crack widths are to be measured in 1/16-inch increments. For wider cracks, a tape measure will provide sufficiently accurate values.

Visual observation of concrete and masonry walls can generally identify most of the earthquake damage to those elements. In some cases, the presence of finishes on the walls can prevent an accurate assessment of the damage. Brittle finishes such as plaster can indicate damage that may not be present in the underlying substrate. Soft finishes such as partitions isolated from the structural walls can obscure minor amounts of damage.

References

ACI Committee 201, 1994a, "Guide for Making a Condition Survey of Concrete in Service", ACI 201.1R-92, *Manual of Concrete Practice*, American Concrete Institute, Detroit, Michigan.

ACI Committee 224, 1994b, Causes, Evaluation and Repair of Cracks in Concrete Structures", ACI Committee 224.1R-93, *Manual of Concrete Practice*, American Concrete Institute, Detroit, Michigan.

ACI Committee 364, 1994c, "Evaluation of Structures Prior to Rehabilitation", ACI 364.1R, *ACI Manual of Concrete Practice*, American Concrete Institute, Detroit, Michigan.

Component Damage Record (Example)		
Building Name:	**Project ID:**	**Prepared by:**
Concrete Shear Wall Building	ATC 43 Example	ATC
Location Within Building:		Date:
Floor: $1^{st}/2^{nd}$ Column Line: B Component Type:		24-Sep-97

Sketch and Description of Damage:

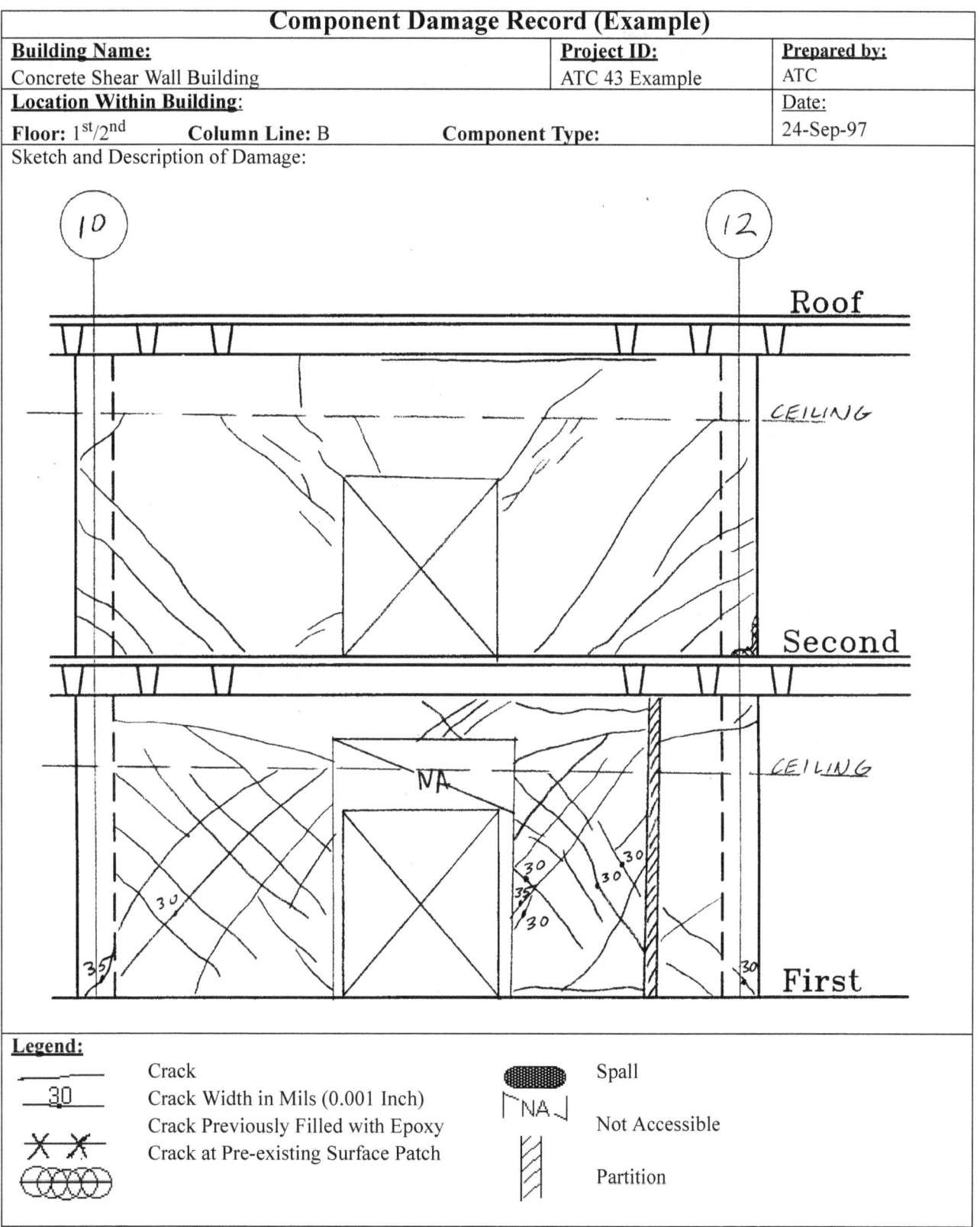

Legend:

—— 30 ——	Crack
30	Crack Width in Mils (0.001 Inch)
X X	Crack Previously Filled with Epoxy
⊕⊕⊕⊕	Crack at Pre-existing Surface Patch

⬬ (Spall)	Spall
⌐NA⌐	Not Accessible
▨	Partition

TEST AND INVESTIGATION GUIDE	Test Type:	**Nondestructive**
SOUNDING	Materials:	**Concrete, Reinforced Masonry**

NDE 2

Description

Tapping on a wall with a dense object, such as a hammer, and listening to the vibrations emitted from the wall can be useful for identifying voids or delaminations in concrete walls. The sound produced from a solid wall will be different from that from a wall with voids or delaminations close to the surface. In concrete block masonry walls, sounding can be used to verify that the cells in the blocks have been grouted.

Equipment

The typical equipment required for sounding is a hammer. However, any hard, dense object can be used.

Execution

In areas where the visual observations indicate that the wall may have delaminations, the wall can be sounded by tapping with a hammer. Delaminations and spalls will generally produce a hollow sound (ACI, 1994) when compared with solid material. The wall should be tapped several times in the suspect area and away from the suspect area, and the sounds compared. It is important to test an area that is undamaged, and of the same material and thickness to use as a baseline comparison. For a valid comparison, the force exerted by the tapping should be similar for both the suspect and baseline areas.

In reinforced masonry construction, sounding can be used to assess whether the cells in the wall have been grouted. Near the ends of a block, the unit is solid for the full thickness of the wall. For most of the length of the block, it is relatively thin at the faces. If the sound near the end of the block is substantially different than at the middle of the cell, the cell is probably not grouted.

Personnel Qualifications

Sounding of concrete and masonry walls should be performed by an engineer or trained technician. Engineers and technicians should have previous experience in identifying damage to concrete and masonry structures. Engineers and technicians should also be able to distinguish between sounds emitted from a hammer strike. Prior experience is necessary for proper interpretation of results.

Reporting Requirements

The personnel conducting the tests should provide sketches of the wall indicating the location of the tests and the findings. The sketch should include the following information:

- Mark the location of the test on either a floor plan or wall elevation.

- Report the results of the test, indicating the extent of delamination.

- Report the date of the test.

- List the responsible engineer overseeing the test and the name of the company conducting the test.

TEST AND INVESTIGATION GUIDE
continued

$\boxed{\text{NDE 2}}$

Limitations

The properties of the wall can influence the usefulness of sounding. The geometry of the wall and the thickness of the wall will affect the results (ASCE, 1990). Sounding is best used away from the perimeter of the wall and on a wall of uniform thickness.

The accuracy of information from sounding with a hammer also depends on the skill of the engineer or technician performing the test and on the depth of damage within the thickness of the wall. Delaminations up to the depth of the cover for the reinforcing bars (usually about 1 to 2 inches) can usually be detected. Detection of deeper spalls or delamination requires the use of other NDE techniques. Sounding cannot determine the depth of the spall or delamination (Poston et al., 1995).

Tapping on a loose section of material can cause the piece to become dislodged and fall. Avoid sounding overhead. A ladder, scaffold, or other lift device should be used to reach higher elevations of a wall.

References

ACI Committee 224, 1994, Causes, Evaluation and Repair of Cracks in Concrete Structures", ACI Committee 224.1R-93, *Manual of Concrete Practice*, American Concrete Institute, Detroit, Michigan.

ASCE, 1990, *Guideline for Structural Condition Assessment of Existing Buildings*, ASCE 11-90, American Society of Civil Engineers, New York, New York.

Poston, R.W, A.R. Whitlock, and K.E. Kesner, 1995, "Condition Assessment Using Nondestructive Evaluation" *Concrete International*, July, 1995, American Concrete Institute, Detroit, Michigan, pp 36-42.

	TEST AND INVESTIGATION GUIDE	Test Type:	**Nondestructive**
NDE 3	**REBOUND HAMMER**	Materials:	**Concrete, Unreinforced Masonry**

Description

A rebound hammer provides a method for assessing the in-situ compressive strength of concrete. In this test, a calibrated hammer impact is applied to the surface of the concrete. The amount of rebound of the hammer is measured and correlated with the manufacturer's data to estimate the strength of the concrete. The method has also been used to evaluate the strength of masonry.

Equipment

A calibrated rebound hammer is a single piece of equipment that is hand operated

Execution

ASTM C805 (ASTM, 1995) provides a standard on the use of a rebound hammer. The person operating the equipment places the impact plunger of the hammer against the concrete and then presses the hammer until the hammer releases. The operator then records the value on the scale of the hammer. Typically three or more tests are conducted at a location. If the values from the tests are consistent, record the average value. If the values vary significantly, additional readings should be taken until a consistent pattern of results is obtained.

Since the test is relatively rapid, a number of test locations can be chosen for each wall. The values from the tests are converted into compressive strength using tables prepared by the manufacturer of the rebound hammer.

Personnel Qualifications

A technician with minimal training can operate the rebound hammer. An engineer experienced with trebound hammer data should be available to supervise to verify that any anomalous values can be explained.

Reporting Requirements

The personnel conducting the tests should provide sketches of the wall, indicating the location of the tests and the findings. The sketch should include the following information:

- Mark the location of the test marked on either a floor plan or wall elevation.

- Record the number of tests conducted at a given location.

- Report either the average of actual readings or the average values converted into compressive strength along with the method used to convert the values into compressive strength.

- Report the type of rebound hammer used along with the date of last calibration.

- Record the date of the test.

- List the responsible engineer overseeing the test and the name of the company conducting the test.

TEST AND INVESTIGATION GUIDE
continued

Limitations

The rebound hammer does not give a precise value of compressive strength, but rather an estimate of strength that can be used for comparison. Frequent calibration of the unit is required (ACI, 1994). Although manufacturers' tables can be used to estimate the concrete strength, better estimates can be obtained by removing core samples at selected locations where the rebound testing has been performed. The core samples are then subjected to compression tests. The rebound values from other areas can be compared with the rebound balues that correspond to the measured core compressive strength.

The results of the rebound hammer tests are sensitive to the quality of the concrete on the outer several inches of the wall (Krauss, 1994). More reproducible results can be obtained from formed surfaces rather than from finished surfaces.

Surface moisture and roughness can also affect the readings. The impact from the rebound hammer can produce a slight dimple in the surface of the wall. Do not take more than one reading at the same spot, since the first impact can affect the surface, and thus affect the results of a subsequent test.

When using the rebound hammer on masonry, the hammer should be placed at the center of the masonry unit. The values of the tests on masonry reflect the strength of the masonry unit and the mortar (Noland et al., 1982). This method is only useful in assessing the strength of the outer wythe of a multi-wythe wall.

References

ACI Committee 364, 1994, "Evaluation of Structures Prior to Rehabilitation", ACI 364.1R, *ACI Manual of Concrete Practice*, American Concrete Institute, Detroit, Michigan

ASTM, 1995, *Test for Rebound Number of Hardened Concrete*, ASTM C805, American Society for Testing and Materials, Washington, DC

Krauss, P.D., 1994, *Repair Materials and Techniques for Concrete Structures in Nuclear Power Plants*, NRC JNC No. B8045, US Nuclear Regulatory Commission, Washington, DC.

Noland, J.L. et al., 1982, *An Investigation into Methods of Nondestructive Evaluation of Masonry Structures*, Report to the National Science Foundation, Atkinson-Noland & Associates, Boulder, Colorado

TEST AND INVESTIGATION GUIDE	Test Type:	**Nondestructive**
NDE 4 **REBAR DETECTOR**	Materials:	**Concrete, Reinforced Masonry**

Description

Covermeter is the general term for a rebar detector used to determine the location and size of reinforcing steel in a concrete or masonry wall. The basic principle of most rebar detectors is the interaction between the reinforcing bar and a low frequency magnetic field. If used properly, many types of rebar detectors can also identify the amount of cover for the bar and/or the size of the bar. Rebar detection is useful for verifying the construction of the wall, if drawings are available, and in preparing as-built data if no previous construction information is available.

Equipment

Several types and brands of rebar detectors are commercially available. The two general classes are those based on the principle of magnetic reluctance and those based on the principle of eddy (Carino, 1992). The various models can have a variety of features including analog or digital readout, audible signal, one-handed operation, and readings for reinforcing bars and prestressing tendons. Some models can store the data on floppy disks to be imported into computer programs for plotting results.

Execution

The unit is held away from metallic objects and calibrated to zero reading. After calibration, the unit is placed against the surface of the wall. The orientation of the probe should be in the direction of the rebar that is being detected. The probe is slid slowly along the wall, perpendicular to the orientation of the probe, until an audible or visual spike in the readout is encountered. The probe is passed back and forth over the region of the spike to find the location of the maximum reading, which should correspond to the location of the rebar. This location is then marked on the wall. The procedure is repeated for the perpendicular direction of reinforcing.

If size of the bar is known, the covermeter readout can be used to determine the depth of the reinforcing bar. If the depth of the bar is known, the readout can be used to determine the size of the bar. If neither quantity is known, most rebar detectors can be used to determine both the size and the depth using a spacer technique. The process involves recording the peak reading at a bar and then introducing a spacer of known thickness between the probe and the surface of the wall. A second reading is then taken. The two readings are compared to estimate the bar size and depth.

Intrusive testing can be used to help interpret the data from the detector readings. Selective removal of portions of the wall can be performed to expose the reinforcing bars. The rebar detector can be used adjacent to the area of removal to verify the accuracy of the readings.

Personnel Qualifications

The personnel operating the equipment should be trained and experienced with the use of the particular model of covermeter being used and should understand the limitations of the unit.

TEST AND INVESTIGATION GUIDE
Continued

$$\boxed{\text{NDE 4}}$$

Reporting Requirements
The personnel conducting the tests should provide a sketch of the wall indicating the location of the testing and the findings. The sketch should include the following information:

• Mark the locations of the test on either a floor plan or wall elevation.

• Report the results of the test, including bar size and spacing and whether the size was verified.

• List the type of rebar detector used.

• Report the date of the test.

• List the responsible engineer overseeing the test and the name of the company conducting the test.

Limitations
The readings can be difficult to interpret if the depth of the reinforcement is too great or if there is heavy congestion of reinforcement, such as at splices or boundaries (ACI, 1994). The accuracy will vary between units and manufacturers. Except at the boundaries of the wall, the spacing of bars is generally wide enough that the influence of adjacent bars should not affect the readings. Other embedded metals, such as metallic conduits or pipes will be detected and may give false readings.

For walls with two layers of reinforcing steel, the rebar detector can only be used to detect the reinforcing bars closest to the face on which the probe is used. The unit should be used on both faces to detect bars in a wall with two layers of reinforcement. When two layers are present, the second layer of reinforcement can affect the readings by producing a stronger signal for a bar of

given size and depth than the bar would produce in the absence of a second layer (ACI, 1997).

Some rebar detectors require recalibration at regular intervals during use. Therefore, the user should frequently check the readings to verify that the readings are still reproducible. The spacer technique to determine the size and depth of reinforcement is only accurate to within 1 or 2 bar diameters (Krauss, 1994 and Bungey, 1989).

When measuring the cover depth, many units actually measure the distance to the center of the reinforcing steel. The manufacturer's literature should be reviewed to determine the meaning of the depth reading.

References
ACI Committee 228, 1997, *Nondestructive Tests Methods for Evaluation of Concrete in Structures*, ACI 228.2R - Draft, American Concrete Institute, Detroit, Michigan.

ACI Committee 364, 1994, Guide for Evaluation of Concrete Structures Prior to Rehabilitation, ACI 364.1R, *ACI Manual of Concrete Practice*, American Concrete Institute, Detroit, Michigan.

Bungey, J.H., 1989, *Testing Concrete in Structures*, 2nd Edition, Chapman and Hall, New York, New York.

Carino, N.J, 1992, *Performance of Electromagnetic Covermeters for Nondestructive Assessment of Steel Reinforcement*, NISTR 4988, National Institute of Standards and Technology, Gaithersburg, Maryland.

Krauss, P.D., 1994, *Repair Materials and Techniques for Concrete Structures in Nuclear Power Plants*, NRC JNC No. B8045, US Nuclear Regulatory Commission, Washington, DC.

TEST AND INVESTIGATION GUIDE		Test Type:	**Nondestructive**
NDE 5	**ULTRASONIC PULSE VELOCITY**	Materials:	**Concrete, Reinforced Masonry, Unreinforced Masonry**

Description

The ultrasonic pulse velocity method measures the travel time of an ultrasonic pulse through the thickness of the wall. The velocity at which the pulse travels through the wall is affected by the quality of the material, including the presence of cracking or damage. By comparing the relative travel time at various sections of known thickness, the ultrasonic pulse velocity can be used to assess relative strength of concrete or masonry and to indicate the presence of cracking or delamination.

Equipment

The equipment and calibration procedures are described in ASTM C 597, *Standard Test Method for Pulse Velocity Through Concrete*. Portable equipment is available from several manufacturers. Transmitting and receiving transducers are required.

The frequency of the transducers is typically about 50 kHz, which is adequate for walls that are at least four inches thick, corresponding to the wave length of the pulse (Krauss, 1994).

A time-measuring meter with either a digital time display or a digital storage oscilloscope is also required.

Execution

The area of the wall to be examined should be laid out with a grid. The location of the grid should be coordinated so that the intersection of the grid lines will be at the same location on both sides of the wall. The spacing of the grid will vary, depending on the size of the wall and the extent of the expected damage. A grid spacing of one-foot centers should provide a reasonably fine spacing to capture potential damage.

The transmitting and receiving transducers are mounted on opposite sides of the wall, using a couplant between the transducer and the surface of the wall. For masonry walls, the transducer should be mounted to the masonry units, not to the mortar joints. The meter sends a series of pulses through the wall and measures the transmission time, which is recorded. The transducers are moved to the next location, and the test is repeated. If high readings are encountered, indicating possible discontinuities, a finer grid should be laid out in the vicinity of the possible damage to eastablish the extent of the discontinuity.

The results are displayed as travel time. The travel time needs to be converted into velocity using the thickness of the wall. The pulse velocity, not the travel time, should be used to compare results at various locations.

To establish the relative strength of the wall, samples should be taken at representative test locations. Correlate the strength of the extracted material samples and the pulse velocity readings at those locations, then use this correlation to estimate the strength at other sections of the wall.

Personnel Qualifications

A technician with training in the use of the equipment can carry out the test. An engineer or technician with extensive experience in the use and limitations of the equipment should be responsible for overseeing the tests and interpreting the results.

TEST AND INVESTIGATION GUIDE
Continued

NDE 5

Reporting Requirements

The personnel conducting the tests should provide a sketch of the wall indicating the location of the testing and the findings. The sketch should include the following information:

• Mark the location of the test on either a floor plan or wall elevation.

• Report the test results as either actual velocity measurements or interpreted results.

• List the type of pulse-velocity equipment used, including the date of last calibration.

• Record the date of the test.

• List the responsible engineer overseeing the test and the name of the company conducting the test.

Limitations

Pulse-velocity measurements require access to both sides of the wall. The wall surfaces need to be relatively smooth. Rough areas can be ground smooth to improve the acoustic coupling. Couplant must be used to fill the air space between the transducer and the surface of the wall. If air voids exist between the transducer and the surface, the travel time of the pulse will increase, causing incorrect readings.

Some couplant materials can stain the wall surface. Non-staining gels are available, but should be checked in an inconspicuous area to verify that it will not disturb the appearance.

Embedded reinforcing bars, oriented in the direction of travel of the pulse, can affect the results, since the ultrasonic pulses travel through steel at a faster rate than

through concrete. Bars larger than 3/8-inch diameter will significantly affect the results (Chung and Law, 1983). The moisture content of the concrete also has a slight effect (up to about 2 percent) on the pulse velocity.

Pulse-velocity measurements can detect the presence of voids or discontinuities within a wall; however, these measurements cannot determine the depth of the voids (ACI, 1997).

References

ACI Committee 228, 1997, *Nondestructive Tests Methods for Evaluation of Concrete in Structures*, ACI 228.2R - Draft, American Concrete Institute, Detroit, Michigan.

ACI Committee 364, 1994, Guide for Evaluation of Concrete Structures Prior to Rehabilitation, ACI 364.1R, *ACI Manual of Concrete Practice*, American Concrete Institute, Detroit, Michigan.

Berra, M., L. Binda, G. Baronio, and A. Fatticcioni, 1987, "Ultrasonic Pulse Transmission: A Proposal to Evaluate the Efficiency of Masonry Strengthened by Grouting*", Evaluation and Retrofit of Masonry Structures*, Proceedings of the Second Joint USA-Italy Workshop on Evaluation and Retrofit of Masonry Structures, pp 93-110.

Chung, H.W. and K.S. Law, 1983, "Diagnosing In Situ Concrete by Ultrasonic Pulse Technique", *Concrete International*, October 1983, American Concrete Institute, Detroit, Michigan, pp 42-49.

Krauss, P.D., 1994, *Repair Materials and Techniques for Concrete Structures in Nuclear Power Plants*, NRC JNC No. B8045, US Nuclear Regulatory Commission, Washington, DC.

TEST AND INVESTIGATION GUIDE		Test Type:	**Nondestructive**
NDE 6	**IMPACT ECHO**	Materials:	**Concrete, Reinforced Masonry, Unreinforced Masonry**

Description

Impact echo is a method for detecting discontinuities within the thickness of a concrete or masonry wall. The surface of the material is struck with an impactor, a small hammer, which introduces an energy pulse into the material. The energy pulse is reflected off of wave-speed discontinuities within the material. The discontinuities can be cracks, the back surface of the material, side surfaces, delaminations, or voids. A transducer mounted to the striking surface records the reflection of the energy. The transducer is connected to a Fast Fourier Transform (FFT) analyzer, which converts the time history signal from the transducer into the frequency domain. The frequency results and the raw time history can be interpreted to assess the thickness of the material and the size and location of discontinuities within the wall, such as voids, cracks, and delaminations.

Equipment

The typical impact echo equipment consists of a small impactor (hammer) to strike the surface of the material, a transducer to measure the surface response of the material, and an FFT analyzer to analyze the measurements made by the transducer. Some FFT analyzers are extremely sophisticated portable computers that allow for extensive manipulation of the output, while others simply provide a graph showing the frequency content of the output.

The equipment can be assembled from available components. Complete systems are also commercially available (ACI, 1997).

Execution

The transducer is placed on the surface of the material. Good contact must be developed between the transducer and the material, or the transducer will not be able to get a clean signal from the energy pulse. Strike the material with the impactor to introduce an energy pulse into the material. The FFT analyzer produces a frequency content analysis of the transducer's signal, but the final analysis of the data rests with the operator. If the impactor and transducer are in the middle of a solid wall, the energy pulse bounces back and forth between the front and back surfaces, typically giving an FFT frequency content with one major frequency peak. This peak corresponds to how quickly the energy pulse bounces between the front and back surfaces of the material. If the impactor and transducer are in the middle of a wall with a delamination, the first peak should be at a higher frequency, since the energy pulse will tend to bounce between the front surface of the wall and the surface of the delamination, a shorter distance requiring less travel time. The presence of side boundaries, voids, large concentrated amounts of reinforcement, and cracks will complicate the signal.

Personnel Qualifications

Impact echo testing should be performed by an engineer or technician well-trained and experienced in using this technique.

TEST AND INVESTIGATION GUIDE

continued

NDE 6

Reporting Requirements

The personnel conducting the tests should provide sketches of the wall indicating the location of the tests and the findings. The sketch should include the following information:

- Mark location of the test on either a floor plan or wall elevation.

- Record the number of tests conducted at a given location.

- Report the results of the test using either the actual readings or the interpreted results, including the peak frequency values.

- Describe the type of impact echo equipment used, along with the date of last calibration.

- Report the date of the test.

- List the responsible engineer overseeing the test and the name of the company conducting the test.

Limitations

The accuracy of impact echo testing is typically highly dependent on the skill of the engineer or technician in understanding the testing method and interpreting the results. Incompletely trained or untrained persons using impact echo methods have a high probability of interpreting the results incorrectly. The physical limitations and accuracy of impact echo are governed in part by the size of the impactor, the type, sensitivity, and natural frequency of the transducer, the uniformity of the concrete, and the ability of the FFT analyzer to manipulate the data into useful information.

The impact echo technique has been applied extensively to concrete structures. However, there is little experience with applying the technique to reinforced or unreinforced masonry components.

References

ACI Committee 228, 1997, *Nondestructive Test Methods for Evaluations of Concrete in Structures*, ACI 228.2R - Draft, American Concrete Institute, Detroit, Michigan.

Poston, R.W, et al., 1995, "Condition Assessment Using Nondestructive Evaluation*" Concrete International*, July, 1995, American Concrete Institute, Detroit, Michigan, pp 36-42.

	TEST AND INVESTIGATION GUIDE	Test Type: **Nondestructive**
NDE 7	**SPECTRAL ANALYSIS OF SURFACE WAVES (SASW)**	Materials: **Concrete**

Description

Spectral analysis of surface waves (SASW) is a method of measuring the propagation of surface waves over a wide range of wavelengths. The propagation velocities are measured using accelerometers and a Fast Fourier Transform (FFT) analyzer. The results can be interpreted to assess the thickness of the material and the size and location of discontinuities within the wall, such as voids, large cracks, and delaminations.

Equipment

The SASW tests require the following equipment:

- An impactor, which is usually a hammer

- Two or more receivers, which could be accelerometers or velocity transducers

- An FFT spectrum analyzer for recording and analyzing the input signal from each receiver

Execution

Mount the receivers on the surface of the wall using a removable adhesive. The spacing between the receivers will depend on the thickness of the wall. Strike the surface of the wall with the hammer away from the receivers, producing a surface R-wave that propagates along the surface (ACI, 1997). The surface velocity or acceleration is recorded by the receivers and processed.

The processed results can then be interpreted to assess the condition of the concrete.

Personnel Qualifications

Use of the SASW equipment should be limited to those with extensive training in the use of the equipment. Specialized experience is required to interpret the results.

Reporting Requirements

The personnel conducting the tests should provide sketches of the wall indicating the location of the tests and the findings. The sketch should include the following information:

- Mark the location of the test on either a floor plan or wall elevation.

- Record the number of tests conducted at a given location.

- Report the results of the test using either the actual readings or the interpreted results.

- Describe the type of equipment used.

- Report the date of the test.

- List the responsible engineer overseeing the test and the name of the company conducting the test.

TEST AND INVESTIGATION GUIDE
continued

NDE 7

Limitations

The signal processing equipment used for the interpretation of the results is very complex and not readily available. The SASW process has been used mainly on pavement, slabs, and other horizontal surfaces. Its use on walls has not been documented extensively.

References

ACI Committee 228, 1997, *Nondestructive Tests Methods for Evaluation of Concrete in Structures*, ACI 228.2R - Draft, American Concrete Institute, Detroit, Michigan.

TEST AND INVESTIGATION GUIDE	Test Type:	**Nondestructive**
RADIOGRAPHY	Materials:	**Concrete, Reinforced Masonry**

NDE 8

Description

Radiography can be used to determine the location of reinforcing steel within a concrete or masonry wall. The process involves transmitting x-rays through the concrete. A radiographic film on the opposite side from the x-ray source records the intensity of the x-rays that exit the wall. The processed film presents an image of the locations of reinforcing bars and other discontinuities.

Equipment

A portable X-ray tube with a radioactive isotope is required. For wall thickness less than six inches, iridium-192 or cesium-137 can be used (see, for example, Mitchell et al., 1979). For thicker materials, more intense isotopes are needed. Special photographic film is used to capture the X-rays.

Execution

An x-ray technician mounts the photographic film on the surface of the wall. The location of the film is marked on the wall for future reference. The X-ray tube is mounted or placed on the opposite side of the wall from the film. The x-ray technician exposes the wall to the radioactive isotope. The length of time for the exposure will depend on the thickness of the wall, the size of the film, and the amount of reinforcement in the wall. Thicker walls and areas with a high concentration of reinforcing bars require longer exposure times.

The film is then processed. The processed image is interpreted to assess the locations of reinforcing bars and discontinuities. Reinforcing bars, which are denser than concrete, show up on the image as light areas. Voids are seen on the film as relatively darker areas (ACI, 1997).

Personnel Qualifications

The personnel operating the x-ray equipment require highly specialized training on the handling of radioactive material. The technicians who interpret the images should be experienced in viewing x-rays from concrete or masonry structures.

Reporting Requirements

The personnel conducting the tests should provide sketches of the wall indicating the location of the tests and the findings. The sketch should include the following information:

- Mark the location of the test on either a floor plan or wall elevation.

- Report the results of the test along with a sketch of the findings.

- Describe the type of X-ray equipment used, along with the date of last calibration.

- Report the date of the test.

- List the responsible engineer overseeing the test and the name of the company conducting the test.

TEST AND INVESTIGATION GUIDE
continued

$$\boxed{\text{NDE 8}}$$

Limitations

Because radiography involves the release of radiation, the vicinity of the testing needs to be evacuated, except for the personnel conducting the tests. The size of the area to be evacuated depends on the type of radioactive isotope used and the thickness of the wall. Thicker walls require longer exposure times, and therefore more radiation is released. The time, expense, and logistics of the evacuation must be considered in planning the tests. Most commercially available x-ray equipment is capable of penetrating walls up to 12 inches thick. For thicker walls, the expense of the highly specialized equipment needed is generally not cost-effective for commercial buildings.

The presence of steel within the concrete will produce a shadow on the film to indicate its location. Since the radiation emits from a point source onto the photographic film, the amount of shadow will depend on the depth of the reinforcing bar from the face where the x-ray source is placed. Therefore, it is usually not possible to determine the size of the reinforcing bars based on the photographic image.

References

ACI Committee 228, 1997, *Nondestructive Test Methods for Evaluation of Concrete Structures*, ACI 228.2R - Draft, American Concrete Institute, Detroit, Michigan.

Mitchell, T.M., P.L. Lee, and G.J. Eggert, 1979, "The CMD: A Device for the Continuous Monitoring of the Consolidation of Plastic Concrete, *Public Roads*, Vol. 42, No. 148.

TEST AND INVESTIGATION GUIDE	Test Type:	Nondestructive
NDE 9 PENETRATING RADAR	Materials:	Concrete, **Reinforced Masonry,** **Unreinforced Masonry**

Description

Penetrating radar transmits electromagnetic waves, which are received by an antenna. The propagation of the waves through the material is influenced by the dielectric constant and the conductivity of the material. The signal received can be interpreted to discover discontinuities and variations in the material properties. The interpreted data can be used to detect the location of reinforcing bars, cracks, voids, or other material discontinuities.

Equipment

The penetrating radar instrumentation consists of several components including:

- An antenna that emits an electromagnetic pulse of various frequencies

- A receiving antenna

- A control unit that provides power to the transmitting antenna and acquires the signal from the receiving antenna

- A data recording device such as a printed display or digital storage device

Execution

Place the antenna of the radar unit on the surface of the wall and move along the surface while data are being recorded. The antenna produces an electromagnetic pulse that passes through the wall. Some of the pulse is reflected back to the receiving antenna. The received signal is printed on a strip-recording chart or stored for later analysis (Mellett, 1992). The recorded data represent the condition along the length of the wall for the width of the antenna. Multiple passes are required to obtain data for widths greater than the width of the antenna.

The data can then be interpreted to evaluate the location and depth of reinforcing steel, the thickness of the wall, and the location of delaminations or voids.

Personnel Qualifications

Use of the penetrating radar equipment should be limited to those with the extensive training required to correctly interpret the results (ACI, 1994).

Reporting Requirements

The personnel conducting the tests should provide sketches of the wall indicating the location of the tests and the findings. The sketch should include the following information:

- Mark the location of the test on either a floor plan or wall elevation.

- Report the results of the test using either the actual recorded data with the interpretations marked on the printed data or the interpreted results only.

- List the type of radar equipment used, including the type of antenna.

- Report the date of the test.

- List the engineer responsible for interpreting the test results and the name of the company conducting the test.

TEST AND INVESTIGATION GUIDE
Continued

NDE 9

Limitations

Although penetrating radar units are commercially available, very few units are in service for use with concrete and masonry structures. For example, less than five units are in use in California. Penetrating radar has been used primarily on slabs-on-grade for detecting subsurface conditions. Some work has been done to apply the method to concrete columns (Delgado and Heald, 1996) and to unreinforced masonry buildings.

A high-frequency antenna provides high resolution, but has shallow penetration, whereas deeper penetration with reduced resolution can be achieved with lower-frequency antennae (Krauss, 1994). Radar cannot effectively detect small differences in materials because the effective resolution is typically one-half of the wavelength (Candor, 1984).

Although penetrating radar is useful for locating the spacing and depth of reinforcing bars, it is not possible to determine the size of the bars. Closely-spaced bars can make it difficult to discern bar locations and depths. Close spacing make it difficult to detect features below the layer of reinforcing steel (ACI, 1997). Large metallic objects, such as embedded steel members cannot be clearly identified because of the scattering of the electromagnetic pulse.

References

ACI Committee 228, 1997, *Nondestructive Tests Methods for Evaluation of Concrete in Structures*, ACI 228.2R - Draft, American Concrete Institute, Detroit, Michigan.

ACI Committee 364, 1994, Guide for Evaluation of Concrete Structures Prior to Rehabilitation, ACI 364.1R, *ACI Manual of Concrete Practice*, American Concrete Institute, Detroit, Michigan.

Candor, T. R., 1984, "Review of Penetrating Radar as Applied to Nondestructive Evaluation of Concrete", *In Situ/Nondestructive Testing of Concrete*, ACI SP-82, American Concrete Institute, Detroit, Michigan, pp 581-601.

Delgado, M., and S.R. Heald, 1996, "Post-Earthquake Damage Assessed Nondestructively", *Materials Evaluation*, pp 378-382.

Krauss, P.D., 1994, *Repair Materials and Techniques for Concrete Structures in Nuclear Power Plants*, NRC JNC No. B8045, US Nuclear Regulatory Commission, Washington, DC.

Mellet, J, 1992, "Seeing Through Solid Materials", *Building Renovation*, Penton Publishing.

TEST AND INVESTIGATION GUIDE	Test Type:	**Intrusive**
SELECTIVE REMOVAL	Materials:	**Concrete, Reinforced Masonry, Unreinforced Masonry**

IT 1

Description

When information regarding the construction of portions of the concrete or masonry cannot be obtained using nondestructive techniques, selective removal of portions of the wall is sometimes required to allow direct observation of the condition of the reinforcing bars or interior portion of the concrete or masonry. Removal is suggested only when visual observations of the surface indicate that the wall may have hidden damage such as buckled rebar, or when it is necessary to determine the construction of the wall.

Equipment

Light chipping tools, small diameter core drills, or masonry saws are used for creating openings in the wall.

A fiber-optic borescope can be used to view interior spaces through small openings.

Execution

Portions of the wall are removed by chipping, drilling, or sawing to a specified depth of the wall. The inner construction and condition of the wall are then observed visually. A small mirror and flashlight can be used to better view spaces that are difficult to examine, eliminating the need to remove extensive portions of the wall.

When small holes are used, a fiber-optic borescope can be used to view the interior construction and to look for evidence of damage or deterioration (ACI, 1994). Some borescopes can be fitted with a camera to produce photographic documentation of the observations. Some models also have flexible shafts and pivoting viewing heads to allow for multidirectional viewing.

Following the observations, the intrusive openings should be patched with appropriate material.

Personnel Qualifications

Engineers performing selective removal should be experienced with the equipment being used. The engineer should also be familiar with the drawings or other available documentation on the building to understand the expected results of the intrusive observation.

Reporting Requirements

The personnel conducting the tests should provide sketches of the wall indicating the location of the intrusive openings and the findings. The sketch should include the following information:

- Mark the location of the test on either a floor plan or wall elevation.

- Describe the size and type of opening.

- Specify the maximum strength of the core obtained during the test, in terms of force and in pressure.

- Describe the results of the test by a written description and a sketch or photograph.

- Report the date of the test.

- List the responsible engineer overseeing the test and the name of the company conducting the test.

TEST AND INVESTIGATION GUIDE
continued

$\boxed{\text{IT 1}}$

Limitations

If the findings of the intrusive opening are substantially different from what was expected, the engineer should review all of the available information before proceeding.

The use of selected intrusive openings is often performed in conjunction with nondestructive testing procedures. For example, a rebar detector can be used to establish the location of reinforcing bars nonintrusively. Selected locations of reinforcement can then be chipped out to determine the size of the bar and/ or depth of the cover. These data are then used to calibrate the rebar detector.

The information gained from observations at selected locations is only applicable to the surveyed areas. Construction with similar appearance and condition may be different. The amount of variability will depend on several factors, including the era of construction, the amount of gravity and lateral load on the wall, and normal variations in construction quality. The locations of the intrusive openings should be carefully chosen to include sufficient typical and atypical areas so that the engineer has confidence that the data gathered represent most of the walls in the building.

Conversely, it is seldom cost effective or necessary to intrusively test every wall in a building. Intrusive openings may damage reinforcing bars or other hidden structural elements. Large intrusive openings can weaken the wall.

References

ACI Committee 364, 1994, "Evaluation of Structures Prior to Rehabilitation", ACI 364.1R, *ACI Manual of Concrete Practice*, American Concrete Institute, Detroit, Michigan.

TEST AND INVESTIGATION GUIDE	Test Type:	**Intrusive**
IT 2 **PETROGRAPHY**	Materials:	**Concrete, Reinforced Masonry, Unreinforced Masonry**

Description

A petrographic evaluation is a microscopic evaluation of the concrete or masonry material. A sample of the material is removed and sent to a laboratory where the sample is prepared and studied using a high-powered microscope. Petrographic examination can also be used to determine the cause of cracking and the approximate mix design of the concrete or mortar.

Equipment

*   Typical equipment includes:

*   Core drill or other tools for removing concrete or masonry

*   Laboratory equipment including concrete saws for sectioning, grinding wheels for polishing, and stereo microscopes

Execution

Remove samples of the concrete or masonry material from the building using core drilling equipment or other concrete removal tools. The samples are then sent to a laboratory where they are cut, polished and examined under a microscope in accordance with ASTM C 856 (ASTM, 1991) procedures.

The condition of the concrete or masonry located along the edge or within a crack can often be used to determine if the crack formed recently and thus may be earthquake related. Some of the methods used to assess the age of cracks are:

*   Weathering along cracks. Since the edges of the material on either side of the crack tend to become rounded over time due to normal weathering, it may be possible to estimate whether a crack is "relatively young" or "relatively old" by estimating the amount of weathering.

*   Secondary deposits. Secondary deposits within a crack such as mortar, paint, epoxy, or spackling compound indicate that the crack formed before the installation of the material contained within it.

*   Interpretation of carbonation patterns. Carbonation of calcium hydroxide contained in hydrated cement paste is inevitable and typically begins along formed or cracked surfaces. Carbonation penetrates into the cementitous material in a direction perpendicular to the plane of the formed or cracked surface. If an estimate of the carbonation rate can be made, then studies of the pattern of cementitious matrix carbonation adjacent to a crack can be used to estimate the age of the crack.

Petrographic studies can also establish the approximate composition of concrete and mortar. This information can be used to establish an approximate material strength. The material composition is needed to formulate or specify compatible materials that will be used for repairs and modifications.

Personnel Qualifications

Much of the results of a petrographic examination are subject to the personal judgment of the petrographer (Krauss, 1994). Therefore, the petrographer must have extensive experience in the evaluation of the materials being tested.

TEST AND INVESTIGATION GUIDE
continued

$$\boxed{\text{IT 2}}$$

Reporting Requirements

A variety of information is available through petrography. The personnel conducting the tests on the material samples should provide a written report of the findings to the evaluating engineer. The results should contain, at a minimum, the following information for each sample:

- Identify the sample using the description of location or sample number provided by the engineer.

- Specify the length and diameter of the core and the cross-sectional area.

- Describe the tests performed on the sample, along with the appropriate references.

- Describe the results of the examination.

- Report the date the sample was taken and the date of the test.

- List the responsible engineer overseeing the test and the name of the company conducting the test.

Limitations

Although petrographic analysis can reveal considerable information regarding the composition of materials and the cause of damage, the results are subjective.

The exact cause and age of cracks may not be discernable. Cracks can have several causes (ACI, 1994). Walls that have been protected with finishes are not subject to the typical surface deterioration that would allow comparison of crack faces for assessing relative age.

References

ACI Committee 201, 1994, "Guide for Making a Condition Survey of Concrete in Service", ACI 201.1R-92, *ACI Manual of Concrete Practice*, American Concrete Institute, Detroit, Michigan.

ASTM, 1991, *Standard Practice for Petrographic Examination of Hardened Concrete*, ASTM C 856-83, *1991 ASTM Annual Book of Standards*, American Society for Testing and Materials, Philadelphia, PA, pp 416-428.

Krauss, P.D., 1994, *Repair Materials and Techniques for Concrete Structures in Nuclear Power Plants*, NRC JNC No. B8045, US Nuclear Regulatory Commission, Washington, DC.

	TEST AND INVESTIGATION GUIDE	Test Type:	Intrusive
IT 3	MATERIAL EXTRACTION AND TESTING	Materials:	Concrete, Reinforced Masonry, Unreinforced Masonry

Description

Material testing requires removal of a sample of the material, which can be either the reinforcing steel, concrete, or concrete masonry. The removed samples are then tested to determine the tensile or compressive strength for the steel or concrete, respectively.

Equipment

Concrete cores should be taken with diamond-studded core bits.

Reinforcing steel should be extracted with a reciprocating saw or torch.

Execution

Concrete testing requires removal of core samples with a diamond-tipped drill bit. Typical cores are three to six inches in diameter. For compression testing of the concrete, the length of the cores should be at least two times the diameter (ACI, 1994a). The cores should be taken through sections of walls that have no significant cracking. The core should be taken through the thickness of the wall and should avoid reinforcing steel. The cores should then be prepared and tested in accordance with ASTM C42 procedures (ASTM, 1991a).

Rebar testing requires removal of concrete surrounding a length of reinforcing bar. The length of the sample required is dependent of the size of the bar. The rebar is removed. The removed sample is then subjected to tensile testing. It can also be subjected to metallurgical examination to assess the weldability of the steel. The reinforcing steel sample should be prepared and tested in accordance with ASTM A 370 (ASTM, 1991b).

Following removal of either concrete or rebar samples, the openings should be patched.

Personnel Qualifications

Material samples should be obtained by an experienced contractor. Most areas have contractors that specialize in concrete coring and sawing. The contractor should be familiar with the use of the equipment. An engineer should be responsible for specifying the locations of the sampling. Testing of the samples should be accomplished by a qualified laboratory under the direction of a licensed engineer.

Reporting Requirements

The personnel conducting the tests on the material samples should provide a written report of the findings to the evaluating engineer. The results for the concrete core tests should contain, at a minimum, the following information for each sample:

- Identify the sample using the description of location or sample number provided by the engineer.

- Specify the length and diameter of the core, and cross-sectional area.

- Report the maximum strength of the core obtained during the test, in terms of force and stress.

- Specify the correction factor applied to the results due to the ratio of the length to the diameter, and report the corrected results.

- Report the date the sample was taken and the date of the test.

- List the responsible engineer overseeing the test and the name of the company conducting the test.

TEST AND INVESTIGATION GUIDE
continued

| IT 3 |

Testing of reinforcing steel should include the following information:

- Identify the sample using the description of location or sample number provided by the engineer.

- Report the length and diameter (or size) of the bar.

- Report the yield and ultimate strength of the core or reinforcing bar obtained during the test, in terms of force and stress.

- Plot the force-elongation data in stress-versus-strain units.

- Report the date the sample was taken and the date of the test.

- List the responsible engineer overseeing the test and the name of the company conducting the test.

Limitations

Extraction of samples causes damage to the wall, and repairs to those areas may be required (ACI, 1994b). Therefore, samples should be removed from areas of low expected demands.

The values obtained from the concrete core tests should not be expected to exactly equal the anticipated design strength values. Core test values can have an average value of 85 percent of the specified strength (ACI, 1995). The core strength values should be adjusted using the procedure described in FEMA 274 (ATC, 1997b) to obtain the in-place strength of the concrete. If the results indicate high variability in values, additional samples should be taken and tested to reduce the coefficient of variation. However, the cost of the additional tests should be considered against the benefit of the increased precision of the results.

References

American Concrete Institute, 1994a, Strength Evaluation of Existing Concrete Buildings, ACI 437R-91, *ACI Manual of Concrete Practice*, Part 3, Detroit, Michigan.

American Concrete Institute, 1994b, Evaluation of Structures Prior to Rehabilitation, ACI 364.1R, *ACI Manual of Concrete Practice*, Part 3, Detroit, Michigan.

American Concrete Institute, 1995, *Building Code Requirements for Structural Concrete (ACI 318-95) and Commentary (ACI 318R-95)*, Detroit, Michigan, pg 52.

American Society for Testing and Materials, 1991a, *Standard Test Method of Obtaining and Testing Drilled Cores and Sawed Beams of Concrete*, ASTM C 42-90, 1991 ASTM Standards in Building Codes, Philadelphia, Pennsylvania, pp 27-30.

American Society for Testing and Materials, 1991b, *Standard Test Methods for Mechanical Testing of Steel Products*, ASTM A 370-92, 1991 ASTM Standards in Building Codes, Volume 1, Philadelphia, Pennsylvania, pp 484-529.

ATC, 1997a, *NEHRP Guidelines for the Seismic Rehabilitation of Buildings*, FEMA 273, prepared by the Applied Technology Council for the Building Seismic Safety Council, published by the Federal Emergency Management Agency, Washington, DC.

ATC, 1997b, *NEHRP Commentary on the Guidelines for the Seismic Rehabilitation of Buildings*, FEMA 274, prepared by the Applied Technology Council for the Building Seismic Safety Council, published by the Federal Emergency Management Agency, Washington, DC.

TEST AND INVESTIGATION GUIDE	Test Type: **Intrusive**
IT 4 **IN SITU TESTING - IN PLACE SHEAR**	Materials: **Unreinforced Masonry**

Description

The shear strength of unreinforced masonry construction depends largely on the strength of the mortar used in the wall. An in-place shear test is the preferred method for determining the strength of existing mortar. The results of these tests are used to determine the shear strength of the wall.

Equipment

- Chisels and grinders are needed to remove the bricks and mortar adjacent to the test area.

- A hydraulic ram, calibrated and capable of displaying the applied load.

- A dial gauge, calibrated to 0.001 inch.

Execution

Prepare the test location by removing the brick, including the mortar, on one side of the brick to be tested. The head joint on the opposite side of the brick to be tested is also removed. Care must be exercised so that the mortar joint above or below the brick to be tested is not damaged.

The hydraulic ram is inserted in the space where the brick was removed. A steel loading block is placed between the ram and the brick to be tested so that the ram will distribute its load over the end face of the brick. The dial gauge can also be inserted in the space.

The brick is then loaded with the ram until the first indication of cracking or movement of the brick. The ram force and associated deflection on the dial gage are recorded to develop a force-deflection plot on which the first cracking or movement should be indicated. A dial gauge can be used to calculate a rough estimate of shear stiffness (Eilbeck et al., 1996).

Inspect the collar joint and estimate the percentage of the collar joint that was effective in resisting the force from the ram. The brick that was removed should then be replaced and the joints repointed.

Personnel Qualifications

The technician conducting this test should have previous experience with the technique and should be familiar with the operation of the equipment. Having a second technician at the site is useful for recording the data and watching for the first indication of cracking or movement. The structural engineer or designee should choose test locations that provide a representative sampling of conditions.

TEST AND INVESTIGATION GUIDE
continued

IT 4

Reporting Results

The personnel conducting the tests should provide a written report of the findings to the evaluating engineer. The results for the in-place shear tests should contain, at a minimum, the following information for each test location:

- Describe test location or give the identification number provided by the engineer.

- Specify the length and width of the brick that was tested, and its cross-sectional area.

- Give the maximum mortar strength value measured during the test, in terms of force and stress.

- Estimate the effective area of the bond between the brick and the grout at the collar joint.

- Record the deflection of the brick at the point of peak applied force.

- Record the date of the test.

- List the responsible engineer overseeing the test and the name of the company conducting the test.

Limitations

This test procedure is only capable of measuring the shear strength of the mortar in the outer wythe of a multi-wythe wall. The engineer should verify that the exterior wythe being tested is a part of the structural wall, by checking for the presence of header courses. This test should not be conducted on veneer wythes.

Test values from exterior wythes may produce lower values when compared with tests conducted on inner wythes. The difference can be due to weathering of the mortar on the exterior wythes. The exterior brick may also have a reduced depth of mortar for aesthetic purposes.

The test results can only be qualitatively adjusted to account for the presence of mortar in the collar joints. If mortar is present in the collar joint, the engineer or technician conducting the test is not able to discern how much of that mortar actually resisted the force from the ram.

The personnel conducting the tests must carefully watch the brick during the test to accurately determine the ram force at which first cracking or movement occurs. First cracking or movement indicates the maximum force, and thus the maximum shear strength. If this peak is missed, the values obtained will be based only on the sliding friction contribution of the mortar, which will be less than the bond strength contribution.

References

Eilbeck, D.E, J.D. Lesak, and J.N. Chiropolos, 1996, "Seismic Considerations for Repair of Terra Cotta Cladding", *Proceedings, Seventh North American Masonry Conference,* June, pp 847-858.

ICBO, 1994, In-Place Masonry Shear Tests, UBC Standard 21-6, International Conference of Building Officials, Whittier, California, pg 3-614.

TEST AND INVESTIGATION GUIDE	Test Type: **Intrusive**
IT 5 IN SITU TESTING - FLAT JACK	Materials: **Unreinforced Masonry**

Description

Flat jacks are thin hydraulic jacks that are inserted into the mortar joints of masonry walls. Flat jacks can be used to measure the state of stress of a masonry wall, the modulus of elasticity of the masonry, and the compressive strength of the masonry.

Equipment

- One or two flat jacks, 1/4- to 3/8-inch thick, with a hydraulic pump and pressure gauge

- Measuring points that are secured to the masonry wall

- Dial calipers or other instruments for measuring the distance between points to within 0.001 inch

- Chipping tools or masonry saws to remove mortar

Execution

A single flat jack is used to determine the state of stress in the masonry. A set of measuring points are attached with epoxy above and below the section of masonry to be tested. The distance between the points is measured and then one horizontal mortar joint is removed with chipping tools or saws. A flat jack is inserted in the mortar joint and pressurized to fill the void. The pressure is increased incrementally while measuring the distance between the measuring points. When the distance between the points returns to the original value, the pressure is recorded and then converted into compressive stress (σ) using the following equation (Rossi, 1987):

$$\sigma = p\,K_m\,K_a$$

Where:

p is the gauge pressure

K_m is a jack constant determined by laboratory calibration

K_a is the ratio of the surface area of the jack to the surface area of mortar removed

A "double flat jack test" is used to evaluate the compressive modulus of elasticity and the compressive strength. A section of the masonry construction to be tested is isolated by cutting two parallel sections of mortar joint. The joints should be separated by approximately the length of the flat jack, typically about 14 inches. Prior to cutting, install measuring points within the section of masonry to be tested. Once the test section has been prepared, insert the flat jacks into the mortar joints and pressurize them to fill the voids. Apply loads to both jacks equally in increments, and measure the distance between the points. If the compressive strength is to be determined, the pressure should be increased until cracking is observed (Kingsley and Noland, 1987a). The pressure and deflection values are then converted into a stress-strain plot for the masonry.

Following the tests, the mortar should be replaced.

Personnel Qualifications

The engineers or technicians conducting the tests should be thoroughly familiar with the use of the equipment and should have experience conducting similar tests.

TEST AND INVESTIGATION GUIDE
continued

IT 5

Reporting Requirements

The personnel conducting the tests should provide a written report of the findings to the evaluating engineer. The report for the flat jack tests should contain, at a minimum, the following information for each test location:

• Describe the test location or use the sample number provided by the engineer.

• For a single flat jack test, report the stress state. For a double flat jack test, report the maximum value for masonry strength, that was measured in the test, in terms of force and pressure.

• Provide the load-deflection curve obtained during the test.

• Report the value of K_m determined by calibration tests.

• Report the date of the test.

• List the responsible licensed engineer overseeing the test and the name of the company conducting the test.

Flat jack tests can be expensive and are prone to several problems.

Limitations

Flat jack tests should be performed in areas that are undamaged. Care should be exercised when removing the mortar for the tests to avoid damaging the mortar or masonry in the test area.

The measuring points must be securely fastened to the masonry units to avoid being dislodged during the testing. The gauge length between the measuring points should be as large as possible so that small changes in movement are easier to detect. The measuring points should be located at the midpoint of the joint and the jack so that the maximum deflections are measured.

If the wall is composed of masonry wythes of differing stiffness (for example, a terra-cotta veneer with brick backup), applying a uniform compressive load using the flat jack may cause out-of-plane bending of the wall. If a flat jack test of the entire wall thickness is performed, measuring points can be attached to both faces of the wall.

Often a flat jack test is performed only on the outer wythe. However, header bricks and mortar in the collar joints may prevent accurate results by restraining the outer wythe. If the masonry units have uneven surfaces, the flat jack will deform into the voids. This causes the pressure to be nonuniformly distributed. An uneven surface also makes it difficult to remove the flat jack from the mortar joint. Also, flat jacks may not have the capacity to fail the masonry in all cases.

References

Kingsley, G.R. and J.L. Noland, 1987a, A Note on Obtaining In-Situ Load-Deformation Properties of Unreinforced Brick Masonry in the United States Using Flatjacks, *Evaluation and Retrofit of Masonry Structures*, Proceedings of the Second Joint USA-Italy Workshop on Evaluation and Retrofit of Masonry Structures, pp 215-223.

Kingsley, G.R. and J.L. Noland 1987b, An Overview of Nondestructive Techniques For Structural Properties of Brick Masonry, *Evaluation and Retrofit of Masonry Structures*, Proceedings of the Second Joint USA-Italy Workshop on Evaluation and Retrofit of Masonry Structures, pp 225-237.

Rossi, P.P., 1987, Recent Developments of the Flat-Jack Test on Masonry Structures, *Evaluation and Retrofit of Masonry Structures*, Proceedings of the Second Joint USA-Italy Workshop on Evaluation and Retrofit of Masonry Structures, pp 257-285.

4. Evaluation of Earthquake Damage

4.1 Basis of Evaluation

The quantitative evaluation of the effects of earthquake damage on structures requires the selection of a measurement parameter. Procedures in this document use change in the anticipated performance of the building during future earthquakes as the measurement parameter. This is the change due directly to effects of earthquake damage on the basic structural properties that control seismic performance. If the structural property changes are estimated, the corresponding change in future performance can also be estimated. The total cost to restore the anticipated performance to approximately that of the building before the damaging earthquake quantifies the effects of the observed damage. These hypothetical "repairs" to structural components are referred to as performance restoration measures.

4.2 Seismic Performance Objectives

The damage evaluation procedures in this document are performance-based; that is, they assess the acceptability of the structural system (and the significance of changes in the structural system) on the basis of the degree to which the structure achieves one or more performance levels for the hazard posed by one or more hypothetical future earthquakes. A performance level typically is defined by a particular damage state for a building. The performance levels defined in FEMA 273, in order of decreasing amounts of damage, are collapse prevention, life safety, and immediate occupancy. Hazards associated with future hypothetical earthquakes are usually defined in terms of ground shaking intensity with a certain likelihood of being exceeded over a defined time period or in terms of a characteristic earthquake likely to occur on a given fault. The combination of a performance level and a hazard defines a performance objective. For example, a common performance objective for a building is that it maintain life safety when subjected to ground motion with a ten-percent chance of exceedance in fifty years.

The damage evaluation begins with the selection of an appropriate performance objective. The performance objective serves as a benchmark for measuring the difference between the anticipated performance of the building in its damaged and pre-event states, that is, relative performance analysis. The absolute performance acceptability of the damaged or pre-event building does not affect the quantification of loss. The quantification of performance loss is affected by the choice of performance objective, as illustrated in the following paragraph. Consequently, the selection of objectives is a matter of policy that depends on the occupancy and use of the facility. Guidance may be found in ATC-40, FEMA 273/274, and FEMA 308.

It is important to note that the damage evaluation procedure can be used to investigate changes in performance characteristics for either single or multiple performance objectives. For example, a hospital might be expected to remain functional (immediate occupancy) after a rare event. For a very rare event, the life safety performance level might be acceptable. The damage evaluation procedure may be used with either or both performance objectives, and the loss associated with the damage may be different for the two objectives. The example hospital might have suffered a $1,000,000 loss, based on the cost of restoration measures, with respect to its ability to remain functional after a rare event. The same level of damage might not have resulted in any loss in its ability to preserve life safety in the very rare event. In summary, the effects of damage can depend greatly on the chosen performance objective.

4.3 Seismic Performance Parameters

Recent research and development activities have resulted in the introduction of structural analysis methodologies based on the inelastic behavior of structures (FEMA 273/274, ATC-40). These techniques generate a plot, called a capacity curve, that relates a global displacement parameter (at the roof level, for example) to the lateral force imposed on the structure. The magnitude of the maximum global displacement to occur during an earthquake depends on elastic and inelastic deformations of the individual components of the structure and their combination into the system response.

For a given global displacement of a structure subject to a given lateral load pattern, there is an associated deformation of each structural component of the building. Since inelastic deformation indicates component damage, the maximum global displacement to occur during an earthquake defines a structural damage state for the building in terms of inelastic deformations for each of its components. The capacity of the structure is represented by the maximum global displacement, d_c, at which the component damage is on the verge of exceeding the tolerable limit for a specific performance level. For example, the collapse prevention capacity of a building might be the roof displacement just short of that at which the associated damage would result in collapse of one or more of the column components. Displacement limits for components are tabulated in FEMA 273 and ATC-40.

The analysis methodologies also include techniques to estimate the maximum global displacement demand, d_d, for a specific earthquake ground motion. The ratio of the displacement capacity, d_c, of the building for a specific performance level to the displacement demand, d_d, for a specific hazard is a measure of the degree to which the building meets the performance objective. If the ratio is less than 1.0 the performance objective is not met. If it is equal to one the objective is just met. If it is greater than 1.0, performance exceeds the objective.

4.4 Relative Performance Analysis

The results of the damage investigation include two related categories of information on the structural damage consequences of the earthquake on the building. First, they comprise a compilation of the physical effects on all of the structural components. These typically consist of cracks in concrete or masonry, spalling or crushing of concrete or masonry, and fracture or buckling of reinforcement. Second, the damage is classified according to component type, behavior mode, and severity. Using these data it is possible for the engineer to quantify the changes attributable to the damage with respect to basic structural properties of the components of the building. These properties include stiffness, strength, and deformation limits.

Damage caused by an earthquake can affect the ability of a structure to meet performance objectives for future earthquakes in two fundamental ways. First, the damage may cause the displacement demand for the future, d_d', event to differ from that for the pre-event structure, d_d. This is due to changes in the global stiffness, strength, and damping of the structure, which in turn affect the

maximum dynamic response of the structure by changing its global stiffness, strength, and damping. Also, the displacement capacity of the damaged structure, d_c', may differ from that of the pre-event structure, d_c. Damage to the structural components can change the magnitude of acceptable deformation for a component in future earthquakes.

The analysis procedure described in the following sections uses the change in the ability of the damaged building to meet performance objectives in future earthquakes to measure the effects of the damage. The same basic analysis procedure is also used to formulate performance restoration measures that quantify the loss of seismic performance.

4.4.1 Overview

This section summarizes the basic steps of a seismic relative performance analysis for concrete and masonry wall buildings. This is a quantitative procedure that uses nonlinear static techniques to estimate the performance of the building in future events in both its pre-event and damaged states. The procedure is also used to investigate the effectiveness of potential performance restoration measures. This procedure requires the selection of one or more performance objectives for the building as discussed in Section 4.2. The analysis compares the degree to which the pre-event and damaged buildings meet the specified objective. Figure 4-1 illustrates a generalized relationship between lateral seismic forces (base shear or spectral acceleration) and global structural displacements (roof or spectral displacement).

This plot of structural capacity is characteristic of nonlinear static procedures (FEMA 273/274, ATC-40). A point on the curve defines a specific damage state for the structure, since the deformation of all of its components can be related to the global displacement of the structure. Figure 4-2 illustrates the basic idealization of force-deformation characteristics for individual components.

The nonlinear static procedures estimate the maximum global displacement of a stucture to shaking at its base. These procedures are easier to implement and interpret than nonlinear dynamic time history analyses, but they are relatively new and subject to further development. In their present form, they have limitations (Krawinkler, 1996), particularly for buildings that tend to respond in their higher modes of vibration. This limitation,

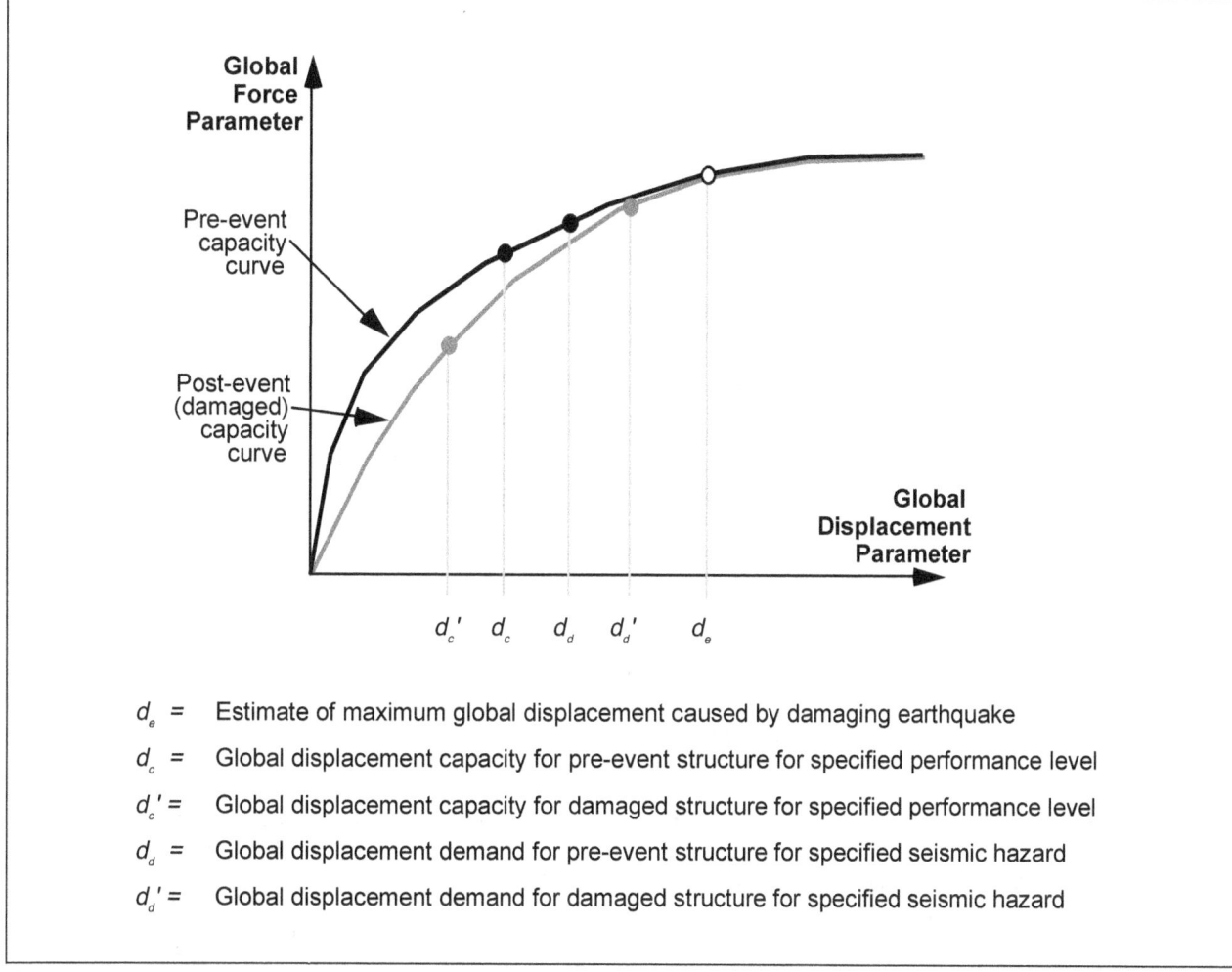

d_e = Estimate of maximum global displacement caused by damaging earthquake

d_c = Global displacement capacity for pre-event structure for specified performance level

d_c' = Global displacement capacity for damaged structure for specified performance level

d_d = Global displacement demand for pre-event structure for specified seismic hazard

d_d' = Global displacement demand for damaged structure for specified seismic hazard

Figure 4-1   ***Displacement Parameters for Damage Evaluation***

however, is relatively less restrictive for concrete and masonry wall buildings because of their tendency to repond in the fundamental mode. Future development of the procedures may also allow improved treatment for higher modes (Paret et al., 1996). Nonlinear static procedures must be carefully applied to buildings with flexible diaphragms.

The basic steps for using the procedure to measure the effect of damage caused by the damaging ground motion on future performance during the performance ground motion is outlined as follows:

1. Using the properties (strength, stiffness, energy dissipation) of all of the lateral-force-resisting components and elements of the pre-event structure, formulate a capacity curve relating global lateral force to global displacement.

2. Determine the global displacement limit, d_c, at which the pre-event structure would just reach the performance level specified for the performance objective under consideration.

3. For the specified performance ground motion, determine the hypothetical maximum displacement for the pre-event structure, d_d. The ratio of d_c to d_d indicates the degree to which the pre-event structure satisfies the specified performance objective.

4. Using the results of the investigation of the effects of the damaging ground motion, modify the component force-deformation relationships using the

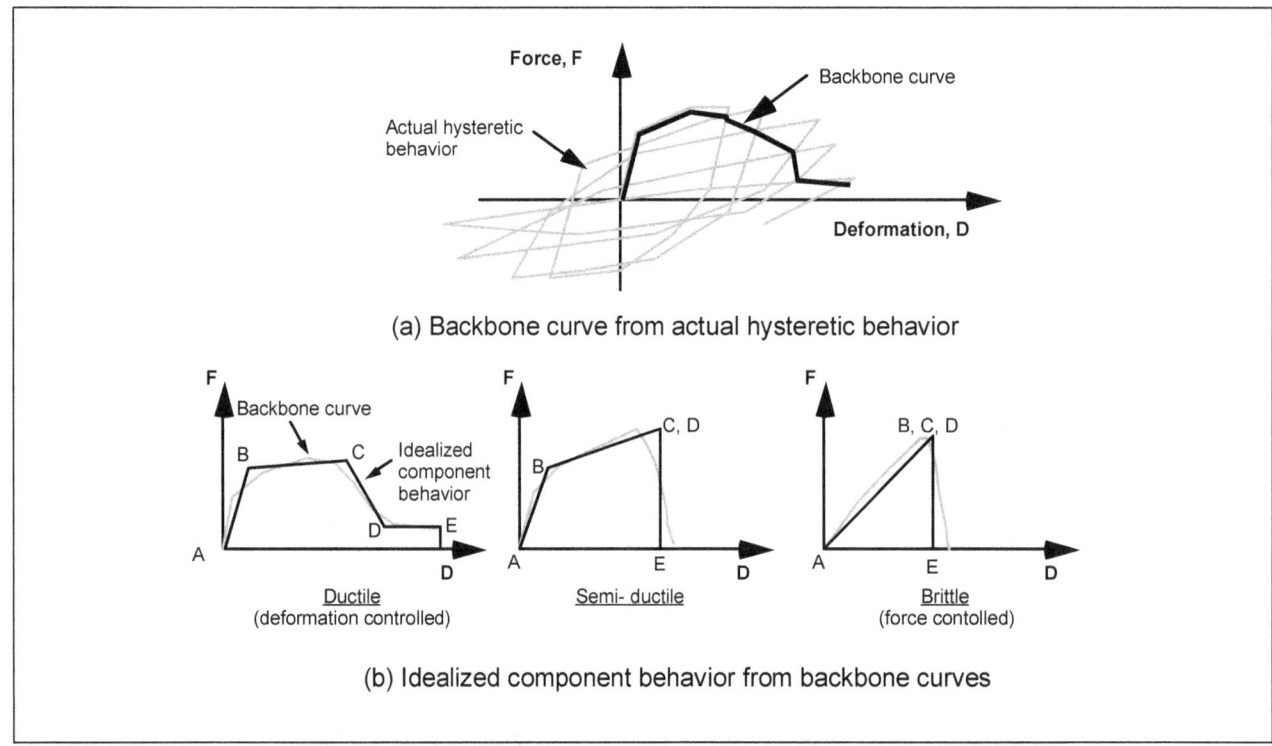

(a) Backbone curve from actual hysteretic behavior

(b) Idealized component behavior from backbone curves

Figure 4-2   *Idealized Component Force-Deformation Relationship*

Component Damage Classification Guides in Chapters 5 through 8. Using the revised component properties, reformulate the capacity curve for the damaged building and repeat steps 2 and 3 to determine d_c' and d_d'. The ratio of d_c' to d_d' indicates the degree to which the damaged structure satisfies the specified performance objective.

5. If the ratio of d_c' to d_d' is the same, or nearly the same, as the ratio of d_c to d_d, the damage caused by the damaging ground motion has not significantly degraded future performance for the performance objective under consideration.

6. If the ratio of d_c' to d_d' is less than the ratio of d_c to d_d, the effects of the damage caused by the damaging ground motion has diminished the future performance characteristics of the structure. Develop hypothetical actions in accordance with Section 4.5, to restore or augment element and component properties so that the ratio of d_c^* to d_d^* (where the * designates the restored condition) is the same, or nearly the same, as the ratio of d_c to d_d.

4.4.2 Global Displacement Performance Limits

The global displacement performance limits (d_c, d_c', d_c^*) are a function of the acceptability of the deformation of the individual components of the structure as it is subjected to appropriate vertical loads and to a monotonically increasing static lateral load distributed to each floor and roof level in an assumed pattern. The deformation of the components depends on both their geometric configuration in the model and their individual force-deformation chacteristics (see Section 2.4) compared to those of other components. The plot of the total lateral load parameter versus global displacement parameter represents the capacity curve for the building for the assumed load pattern. Thus, the capacity curve is characteristic of the global assembly of individual components and the assumed load pattern.

The current provisions of FEMA 273 limit global displacements for the performance level under consideration (e.g., Immediate Occupancy, Life Safety, Collapse Prevention) to that at which any single component reaches its acceptability limit (see

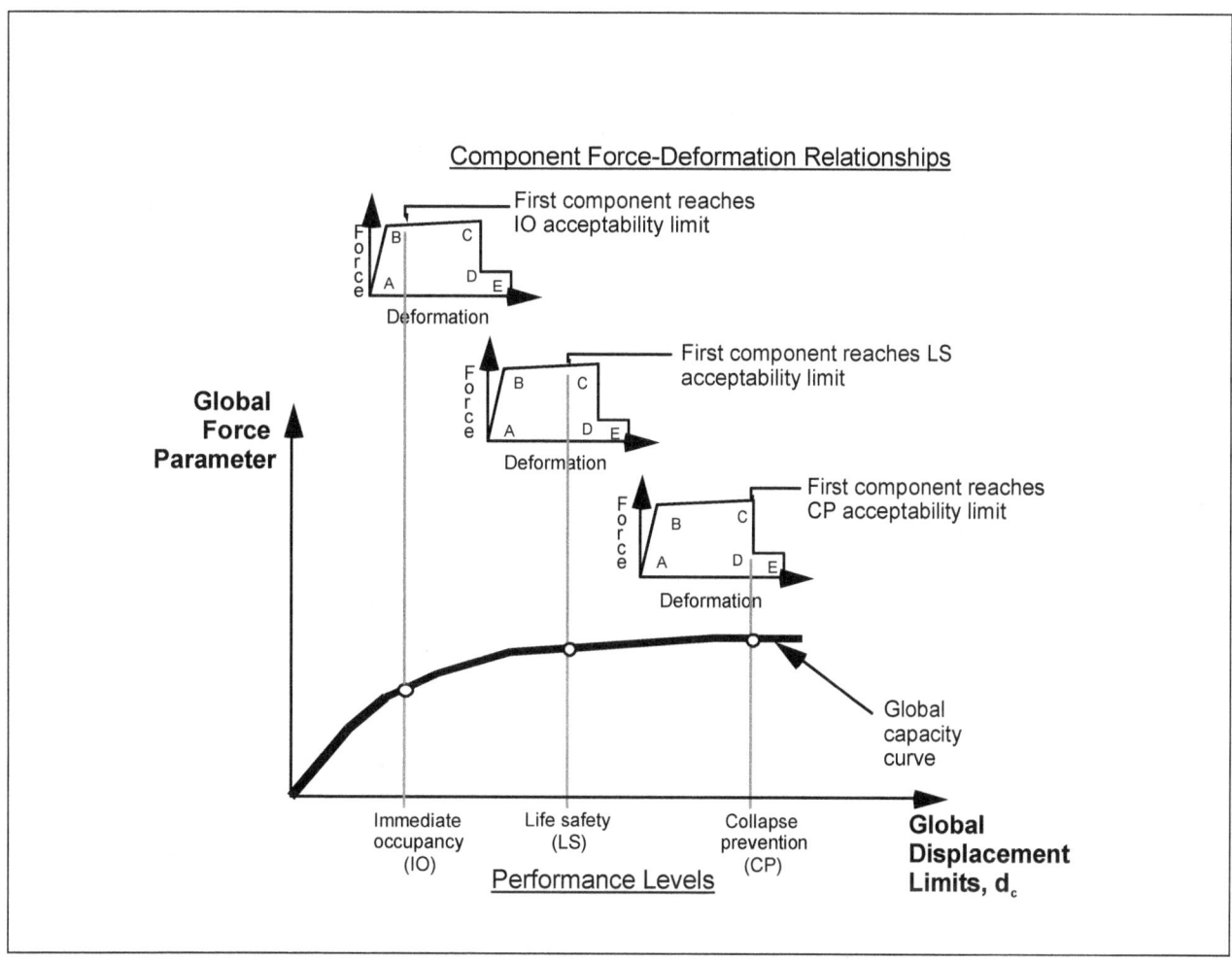

Figure 4-3   Global Displacement Limits and Component Acceptability used in FEMA 273/274

Figure 4-3). The provisions of FEMA 273/274 allow for the re-designation of such components as "secondary". Secondary components have higher deformation acceptability limits but the remaining primary lateral load resisting system components must be capable of meeting acceptability criteria without them. The same allowance may be made for relative performance analysis of earthquake damaged buildings as long as it is applied appropriately to both the pre-event and damaged models.

The acceptability limits were developed for FEMA 273 to identify and mitigate specific seismic deficiencies in buildings to improve anticipated performance. As such, they are intended to be conservative. In a relative performance analysis, the degree of conservatism should be same for both the pre-event and damaged models to give reliable results to estimate the scope of

restoration repairs. In an actual earthquake, some "unacceptable" component behavior may not result necessarily in unacceptable global performance. In the future, it is possible that alternative procedures for better estimating global displacement limits will emerge. These also may be suitable for relative performance analyses provided that they are applied consistently and appropriately to both the pre-event and the damaged models.

4.4.3   Component Modeling and Acceptability Criteria

4.4.3.1   Pre-Event Building

In determining the capacity curve for the pre-event building, component properties are generated using the procedures of FEMA 273/274 or ATC-40, modified, if necessary, to reflect the results of the damage

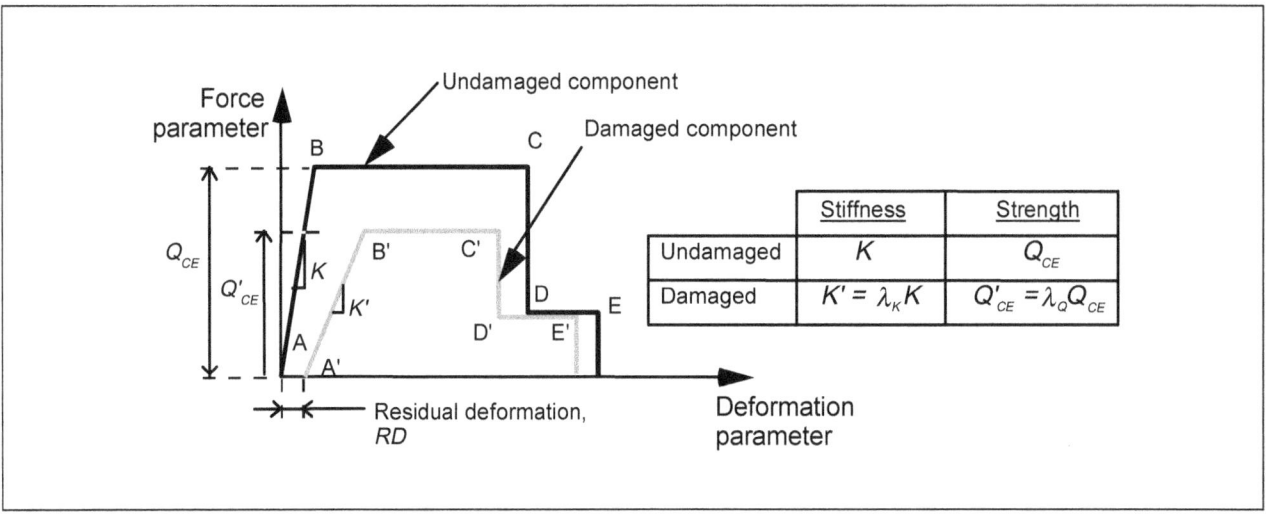

Figure 4-4 *Component Modeling Criteria*

investigation. Modifications may be warranted for two reasons:

1. The procedures assume a normal, relatively minor, degree of deterioration of the building due to service conditions. If the investigation reveals pre-existing conditions (see Section 3.4) that affect component properties beyond these normal conditions, then the "pre-event" component properties must be modified to reflect the condition of the structure just before the earthquake.

2. If the verification process (see Section 3.6) indicates component types or behavior modes inconsistent with the FEMA 273/274 or ATC-40 predicted properties, then the pre-event component properties are modified to reflect the observed conditions.

4.4.3.2 Damaged Building

The effects of damage on component behavior are modeled as shown generically in Figure 4-4. Acceptability criteria for components are illustrated in Figure 4-5. The factors used to modify component properties are defined as follows:

λ_K = modification factor for idealized component force-deformation curve accounting for change in effective initial stiffness resulting from earthquake damage.

λ_Q = modification factor for idealized component force-deformation curve accounting for

change in expected strength resulting from earthquake damage.

λ_D = modification factor applied to component deformation acceptability limits accounting for earthquake damage.

RD = absolute value of the residual deformation in a structural component, resulting from earthquake damage.

The values of the modification factors depend on the behavior mode and the severity of damage to the individual component. They are tabulated in the Component Guides in Chapters 5 through 8. The notation λ^* is used to denote modifications to pre-event properties for restored components. These also vary by behavior mode, damage severity, and type of restoration measure, in accordance with the recommendations of Chapters 5 through 8. Figure 4-6 illustrates the general relationship between damage severity and the modification factors. Component stiffness is most sensitive to damage, so this parameter must be modified even when damage is slight. Reduction in strength implies more significant damage. After relatively severe damage, the magnitudes of acceptable displacements are reduced.

4.4.3.3 Establishing λ Factors by Structural Testing

The component modification factors (λ factors) for an earthquake damaged building can be established by

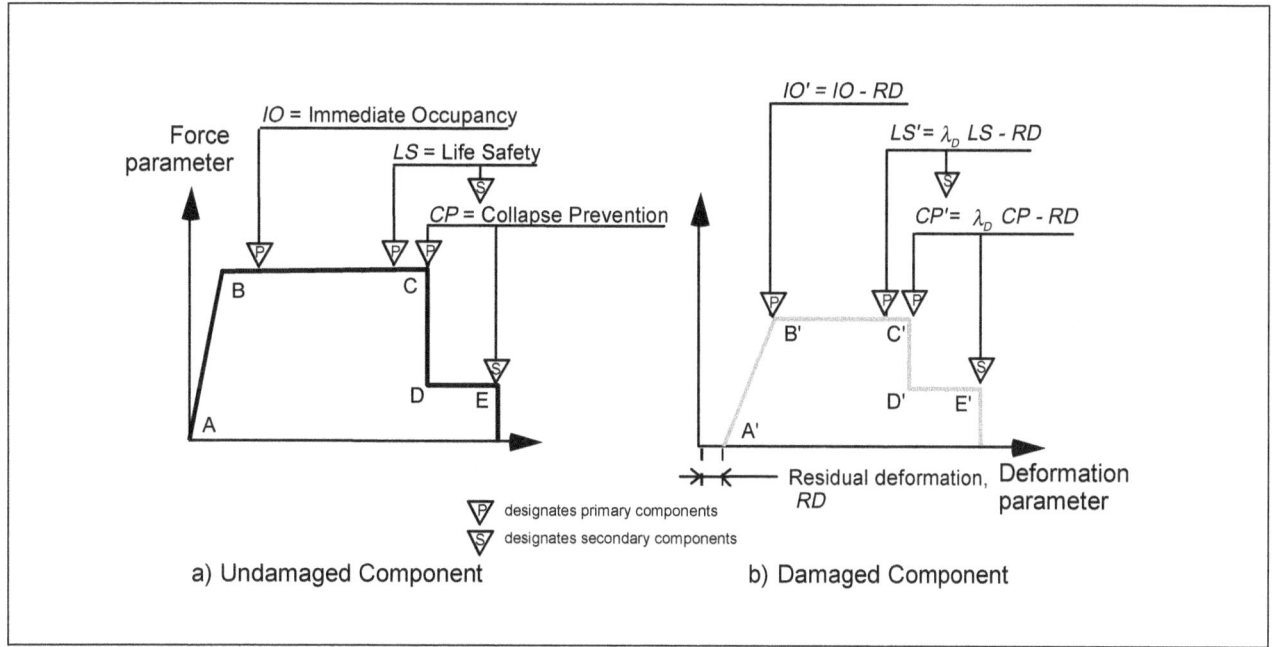

Figure 4-5   *Component Acceptability Criteria*

laboratory structural testing of critical components, rather than using the values given in the Component Guides of Chapters 5, 6, 7, and 8. The testing must be directly applicable to the specific structural details of the building, and to the damaging and performance earthquakes considered. Typically this would mean that a project-specific test program must be carried out. In certain circumstances, the expense of such a test program may be justified.

If testing is carried out to establish λ values, the test program should conform to the following guidelines:

A. Test Procedure

Two identical test specimens are required for each structural component of interest. One specimen is tested to represent the component in its *post-event* condition subjected to the performance earthquake; the second specimen is tested to represent the component in its *pre-event* condition subjected to the performance earthquake. The λ values are derived from the differences in the force-displacement response between the two specimens.

Figure 4-7 schematically illustrates the required testing. Specimen A is tested first by a load-displacement sequence representative of the damaging earthquake. After this testing, the damage in the specimen should be

similar in type and severity to that observed in the actual building after the damaging earthquake.. Specimen A is then tested by a load-displacement sequence representative of the performance level earthquake. From the resulting force-displacement hysteresis data, a backbone curve is drawn, according to Section 2.13.3 of FEMA 273.

A similar testing process is carried out on Specimen B, except that the initial test sequence representing the damaging earthquake is not applied. The λ values are derived from a direct comparison of the backbone curve of Specimen B with the backbone curve of Specimen A.

B. Test Specimens

Each pair of test specimens are to be identical in all details of construction, and in material strengths.

Scale. The scale of the components should be as near to full-scale as is practical. Generally, reinforced concrete specimens should not be tested below 1/4 to 1/3 scale.

Materials. Material strengths for the actual building should be established by testing, and test specimen materials shall be used that match the actual strengths as closely as possible. Material strengths should be identical between Specimen A and Specimen B.

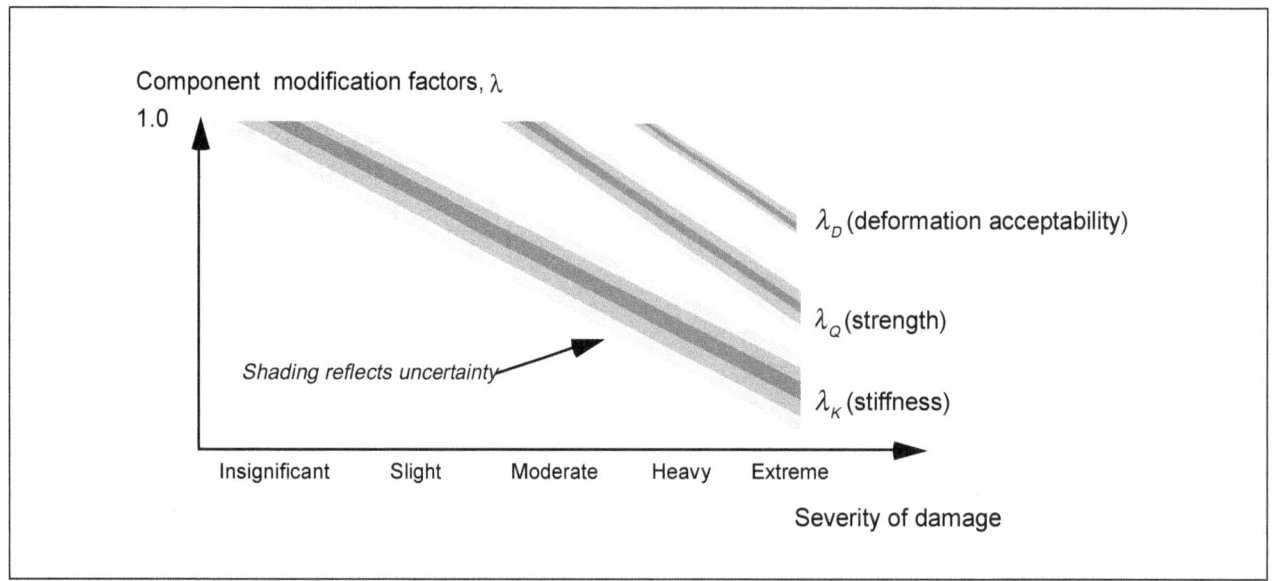

Figure 4-6   Component Modification Factors and Damage Severity

For reinforced concrete structures, it may be preferable to cast Specimens A and B at the same time from the same batch of concrete. Cylinders should be tested to establish the concrete strength at the time each specimen is tested.

Reinforcing steel used in the test specimens should be tested, and the yield strength should be close to that for the actual building. If it is not possible to match yield strength, the area of reinforcement can be adjusted to compensate for the difference in yield strength.

Pre-existing Damage. If the component of the actual building contains damage that is identified as pre-existing (e.g. cracks from shrinkage or a previous earthquake) then, to the extent possible, this damage should be induced in both Specimen A and Specimen B.

Number of Specimens. For simplicity of presentation, these guidelines refer to testing only one pair of specimens. For behavior modes or seismic response that shows substantial variability, more specimens would need to be tested so that results are based on a statistically significant sample.

C. Loading

Cyclic-static loading would typically be used for the testing. The test set-up and applied loading should be designed so that moment and shear diagrams and axial stress levels are representative of those occurring in the actual structure. Established test protocols, e.g. ATC (1992), should be followed to the extent applicable. A pseudo-dynamic loading sequence may be used, with consideration of the issues identified below.

Representing the Damaging Earthquake. To represent the damaging earthquake, the load sequence should recreate the displacement amplitudes and number of cycles undergone by the actual component in the damaging earthquake. Ground motion parameters for the damaging earthquake, from contour maps and recording stations near to the building site, should be reviewed for preliminary estimates of these parameters. The final determination of appropriate displacement amplitudes should be made during testing, so that the loading produces damage that is similar in type and severity to that observed in the actual building.

If a pseudo-dynamic loading sequence is used, it must be designed to allow adjustments during testing to better represent the actual input level of the damaging earthquake.

Representing the Performance Earthquake. To represent the performance earthquake, the load sequence should reflect the demands associated with the selected seismic hazard level. The load sequence should also contain enough cycles at different displacement levels to allow the construction of the backbone curve per Section 2.13.3 of FEMA 273.

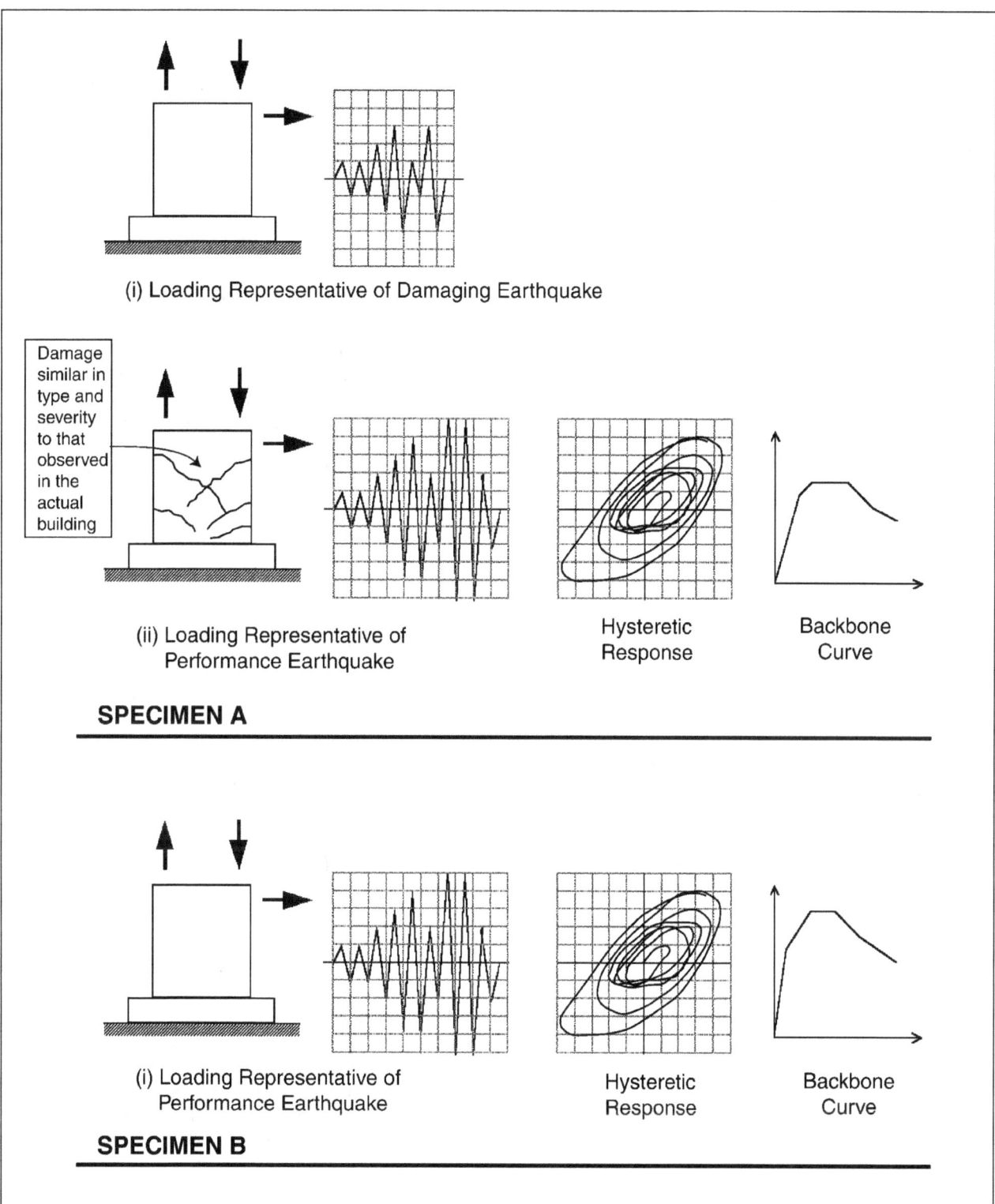

(i) Loading Representative of Damaging Earthquake

Damage similar in type and severity to that observed in the actual building

(ii) Loading Representative of Performance Earthquake

Hysteretic Response

Backbone Curve

SPECIMEN A

(i) Loading Representative of Performance Earthquake

Hysteretic Response

Backbone Curve

SPECIMEN B

Figure 4-7   Determining λ values from structural testing

Pseudo-dynamic test sequences may need to be carefully selected to produce enough cycles in each direction.

4.4.4 Global Displacement Demand

Prior earthquake damage may alter the future seismic response of a building by affecting the displacement demand and the displacement capacity. Effects of prior damage on the future displacement demands may be evaluated according to methods described in this section. Effects of prior damage on displacement capacity are described in Section 4.4.3.

FEMA 307 describes analytical and experimental studies of effects of prior damage on future earthquake response demands. A primary conclusion is that prior earthquake damage often does not cause a statistically significant change in maximum displacement demand for the overall structural system in future earthquakes under the following circumstances:

a. there is not rapid degradation of resistance with repeated cycles.

b. the performance ground motion associated with the future event produces a maximum displacement, d_d, larger than that produced by the damaging ground motion, d_e.

c. the residual drift of the damaged or repaired structure is small relative to d_e.

If the performance ground motion produces a maximum displacement, d'_d less than that produced by the damaging ground motion, d_e. the response of the damaged structure is more likely to differ from that of the pre-event structure, d_d (see Figure 4-8).

There are several alternatives for estimating the displacement demand for a given earthquake motion. FEMA 273 relies primarily on the displacement coefficient method. This approach uses a series of coefficients to modify the hypothetical linear-elastic response of a building to estimate its nonlinear-inelastic displacement demand. The capacity spectrum method (ATC 40) characterizes seismic demand initially using a 5% damped linear-elastic response spectrum and reduces the spectrum to reflect the effects of energy dissipation in an iterative process to estimate the inelastic displacement demand. The secant stiffness method (Kariotis et al., 1994), although formatted differently, is fundamentally similar to the capacity

spectrum method. Both these latter two methods can be related to the substitute structure method (Shibata and Sozen, 1976). The use of each of these approaches to generate estimates of global displacement demand (d_d, $d_{d'}$, and d_{d*}) is summarized in the following sections. Generally, any of the methods may be used for the evaluation of the effects of damage; however, the same method should be used to calculate each of the global displacement demands (d_d, $d_{d'}$, and d_{d*}) when making relative comparisons using these parameters.

4.4.4.1 Displacement Coefficient Method

The displacement coefficient method refers to the nonlinear static procedure described in Chapter 3 of FEMA 273. The method also is described in Section 8.2.2.2 of ATC 40. The reader is referred to those documents for details in application of the procedure. A general overview and a description of the application of the method to damaged buildings are presented below.

The displacement coefficient method estimates the earthquake displacement demand for the building using a linear-elastic response spectrum. The response spectrum is plotted for a fixed value of equivalent damping, and the spectral response acceleration, S_a, is read from the spectrum for a period equal to the effective period, T_e. The effective period is defined by the following:

$$T_e = T_i \sqrt{\frac{K_i}{K_e}} \qquad (4\text{-}1)$$

where T_i is the elastic fundamental period (in seconds) in the direction under consideration calculated by elastic dynamic analysis, K_i is the elastic lateral stiffness of the building in the direction under consideration (refer to Figure 4-9), and K_e is the effective lateral stiffness of the building in the direction under consideration (refer to Figure 4-9). As described in FEMA 273, the effective lateral stiffness is taken as a secant to the capacity curve at base shear equal to $0.6V_y$. For a concrete or masonry wall building that has not been damaged previously by an earthquake, the effective damping is taken equal to 5% of critical damping.

The target displacement, δ_t, is calculated as:

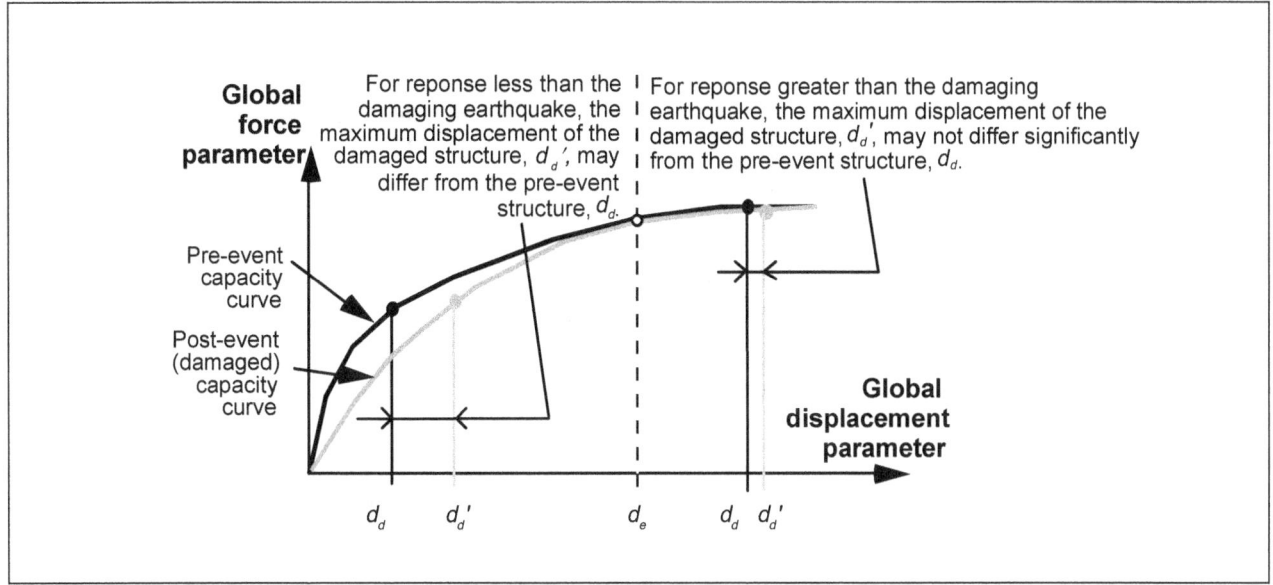

Figure 4-8   Maximum Displacement Dependency on Damaging Earthquake

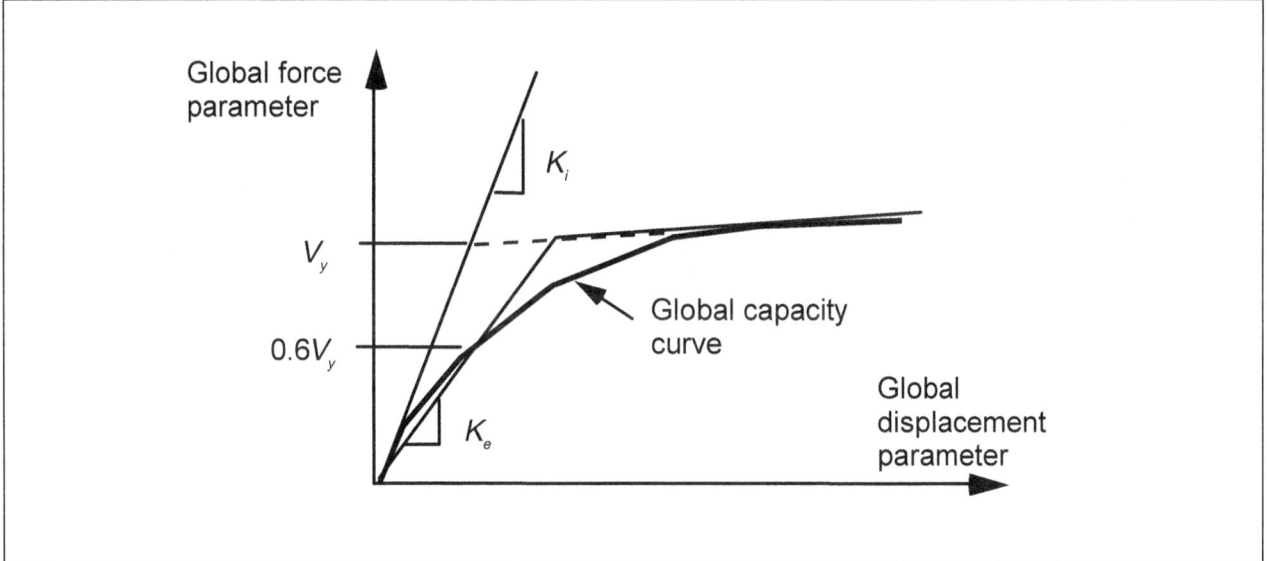

Figure 4-9   Global Capacity Dependency on Initial and Effective Stiffness

$$\delta_t = C_0 C_1 C_2 C_3 S_a \frac{T_e^2}{4\pi^2} \qquad (4\text{-}2)$$

where C_0, C_1, C_2, C_3 are modification factors defined in FEMA 273, and all other terms are as defined previously.

The maximum displacement, d_d, of the building in its pre-event condition for a performance ground motion is estimated by applying the displacement coefficient method using component properties representative of the pre-event conditions. To use the displacement coefficient method to estimate the maximum displacement demand, d_d', during a performance

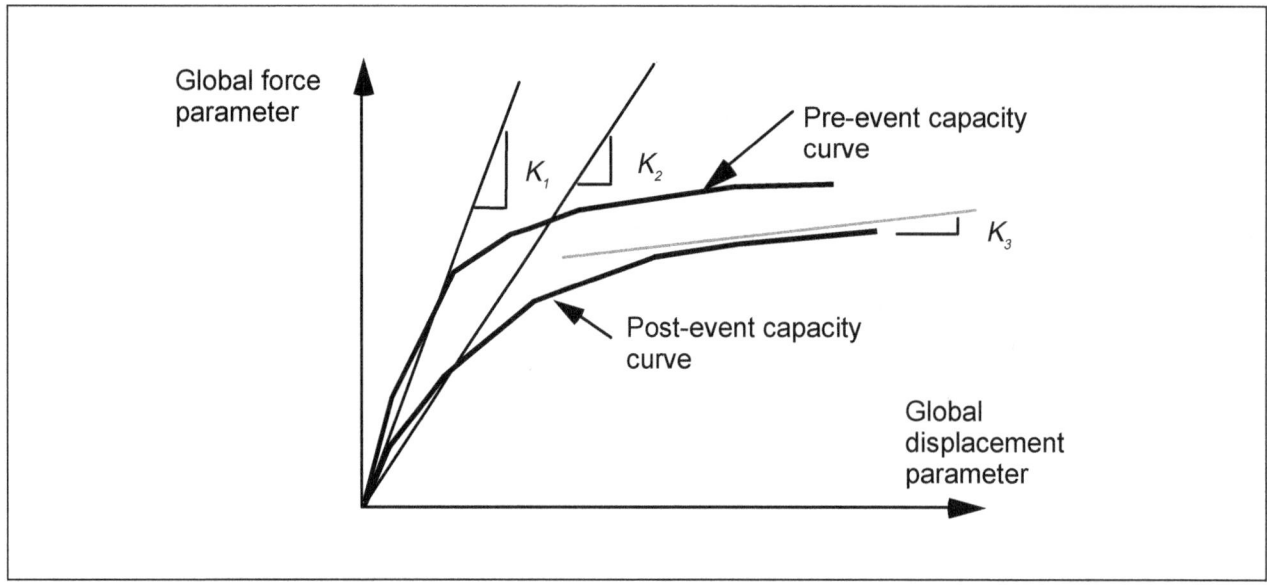

Figure 4-10   Pre- and Post-Event Capacity Curves with Associated Stiffnesses

ground motion for a building damaged by a previous earthquake use the following steps: (See Figure 4-10.)

1. Construct the relation between lateral seismic force (base shear) and global structural displacement (roof displacement) for the pre-event structure. Refer to this curve as the pre-event capacity curve. Pre-event force-displacement relations should reflect response characteristics observed in the damaging earthquake, as discussed in Section 3.6.

2. Construct a similar relationship between lateral seismic force and global structural displacement for the structure based on the damaged condition of the structure, using component modeling parameters defined in Section 4.4.3. Refer to this curve as the post-event capacity curve.

3. Define effective stiffnesses K_1, K_2, and K_3 as shown in Figure 4-10. K_1 is K_e (see Figure 4-9) calculated from the pre-event capacity curve. K_2 is K_e (see Figure 4-9) calculated from the post-event capacity curve. K_3 is the effective post-yield stiffness from the post-event capacity curve.

4. Apply the displacement coefficient method as defined in FEMA 273 with the effective stiffness taken as $K_e = K_1$, effective damping equal to 5% of critical damping, post-yield stiffness defined by stiffness K_3, and effective yield strength defined by the intersection of the lines having slopes K_1 and K_3

to calculate δ_t using Equation 4-2. Assign the displacement parameter d'_{d1} the value calculated for δ_t.

5. Apply the displacement coefficient method as defined in FEMA 273 with the effective stiffness taken as $K_e = K_2$, effective damping as defined by Equation 4-3, post-yield stiffness defined by stiffness K_3, and effective yield strength defined by the intersection of the lines having slopes K_2 and K_3 to calculate the displacement parameter d'_{d2}.

6. Using the displacement parameters d'_{d1} and d'_{d2}, estimate the displacement demand, d'_d, for the structure in its damaged condition as follows:

   a. If d'_{d1} is greater than d_e, then $d'_d = d'_{d1}$

   b. If d'_{d1} is less than d_e, then $d'_d = d'_{d2}$

The effective damping as defined by Equation 4-3 is consistent with experimental results obtained by Gulkan and Sozen (1974),

$$\beta = 0.05 + 0.2\left[1 - \left(\frac{K_2}{K_1}\right)^{0.5}\right] \qquad (4\text{-}3)$$

For a restored or upgraded structure, the displacement demand, d_d^*, for a performance ground motion may be calculated using the displacement coefficient method with 5% damping using a capacity curve generated using applicable properties for existing components, whether repaired or not, and any supplemental components added to restore or upgrade the structure.

4.4.4.2   Capacity Spectrum Method

The capacity spectrum method is described in Section 8.2.2.1 of ATC 40. The reader is referred to that document for details in application of the procedure. A general overview and a description of the application of the method to damaged buildings are presented below.

The capacity spectrum method estimates the earthquake displacement demand for the building using a linear-elastic response spectrum. The response spectrum is plotted for a value of equivalent damping based on the degree of nonlinear response, and the spectral displacement response is read from the intersection of the capacity curve and the demand curve. In some instances of relatively large ground motion, the curves may not intersect, indicating potential collapse. In these cases the displacement coefficient method could be used as an alternate method for damage evaluation.

The maximum displacement of the building in its pre-event condition, d_d, for a performance ground motion is estimated by applying the capacity spectrum method using component properties representative of the pre-event conditions. To use the capacity spectrum method to estimate the maximum displacement demand, d'_d, during a performance ground motion for a building damaged by a previous earthquake, use the following steps:

1. Construct the relation between lateral seismic force (spectral acceleration) and global structural displacement (spectral displacement) for the structure assuming the damaging ground motion and its resultant damage had not occurred. Pre-event component force-deformation relationships should reflect response characteristics observed in the damaging earthquake as discussed in Section 3.6. Refer to this curve as the pre-event capacity curve.

2. Construct a similar relation between lateral seismic force and global structural displacement for the structure based on the damaged condition of the structure, using component modeling parameters defined in Section 4.4.3. Refer to this curve as the post-event capacity curve.

3. Apply the capacity spectrum method using the pre-event capacity curve to calculate the displacement parameter d'_{d1}.

4. Apply the capacity spectrum method using the post-event capacity curve to calculate the displacement parameter d'_{d2}. For determining the effective damping, the yield strength and displacement for the post-event capacity curve should be taken identically equal to the yield strength and displacement determined for the pre-event capacity curve. (See Equation 4-3.)

5. Using the displacement parameters d'_{d1} and d'_{d2}, estimate the displacement demand, d'_d, for the structure in its damaged condition as follows:

    a. If d'_{d1} is greater than d_e, then $d'_d = d'_{d1}$

    b. If d'_{d1} is less than d_e, then $d'_d = d'_{d2}$

For a restored or upgraded structure the displacement demand for a performance ground motion, d_{d*}, may be calculated using the capacity spectrum method based on a capacity curve using applicable properties for existing components, whether repaired or not, and any supplemental components added to restore or upgrade the structure.

4.4.4.3   Secant Stiffness Method

The secant stiffness method is described in Section 8.4.2.1 of ATC-40, *Seismic Evaluation and Retrofit of Concrete Buildings* (ATC, 1996). The reader is referred to that document for details in application of the procedure. To use the method for damaged buildings, the general procedure should be applied based on the properties of the damaged building.

4.4.4.4   Nonlinear Dynamic Procedure

As an alternative to the nonlinear static procedures described above, nonlinear dynamic response histories may be computed to estimate the displacement demand for the building. This dynamic analysis approach requires that suitable ground motion records be selected for both the damaging event and the performance ground motion. It also requires that representative structural models be prepared for the building in its pre-event (no superscript), damaged ('), and restored or upgraded (*) conditions. Detailed procedures have not been developed for the use of nonlinear dynamic response histories in relative performance analyses. The following sections offer general guidance on nonlinear dynamic procedures consistent with the nonlinear static procedures.

A. Ground Motions

Damaging Event. If available, ground motion time histories at, or near, the site may be used to represent the damaging ground motion. Alternatively, an estimate of spectral response can be generated using the procedures of Section 3.1. Time histories consistent with the estimated spectral response may then be generated to represent the damaging ground motion. The average maximum displacement response, d_d, of the pre-event structural model to the time histories should be near that which is estimated to have actually occurred in the structure under evaluation. This effort may involve some adjustments to both the structural model and the ground motion in a verification process, similar to that outlined in Section 3.6, to calibrate the analysis with the observed damage.

Performance Ground Motions. Three to five ground motion accelerograms might be used to represent potential motions at the site for each performance level considered. Each of the records should be consistent with the response spectra that would be used with the nonlinear static procedures for the performance level under consideration as presented in FEMA 273/274 (ATC, 1997a,b) and ATC-40 (ATC, 1996). The maximum global displacement responses of the structural models can be averaged to generate a best estimate of the response.

B. Structural Modeling

The analysis procedure will vary, depending on the type of model used for the individual structural components.

Non-degrading Component Models. If the components are modeled using force-deformation relations that do not include strength degradation (non-degrading model), a procedure to estimate displacement demands is as follows:

1. Determine the maximum displacement response of the pre-event building, d_d, to the performance ground motion using a structural model with component properties representative of pre-event conditions. This pre-event model should be calibrated, as discussed above, to the damaging ground motion.

2. Modify the component properties to reflect the effects of the observed damage in accordance with the recommendations of Section 4.4.3. Determine the maximum displacement response of the build-

ing in its damaged condition, d_d', to the performance ground motion using the damaged structural model.

3. Modify the model to reflect the effects of restoration or upgrade measures. These may include the modification of existing components or the addition of new components as discussed in Section 4.5. Determine the maximum displacement response of the building in its restored or upgraded condition, d_d*, to the performance ground motion using the restored or upgraded structural model.

Degrading Component Models. If the components are modeled using force-deformation relations that allow strength degradation during the response history (degrading model), a procedure to estimate displacement demands is as follows:

1. Determine the maximum displacement response of the pre-event building, d_d, to the performance ground motion using a structural model with component properties representative of pre-event conditions. This pre-event model should be calibrated, as discussed in Section 3.6, to the damaging ground motion.

2. Subject the pre-event model to a composite ground motion comprised of the damaging ground motion, followed by a quiescent period in which the structure comes to rest, followed by the performance ground motion. Estimate the maximum displacement response, d_d', of the building in its damaged condition as the maximum displacement to occur in the time period after the quiescent period in the record.

3. Modify the model to reflect the effects of both the damage and restoration or upgrade measures. These may include the modification of existing components or the addition of new components as discussed in Section 4.5. Determine the maximum displacement response, d_d*, of the building in its restored or upgraded condition to the performance ground motion using the modified structural model.

4.5 Performance Restoration Measures

If the performance capability of the structure is diminished by the effects of earthquake damage $(d_c'/d_d' < d_c/d_d)$, the magnitude of the loss is quantified by the costs of performance restoration measures. These are hypothetical actions that, if implemented, would result in future performance approximately equivalent to the undamaged building $(d_c^* / d_d^* \approx d_c / d_d)$. Performance restoration measures may take several different forms:

- Component restoration entails the repair of individual components to restore structural properties that were diminished as a result of the earthquake damage. The Component Guides in Chapters 5 through 8 provide guidance based on component type, behavior mode, and severity of damage. They refer to outline specifications for individual restoration techniques that are compiled in FEMA 308: *The Repair of Earthquake Damaged Concrete and Masonry Wall Buildings.*

- An extreme case of component repair is complete replacement. In some cases, this is the only alternative indicated in the Component Guides. In other cases, it may be a more economical alternative than component repair.

- Performance can also be restored by the addition of new supplemental lateral-force-resisting elements or components.

Performance restoration measures are specified at the component level using one or more of the above alternatives. The measures are then tested by analyzing the performance of the modified structure, as outlined in Section 4.4. If necessary, the scope of the measures should be adjusted until the performance is approximately the same as that of the undamaged building. It should be noted that all components need not necessarily be restored individually to restore overall performance. It is advisable to explore several strategies to reach an economical solution.

Once the scope has been determined, the loss associated with the earthquake damage can be calculated as the cost of the performance restoration measures if they were to be actually implemented. The cost should include estimates of direct construction costs, as well as the associated indirect costs. Indirect costs include project costs such as design and management fees. In some cases, the costs of hypothetical temporary relocation of building occupants, loss of revenue, and other indirect costs, should be included. Detailed guidelines on determining both direct and indirect costs are not included in this document.

4.6   An Alternative—The Direct Method

A direct method of determining performance restoration measures may be used to estimate the loss caused by the earthquake. This method assumes that the scope of performance restoration measures is equivalent to restoring the significant structural properties of all of the components of the structure. The method uses individual repair actions for each component addressing the observed damage directly without explicitly considering its effect on seismic performance. These repair actions are summarized in the Component Guides of Chapters 5 through 8. The fundamental assumption is that the restored structure would have equivalent performance capability to the undamaged building for performance ground motions greater than that of the damaging ground motion ($d_d^* > d_e$). For this case the direct method will tend to overestimate losses, because some of the damage will have no effect on seismic performance. If the anticipated performance ground motion is less than that of the damaging ground motion ($d_d^* < d_e$), the direct method may underestimate losses in some cases, depending on individual building characteristics. This is because it neglects some loss of stiffness in components that theoretically could increase displacement response for smaller events. In these cases, relative performance analysis may be necessary to evaluate losses more accurately.

The direct method should be used only if the sole objective is to estimate the loss from the damaging earthquake based on the cost of the performance restoration measures. The direct method provides no information on the actual performance of the building in its damaged or undamaged states and cannot be used for design purposes.

Relative performance analysis is preferred because it determines the seismic performance of the building in its damaged and undamaged states.The scope of performance restoration measures is determined by analyzing their effect on the predicted performance. Since the effect on global performance of damage to individual components is considered, this technique generally provides a more accurate evaluation of the actual loss due to the damaging earthquake. The analysis also provides information that may be used for design purposes to restore or upgrade the building.

5: Reinforced Concrete

5.1 Introduction and Background

This section provides information on reinforced concrete wall components including the Component Damage Classification Guides (Component Guides) for reinforced concrete (RC). Section 5.2 defines and describes the component types and behavior modes that may be encountered in reinforced concrete wall structures. Section 5.3 gives evaluation procedures for assessing component strength and determining likely behavior modes. Section 5.4 defines symbols for studies on reinforced concrete walls.

The Component Guides and evaluation procedures in Section 5.5 are based on a review of the applicable research. Extensive structural testing has been done on reinforced concrete walls and wall components. In-plane tests on concrete walls in the 1950s and early 1960s generally applied monotonic loading and focused on ultimate strength without considering displacement or ductility capacity. Two major programs during this time period were carried out at Stanford University (Benjamin and Williams, 1957 and 1958) and at MIT (Antebi et al., 1960).

From the late 1960s to early 1980s, the Portland Cement Association conducted a comprehensive and pioneering test program on the earthquake resistance of reinforced concrete walls (Cardenas, 1973; Oesterle et al., 1976, 1979, 1983; Shiu et al., 1981; Corley et al., 1981). The tests principally used cyclic-static loading and identified several possible behavior modes in reinforced concrete walls, including flexural behavior, diagonal tension, diagonal compression (web crushing), boundary compression and bar buckling, sliding shear, and out-of-plane wall buckling.

Research on the behavior of coupling beams of walls was begun in the late 1960s at the University of Canterbury in New Zealand (Paulay, 1971a, b; Paulay & Binney, 1974; Paulay and Santhakumar, 1976). Research on numerous aspects of reinforced concrete wall behavior continued at Canterbury through the 1990s, covering ductile flexural behavior, diagonal tension, boundary confinement and bar buckling, sliding shear, out-of-plane stability, and walls with irregular openings (Paulay, 1980, 1986; Paulay et al., 1982; Paulay & Priestley, 1992, 1993). The research findings have contributed to code provisions for reinforced concrete walls in the United States and elsewhere.

As part of a U.S.- Japan cooperative research program, a seven-story full-scale, shear-wall structure was tested (Wight, 1985). This test demonstrated the contribution made by beams and slabs that framed into the wall to the capacity of wall structures.

Wall testing has also been carried out at the University of California, Berkeley (Wang et al., 1975; Vallenas et al., 1979; Iliya & Bertero, 1980) and on small shake-table specimens at the University of Illinois (Aristisabal and Sozen, 1976; Lybas and Sozen, 1977). More recent tests have been done by Wallace and Thomsen (1995). A number of tests have been carried out in Japan, including those by Ogata and Kabeyasawa (1984).

5.2 Reinforced Concrete Component Types and Behavior Modes

5.2.1 Component Types

Five possible component types are defined for reinforced concrete wall structures. The component types are listed and described in Table 5-1. Typically only component types RC1, RC2, and RC3 will suffer earthquake damage. Component types RC4 and RC5 are mentioned for completeness, but since they are not expected to suffer earthquake damage, they are not discussed in detail.

Wall Component types are assigned based on identifying the governing mechanism for nonlinear lateral deformation for the structure, as described in Section 2.4.

5.2.2 Behavior Modes and Damage

The possible modes of nonlinear behavior and damage for reinforced concrete wall components are outlined in Table 5-2, along with their response characteristics. Section 2.2 of FEMA 307 presents typical force-displacement hysteresis loop shapes for the behavior modes. The likelihood of such behavior modes occurring in each of the prevalent component types is shown in Table 5-3.

Table 5-1   Component Types and Descriptions for Reinforced Concrete Walls.

Component Type		Description
RC1	Isolated Wall or Stronger Wall Pier	Stronger than beam or spandrel elements that may frame into it, so that nonlinear behavior (and damage) is generally concentrated at the base, with, for example a flexural plastic hinge or shear failure. Includes isolated (cantilever) walls. If the component has a major setback or cutoff of reinforcement above the base, this section should be also checked for nonlinear behavior.
RC2	Weaker Wall Pier	Weaker than the spandrels to which it connects, characterized, for example, by flexural hinging at top and bottom, or shear failure.
RC3	Weaker Spandrel or Coupling Beam	Weaker than the wall piers to which it connects, characterized, for example, by hinging at each end, shear failure, or sliding shear failure.
RC4	Stronger Spandrel	Should not suffer damage because it is stronger than attached piers. If this component is damaged, it should probably be re-classified as RC3.
RC5	Pier-Spandrel Panel Zone	Typically not a critical area in RC walls.

The behavior modes are described in the following sections according to the ductility categories given in Table 5-2.

This document focuses on structures for which earthquake damage occurs primarily in wall components. Engineers should be aware that damage can also occur in other structural elements such as foundations, columns, beams, and slabs.

5.2.3 Behavior Modes with High Ductility Capacity (Flexural Response)

Adequately designed reinforced concrete walls of various configurations can respond to earthquake shaking in a ductile manner. Ductile wall response usually results from flexural behavior, which requires that the wall components be designed to avoid the following less desirable behavior effects:

- Failures in shear corresponding to diagonal tension, web crushing, or sliding shear

- Buckling of longitudinal bars in boundary regions of plastic hinge zones

- Loss of concrete strength due to high compressive strains in unconfined boundary regions of plastic hinge zones

- Slip of lap splices

- Out-of-plane buckling of thin wall sections

The strength of wall components in flexure is calculated using conventional procedures given in Section 5.3.5. Wall components responding in flexure generally have good displacement capacity, typically in-plane rotations exceeding two percent (or 0.02 radians), or displacements at least eight times yield.

5.2.4 Behavior Modes with Intermediate Ductility Capacity

The following behavior modes can be defined as having intermediate ductility capacity:

- Flexure/Diagonal tension

- Flexure/Web crushing

- Flexure/Sliding shear

- Flexure/Boundary-zone compression

- Flexure/Lap-splice slip

- Flexure/Out-of-plane wall buckling

The earthquake response in these behavior modes is initially governed by flexure, but after some number of cycles, reaching some level of earthquake displacements, a response mode other than flexure predominates. At this point, the component suffers strength degradation.

Table 5-2   *Behavior Modes for Reinforced Concrete Wall Components.*

Behavior Mode	Approach to calculate strength (use expected material values)	Approach to estimate displacement capacity	Ductility Category
A. Ductile flexural response	Conventional calculations per Section 5.3.5.	Good displacement capacity (e.g. 2% drift, or 8x yield displacement).	High ductility capacity
B. Flexure/ Diagonal tension	Moment strength per Section 5.3.5 initially governs strength.	Based on shear strength as a function of ductility. See Section 5.3.6.b.	Ductility capacity varies (Failure only occurs after some degree of flexural yielding and concrete degradation.)
C. Flexure/ Diagonal compression (web crushing)		Based on relationship of web crushing strength to drift per Oesterle et al (1983). See Section 5.3.6.c.	
D. Flexure/ Sliding shear		Shear friction approach per ACI 318, or recommendations of Paulay and Priestley (1992). See Section 5.3.6.d.	
E. Flexure/ Boundary-zone compression		Based on amount of ties required for moderate and high ductility levels, per Paulay and Priestley (1992). See Section 5.3.7.	
F. Flexure/Lap-splice slip		Based on lap strength as a function of ductility. See Section 5.3.8	
G. Flexure/Out-of-plane wall buckling		Based on wall thickness requirements for moderate and high ductility levels. See Section 5.3.9.	
H. Preemptive diagonal tension	Shear strength governs at low ductility levels, per Section 5.3.6.b.	No inelastic displacement capacity.	Little or no ductility capacity (Flexural reinforcement does not yield.)
I. Preemptive web crushing	May occur at shear stresses of $12\sqrt{f'_{ce}} - 15\sqrt{f'_{ce}}$. See Section 5.3.6.c.		
J. Preemptive sliding shear	Shear friction approach per ACI 318. See Section 5.3.6.d.		
K. Preemptive boundary zone compression	Applies only to unusually high axial loads, above the balance point. Moment strength calculation still governs.		
L. Preemptive lap-splice slip	Lap strength, per FEMA 273 and ATC-40, or approach of Priestly et al. (1996) governs. See Section 5.3.8.		
M. Global foundation rocking of wall	See FEMA 273 or ATC-40		Moderate to high ductility capacity
N. Foundation rocking of individual piers			

Table 5-3   *Likelihood of Earthquake Damage to Reinforced Concrete Walls According to Wall Component and Behavior Mode.*

Behavior Mode	Wall Component Type		
	Isolated Wall or Stronger Wall Pier (RC1)	**Weaker Wall Pier** (RC2)	**Weaker Spandrel or Coupling Beam** (RC3)
A. Ductile flexural response	Common in well-designed walls See Guide RC1A	May occur See Guide RC2A	May occur, particularly if diagonally reinforced Similar to Guide RC2A
B. Flexure/Diagonal tension	Common See Guide RC1B	Common Similar to Guide RC1B	Common See Guide RC3B
C. Flexure/Diagonal compression (web crushing)	Common (frequently observed in laboratory tests) See Guide RC1C	May occur	May occur
D. Flexure/Sliding shear	May occur, particularly for squat walls See Guide RC1D	May occur	Common See Guide RC3D
E. Flexure/ Boundary-zone compression	Common See Guide RC1E	May occur	Unlikely
F. Flexure/Lap-splice slip	May occur	May occur	May occur
G. Flexure/Out-of-plane wall buckling	May occur (observed in laboratory tests)	Unlikely	Unlikely
H. Preemptive diagonal tension	Common Similar to Guide RC2H	Common See Guide RC2H	Common Similar to Guide RC2H
I. Preemptive web crushing	May occur in squat walls (observed in laboratory tests)	May occur	May occur
J. Preemptive sliding shear	May occur in very squat walls or at poor construction joints.	May occur in very squat walls or at poor construction joints.	Unlikely
K. Preemptive boundary zone compression	May occur in walls with unsymmetric sections and high axial loads	May occur in walls with unusually high axial load	Unlikely
L. Preemptive lap-splice slip	May occur	May occur	May occur
M. Global foundation rocking of wall	Common	n/a	n/a
N. Foundation rocking of individual piers	May occur	May occur	n/a

Notes:
- Shaded areas of table indicate behavior modes for which a specific Component Damage Classification Guide is provided in Section 5.5. The notation *Similar to Guide...* indicates that the behavior mode can be assessed by using the guide for a different, but similar, component type or behavior mode.
- *Common* indicates that the behavior mode has been evident in postearthquake field observations and/or that experimental evidence supports a high likelihood of occurrence.
- *May occur* indicates that the behavior mode has a theoretical or experimental basis, but that it has not been frequently reported in postearthquake field observations.
- *Unlikely* indicates that the behavior mode has not been observed in either the field or the laboratory.

a. Strength and Displacement Capacity

Flexural behavior governs the maximum strength achieved in behavior modes with intermediate ductility capacity. For these behavior modes, the full flexural strength will not be sustained at high levels of cyclic deformation. Section 5.3.5 gives guidelines for calculating flexural strength.

Displacement capacity may be difficult to assess for behavior modes with intermediate ductility capacity. One approach for estimating displacement capacity is to consider the intersection of the force-displacement curve for flexural response with a degrading strength envelope for the governing failure mechanism. The degrading strength envelope may represent, for example, lap-splice strength or shear strength. Useful research has been carried out using this approach. For example, Priestley et al. (1996) have developed specific recommendations on the degradation of strength as a function of ductility for lap-splice failure and shear failure.

b. Flexure/Diagonal Tension

The flexure/diagonal tension behavior mode occurs in a wall component when the shear strength in diagonal tension initially exceeds the flexural strength, allowing flexural yielding to occur. However, after the cracks open and the concrete in the plastic hinge zone degrades, the shear strength is reduced below the flexural strength, and shear behavior predominates.

At low levels of response, this behavior mode may appear similar to a ductile flexural response, although diagonal cracks due to shear stress may be more prominent. At higher levels of response, diagonal cracking tends to concentrate in one or two wide cracks. Eventually horizontal reinforcement can be strained to the point of fracture, signaling a diagonal tension failure.

c. Flexure/Diagonal Compression (Web Crushing)

For heavily reinforced walls subject to high shear forces, shear-related compression failures may occur rather than diagonal tension failures. This mode of behavior has been commonly observed in laboratory testing, and it may be prevalent in low-rise walls or when shear reinforcement is sufficient to prevent a diagonal tension failure. Higher axial loads also increase the likelihood of web-crushing behavior.

Web crushing generally occurs after some degree of cyclic flexural behavior and degradation. The vulnerability to web crushing can be considered to be proportional to the story drift ratio to which the component is subjected. This behavior mode is characterized by diagonal cracking and spalling in the web region of the wall. Localized web crushing can be initiated by the uneven closing of diagonal cracks under cyclic earthquake forces.

d. Flexure/Sliding Shear

Coupling beams and low-rise walls are particularly vulnerable to failure by sliding shear. Low axial loads and poor construction joint details increase the probability of sliding shear.

In this behavior mode, flexural yielding initially governs the response. Flexural cracks at the critical section tend to join up to form a single crack across the section which becomes a potential sliding plane. Under cyclic forces and displacements, this crack opens more widely so that the aggregate interlock and shear friction resistance on the sliding plane degrade. When the sliding shear strength drops below the shear corresponding to the moment strength, lateral sliding offsets begin to occur.

For many low-rise walls, lateral strength may be governed by the strength of the foundation to resist overturning. Sliding shear behavior is likely to occur only in low-rise walls where the foundations have the capacity to force flexural yielding.

e. Flexure/Boundary-Zone Compression

Taller walls with adequate shear strength but inadequate boundary tie reinforcement tend to be vulnerable to this behavior mode. Under inelastic flexural response, the boundary regions of plastic hinge zones may be subjected to high compression strains, which cause spalling of the cover concrete. If sufficient tie reinforcement is not placed around the longitudinal bars in the wall boundaries, the longitudinal bars are prone to buckling. Additionally, in walls where concrete compressive strains exceed 0.004 or 0.005, the concrete in the boundary regions can rapidly lose compressive strength if it is not confined by adequate boundary ties. In addition to bar-buckling restraint and confinement, ties around the lap splices of boundary longitudinal bars significantly increase lap-splice strength.

f. Flexure/Lap-Splice Slip

Lap splices in the critical plastic hinge regions of walls are commonly encountered in existing buildings. Even when relatively good lap-splice length is provided, lap splices in plastic hinge zones tend to slip when the concrete compressive strain exceeds 0.002, unless ties are provided around the lap splices.

Slipping of lap splices is accompanied by splitting cracks in the concrete, oriented parallel to the spliced reinforcement. The use of tie reinforcement around lap splices, which restrains the opening of the splitting cracks, can prevent or delay the onset of lap-splice slip.

Once lap splices slip, the component strength falls below the full moment strength of the section and the strength is governed by the residual strength of the splices plus the moment capacity due to axial load.

g. Flexure/Out-of-Plane Wall Buckling

Several experimental studies have shown that thin wall sections can experience out-of-plane buckling when subjected to cyclic flexural forces and displacements. For typical wall sections the buckling occurs only at high ductility levels.

Single curtain walls and walls with higher amounts of longitudinal reinforcement tend to be more vulnerable to out-of-plane buckling. Walls with large story heights between floors that brace the wall in the out-of-plane direction are more vulnerable to buckling. T- or L-shaped wall sections with thin stems may also be more vulnerable. Walls with flanges or other enlarged boundary elements are less susceptible.

5.2.5 Behavior Modes with Little or No Ductility Capacity

The following five behavior modes can be considered to have little or no inelastic deformation capacity:

- Preemptive diagonal tension

- Preemptive diagonal compression (web crushing)

- Preemptive sliding shear

- Preemptive boundary-zone compression

- Preemptive lap-splice slip

The term preemptive is used to indicate that a brittle failure mode preempts any flexural yielding of the wall component. These are force-controlled rather than displacement-controlled behavior modes, as defined in FEMA 273.

a. Strength and Displacement Capacity

During preemptive boundary-zone compression, peak strength is equal to the component flexural strength. For the other four behavior modes of this category, strength will be less than the component flexural strength.

Displacement capacity of these force-controlled behavior modes is limited to the elastic displacement corresponding to peak strength. These behavior modes cannot be considered to have any dependable inelastic displacement capacity.

b. Preemptive Diagonal Tension

Preemptive shear failure of wall components in diagonal tension has been commonly observed after earthquakes. This type of failure typically occurs in components with high flexural strength and inadequate shear reinforcement. The failure is characterized by one or more wide diagonal cracks, which can occur suddenly, with little or no early indication of incipient failure.

c. Preemptive Diagonal Compression (Web Crushing)

Preemptive web crushing is a compression failure caused by high shear forces in the web of a wall section. This behavior mode has been observed in laboratory tests of low-rise flanged walls. Walls with flanges or heavy boundary elements are more prone to this type of failure because larger shear stresses are typically generated in the webs of such sections, as compared to rectangular sections. The web crushing begins at small displacement values, preempting any flexural yielding of the wall.

This failure mode has not been reported in actual structures. Typical buildings do not have foundations with enough overturning capacity to sustain the high forces associated with preemptive diagonal compression failures.

d. Preemptive Sliding Shear

Preemptive sliding shear is most likely to occur in low-rise wall piers that have poor construction joints. Before flexural strength can be reached in such walls, sliding occurs along the surfaces of the construction joint.

e. Preemptive Boundary Zone Compression

This behavior mode only occurs in walls with unusually high axial load — above the balance point considering a maximum concrete strain of 0.004 or 0.005. Such conditions typically occur only in T- or L-shaped sections where the stem of the section is in compression and has inadequate boundary confinement.

f. Preemptive Lap-Splice Slip

Wall behavior governed by preemptive lap-splice slip has not been widely reported. Lap-splice lengths may need to be unusually short for splice failures to occur without prior flexural cyclic behavior. Damage in this behavior mode would be characterized by splitting cracks at lap splices and eventual rocking of the wall component on a crack across the lap-spliced section.

5.2.6 Foundation Rocking Response

Foundation rocking of walls and wall piers is usually a ductile mode of behavior. Capacities depend on the foundation type and geometry and the properties of the soil material. Displacement capacity is normally very high, but the effects of large foundation movements on the superstructure must be considered. FEMA 273 and ATC-40 provide detailed recommendations for foundation components.

5.3 Reinforced Concrete Evaluation Procedures

5.3.1 Cracking

The Component Damage Classification Guides require the user to distinguish between flexural cracks and shear cracks, to identify vertical cracking in the compression zone of wall piers, and to identify horizontal cracking in the compression zone of wall spandrels. The guides also require the user to identify cracks that may indicate lap-splice slipping.

The guides require the user to determine crack widths, which is a factor in assessing the severity of earthquake damage in reinforced concrete wall components.

a. Flexural and Shear Cracks

Flexural cracks are those that develop perpendicular to flexural tension stresses. In wall piers, flexural cracks run horizontally; in wall spandrels, the cracks run vertically. Flexural cracks typically initiate at the extreme fiber of a section and propagate towards the section's neutral axis. For components that have

undergone cyclic earthquake displacements in both directions, opposing flexural cracks often join with each other to form a relatively straight crack through the entire section.

Shear cracks are those that result from diagonal tension stresses corresponding to applied shear forces. The cracks run diagonally, typically at an angle of 35° to 70° from the horizontal. The angle of cracking depends on normal forces (e.g., axial load) and on the geometry of the component. For components that have undergone cyclic earthquake displacements of similar magnitude in both directions, the cracks cross each other, forming X patterns.

Flexural cracks often join up with diagonal shear cracks. A typical case is in a wall pier where a horizontal crack at the wall boundary curves downward to become a diagonal shear crack as it approaches the pier centerline. When shear cracks connect to flexural cracks, determine the widths of the flexural portion of the crack and the shear portion of the crack separately.

Cracks initially form perpendicular to the direction of the principal tension stresses in a section. At any point of a component, it is possible to relate the orientation of initial cracking to the applied stresses by considering the stress relationships represented by Mohr's Circle. However, after initial cracking, the orientation of principal stresses will change and crack patterns and stress orientations are affected by the reinforcement.

b. Full-Thickness versus Partial-Thickness Cracking

In investigating reinforced concrete wall components, the engineer should establish whether critical flexural and shear cracks extend through the thickness of the wall. The Component Damage Classification Guides are written under the assumption that the most significant flexural and shear cracks are full-thickness cracks having a similar crack width on each side of the wall.

Laboratory tests on walls have invariably used in-plane loading. Therefore, significant cracks observed in these studies are typically full-thickness. In actual buildings, out-of-plane forces and wall deformations may cause cracks to be partial-thickness, or they may result in cracks that remain open to a measurable width on one wall face, but are completely closed on the opposite wall face. In such cases, the engineer should use judgment in assessing the consequences of the critical cracks. It may be justified to use the average of the measured crack width on each face of the wall. More conservatively, the maximum crack width on either face of the wall can be used in the Component Damage Classification Guides.

c. Cracking as a Precursor to Spalling

In the compression region of wall components, cracks occur as a precursor to concrete spalling. Such cracks form parallel to the principal compression stresses, and they may develop when compressive strains in the concrete exceed 0.003 to 0.005. Such cracking typically signals an increased damage severity in the Component Damage Classification Guides. This type of cracking occurs (1) at the boundary regions of component plastic-hinge zones for flexural behavior, and (2) under a diagonal-compression (web-crushing) type of shear failure.

For wall piers in flexure, this type of cracking is vertical. For wall spandrels in flexure, the cracks are horizontal. In both cases, the cracks occur near the extreme fibers of the section in the plastic hinge zone(s). Such cracking is less likely in spandrels because of the absence of axial load.

The cracking in compression regions of flexural members could appear similar to splitting cracks resulting from lap-splice or bond slip of the reinforcement. Both types of cracking tend to occur in the boundary regions of plastic-hinge zones. Some distinguishing features of the two different types of cracks are described below:

Cracks as a precursor to spalling in the compression region:	Bond or lap-splice splitting cracks:
• Occur under conditions of high compressive strain.	• Occur at the locations of longitudinal reinforcement that is susceptible to bond or lap-splice slip. (Large bar diameters or inadequate lap-splice length.)
• Cracks may be relatively short. Sounding with a hammer (See Section 3.8) may reveal incipient spalling.	
• Cracks occur at the extreme fibers of the section, typically within the cover of the concrete.	• Cracks tend to be relatively long and straight, mirroring rebar locations. The cracks originate at the reinforcement and propagate to the concrete surface.

Diagonal cracking in the web of the wall can be a precursor to a diagonal-compression (web-crushing)

shear behavior. Unlike diagonal tension cracks, these cracks may not open widely, but under increasing damage, the cracks will be followed by spalling of the web concrete. This occurs because the compressive strength of concrete reduces in the presence of transverse tensile strains.

d. Splitting Cracks at Lap Splices

If lap splices are insufficient to develop the required tension forces in the reinforcement, slip occurs at the splices. The visible evidence of lap-splice slip is typically longitudinal cracks (parallel to the splice) that originate at the lap splice and propagate to the concrete surface. Thus, the crack locations reflect the locations of the lap-spliced reinforcement.

e. Crack Widths

Crack widths are to be measured according to the investigation procedures outlined in this document. In the Component Damage Classification Guides, the maximum crack width defines the damage severity. When multiple cracks are present, the widest crack of the type being considered (e.g., shear or flexure) governs the damage severity classification.

The maximum crack width may be significantly larger than the average width of a series of parallel cracks. Although average crack width may be a better indicator of average strain in the reinforcement, maximum crack width is judged to be more indicative of maximum reinforcement strain, and, in general, damage severity. A concentration of strain at one or two wide cracks typically indicates an undesirable behavior mode and more serious damage, whereas an even distribution of strain and crack width among numerous parallel cracks indicates better seismic performance.

The crack width criteria in the Component Damage Classification Guides are based on a comparison to research results, rather than on detailed analyses of crack width versus strain relationships. The criteria recognize that the residual crack width observed after an earthquake may be less than the maximum crack widths occuring during the earthquake.

5.3.2 Expected Strength and Material Properties

a. Expected Strength

The capacity of reinforced concrete components is calculated initially using expected strength values. Expected strength is defined in Section 6.4.2.2 of FEMA 273 and Section 9.5.4.1 of ATC-40 as "the mean maximum resistance expected over the range of deformations to which the component is likely to be subjected."

Expected component strength may be calculated according to the procedures of ACI 318 — or other procedures specified in this document — with a strength-reduction factor, ϕ, taken equal to 1.0. Expected material strength rather than specified minimum material strength is used in the calculations. Material strength values are discussed below.

b. Reinforcing Steel Strength

Tables 6-1 and 6-2 of FEMA 273 gives the specified yield and tensile strength of reinforcing steel that has been used in buildings since the turn of the century. ASTM A432 reinforcing steel, with a specified yield strength of 60 ksi, was introduced in 1959 (CRSI Data Report 11). Prior to this date, reinforcing steel typically had a specified yield strength of 40 ksi or less.

The actual yield strength of reinforcing steel typically exceeds the specified value, as discussed in Section 9.5.4.1 of ATC-40. Tests by Wiss, Janney, Elstner Associates (1970) on ASTM A432 Grade-60 bars showed an average tensile stress of 67 ksi for bars at a strain of 0.005 and an average tensile stress of 70 ksi for bars at a strain of 0.008. Stresses were based on actual rather than nominal bar areas, and the standard deviation was about 7 ksi. Similarly, data cited by Park (1996) indicate that actual bar yield strength averages about 1.15 times the specified value. At moderate-to-high ductilities, strain hardening will further increase the stress in yielding reinforcement.

FEMA 273 (Section 6.4.2.2) and ATC-40 prescribe that expected strength values be calculated assuming a strength of yielding reinforcement equal to "at least 1.25 times the nominal yield strength." For this document, in the absence of applicable test data, the initial expected strength of yielding reinforcement, f_{ye}, is assumed equal to 1.25 times the nominal yield strength. A range of reinforcement strength, between 1.1 and 1.4 times the nominal yield strength, can also be considered in the evaluation procedures, should field observation warrant.

Section 6.3.2 of FEMA 273 gives recommendations for establishing reinforcing steel strength by testing.

c. Concrete Strength

Table 6-3 of FEMA 273 gives typical concrete compressive strength values that may be assumed in buildings according to year of construction and structural member type. These values can be considered as specified or nominal values, f_c', rather than expected values, f_{ce}'. If structural drawings are available that indicate specified concrete strengths, use the values from the drawings instead of the assumed values from the table.

The actual concrete strength in existing structures can significantly exceed the specified minimum concrete strength, by factors of up to 2.3 (Park, 1996). For this document, in the absence of applicable test data, the initial expected concrete compressive strength is assumed to be equal to 1.5 times the specified strength. A range of concrete strength, between 1.0 and 2.0 times the specified strength, can also be considered in the evaluation procedures, when based on field observations. FEMA 273 and ATC-40 do not specifically address the relationship between expected and specified concrete strengths.

In the case that concrete compressive strength test results from the existing construction are available, use these results to establish the expected concrete strength. The expected concrete strength considers the likely strength increase of the concrete over time, as is discussed in Section 9.5.2.2 of ATC-40. In the absence of more specific data, the initial expected strength can be taken equal to 1.2 times the tested strength at 28 days after construction.

Section 6.3.2 of FEMA 273 gives recommendations for establishing concrete compressive strength by testing.

Concrete strength seldom has a significant effect on wall flexural strength. It will have a more significant effect on shear strength. one component of which is taken proportional to $\sqrt{f_{ce}'}$.

d. Concrete Modulus of Elasticity

The modulus of elasticity for concrete is calculated according to ACI-318, using the expected concrete strength as defined above.

5.3.3 Plastic-Hinge Location and Length

Plastic hinges occur at the critical flexural regions of wall members where moment demand reaches moment

---

strength. For earthquake-induced forces, plastic hinges typically occur at the face of a supporting member or foundation. Lap splices may force plastic hinges to develop or concentrate at the ends of the lap-splice length.

Potential plastic-hinge locations are to be identified for all wall elements subjected to earthquake forces and displacements. The locations are established as part of identifying the governing mechanism for nonlinear lateral deformation of the structure, as described in Section 2.4.

Isolated walls or stronger wall piers (component type RC1) typically have a single plastic hinge region at the base of the wall. A plastic hinge could also occur above the base of a wall at a location of reduced strength such as (1) a setback of the wall, (2) a level where a substantial amount of the vertical reinforcement is curtailed, or (3) a level above which the number of walls resisting seismic forces is reduced.

The curtailment of reinforcement may need to be investigated in some detail. Plastic hinging may occur at an area of reinforcement curtailment because (1) higher mode effects cause a moment at that level which exceeds the moment diagram assumed in design, and (2) designers may not have extended reinforcement "beyond the point at which it is no longer required to resist flexure for a distance equal to the flexural depth of the member" as is required by ACI 318 (1995).

Weaker wall piers (component type RC2) under flexural behavior develop plastic hinges at the top and bottom regions of the component, typically at the face of the connecting spandrel or foundation component. Similarly, weak spandrels (component type RC3) under flexural behavior develop plastic hinges at each end of the component, typically at the face of the connecting wall piers.

Plastic hinges are developed only for the ductile flexural behavior mode or for behavior modes with intermediate ductility capacity in which flexure initially governs response. Wall components governed by preemptive shear failures or foundation rocking do not exhibit plastic hinging (although, for foundation rocking, the soil beneath the foundation could be treated conceptually as the plastic-hinge region).

Plastic-hinge lengths define the equivalent zones over which nonlinear flexural strain can occur. The length of plastic hinging generally depends on the depth of the member and on the moment-to-shear ratio (M/V). Bond conditions of the reinforcement also affect the length over which yielding occurs and the penetration of reinforcement yielding into the supporting member.

For reinforced concrete, equivalent plastic-hinge length can be roughly estimated as equal to one-half the member depth (Park and Paulay, 1975). A similar estimate is applied to walls in Section 6.8.2.2 of FEMA 273, where l_p is "set equal to one half the flexural depth, but less than one story height." The 1997 UBC (ICBO, 1997) states that l_p "shall be established on the basis of substantiated test data or may be alternatively taken as $0.5l_w$."

Based on research specifically applicable to walls, the equivalent plastic-hinge length, l_p, can be set at 0.2 times the wall length, l_w, plus 0.07 times the moment-to-shear ratio, M/V (Paulay and Priestley, 1993).

Equivalent plastic-hinge length, as calculated above, is used to relate plastic curvature to plastic rotation and displacement. The actual zone of nonlinear behavior may extend beyond the equivalent plastic-hinge zone.

The Component Damage Classification Guides refer to the plastic hinge length to identify the zone over which nonlinear flexural behavior and damage may be observed. The expected zone of inelastic flexural behavior and damage in the Component Damage Classification Guides can be taken as two times l_p.

In short spandrel beams, the plastic zones at the ends of the beam may merge. In diagonally reinforced coupling beams, the entire length of the spandrel will yield.

5.3.4 Ductility Classifications

In these evaluation procedures, ductility capacity and demand are classified as either low, moderate, or high. The following approximate relationship can be assumed:

Classification	Displacement Ductility
Low Ductility	$\mu_\Delta < 2$
Moderate Ductility	$2 \leq \mu_\Delta \leq 5$
High Ductility	$\mu_\Delta > 5$

This is similar to the classification in Table 6-5 of FEMA-273, except that $\mu_\Delta = 5$ is used as the threshold for high ductility, rather than $\mu_\Delta = 4$. The less conservative value of five is considered more appropriate for damage evaluation, as opposed to retrofit design. The value of five also correlates best with the data for the shear strength recommendations of Section 5.3.6.b.

5.3.5   Moment Strength

The moment strength of a reinforced concrete component under flexure and possible axial loads is calculated according to conventional procedures, as defined in ACI-318, Section 10.2 (ACI, 1995), except that expected material strengths are used as discussed in Section 5.3.2 of this document. The moment strength accounts for all reinforcement that contributes to flexural strength. For example, the moment strength for a wall pier (component type RC1 or RC2) includes all well-anchored vertical bars at the section of interest, not just those in the wall boundaries. The axial load present on the wall component is taken into account in the calculation of moment strength.

For wall components that experience significant earthquake axial loads, such as the piers in a coupled wall system, the moment strength in each direction must consider the axial load combination corresponding to moments in that direction.

For sections with an overall reinforcement ratio, ρ_g, less than $0.008 \times (60\text{ksi}/f_y)$ the expected cracking moment strength, M_{cr}, may exceed the expected moment strength M_e. In such a case, both M_e and M_{cr} are considered in determining the governing mechanism and behavior mode.

a.   Uncertainties or discrepancies in strength

Typically, there should be little uncertainty in the calculation of moment strength for a reinforced concrete component if reinforcement sizes, layout, and the steel and concrete material strengths have been established. The possible range of axial load on the component must also be considered.

b.   Effective Flange Width

When wall sections have flanges or returns, the moment strength includes the effective width of flanges that contribute to flexural strength. C-shaped, I-shaped, L-shaped, T-shaped, and box-shaped wall sections fall in this category. The effective flange width is a function of the moment-to-shear ratio (M/V) for the wall component. Moment strength is relatively insensitive to the assumed flange width in compression, but can be quite sensitive to the assumed flange width in tension. Underestimating the effective flange width could lead to a conclusion that a wall is flexure-critical when in reality it is shear-critical. Typically, as displacement (or ductility level) increases, more of the vertical reinforcement in the flange is mobilized to resist flexure, and the effective flange width increases.

For isolated (cantilever) walls, effective flange width can be related to wall height, h_w, as described and illustrated in Section 5.22 of Paulay and Priestley (1992). For wider applicability to different loading patterns, the moment-to-shear ratio (M/V) can be used in place of the wall height.

FEMA 273 and ATC-40 prescribe an effective flange width of one-quarter of the wall height on each side of the wall web, with engineering judgment to be exercised if significant reinforcement is located outside this width. The 1997 UBC prescribes an effective flange width on each side of the wall web of 0.15 times the wall height. The proposed *NEHRP Provisions for New Buildings* (BSSC, 1997) prescribe a maximum effective width on each side of the wall web of 0.15 times the wall height for compression flanges and 0.30 times the wall height for tension flanges.

A more specific estimate of effective flange width is supported by research (Paulay and Preistley, 1992; Wallace and Thomsen, 1995) and is recommended in this document as defined below:

The effective flange width in compression, on each side of the wall web, may be taken as 0.15 times the moment-to-shear ratio (M/V). The effective flange width in tension, on each side of the wall web, may be taken as 0.5 to 1.0 times the moment-to-shear ratio (M/V). The effective width of the flange does not exceed the actual width of the flange, and the assumed flange widths of adjacent parallel walls do not overlap.

The foundation structure should be checked to ensure that the uplift forces in tension flanges can be developed.

c. Contribution of Frame and Slab Coupling to Wall Capacity

Beams and slabs that frame into a wall may contribute to the lateral capacity of the structural system. This was demonstrated in the testing of a full-scale seven-story wall structure in Japan (Wight, 1985). Beams transverse to the wall and in-line with the wall helped resist the lateral displacement of the wall, resulting in a total strength significantly greater than that of the wall alone.

5.3.6 Shear Strength

a. Shear Demand and Capacity

Consistent with the requirement in Section 2.4 to identify the mechanism of inelastic lateral response for the structure, shear demand is based on the expected strength developed at the locations of nonlinear action (e.g., plastic hinge zones). This is also addressed in Section 6.4.1.1 of FEMA 273.

For behavior modes with intermediate ductility capacity such as flexure/diagonal tension, flexure/diagonal compression, and flexure/sliding shear, the shear demand is based on the expected moment strength developed in the plastic hinge regions. The shear demand so derived can be magnified because of inelastic dynamic effects which change the pattern of inertial force in the building from the inverted triangular distribution typically assumed in analysis and design.

For the example of a cantilever wall with a plastic hinge at the base, the shear demand will equal the expected moment strength at the base divided by 2/3 the wall height for an inverted triangular distribution of lateral forces. However, if inelastic dynamic effects cause the pattern of lateral forces to approach a uniform distribution, then the shear demand will increase to a value equal to the expected moment strength at the base (which will still be developed) divided by 1/2 the wall height.

Inelastic dynamic effects have been studied by researchers, and a shear magnification factor, ω_v, taken as a function of the number of stories, is recommended by Paulay and Priestley (1992). The dynamic amplification of shear demand can be considered by use of such a factor or by considering different vertical distributions of lateral forces in the nonlinear static analysis.

Traditional design equations for shear strength tend to reflect the lower bound of test results, but the overall correlation of the equations with the data is not good. While some wall specimens show strength values close to the prediction of design equations, others show strength values five times higher than the predicted values (Cardenas, 1973).

b. Diagonal Tension

FEMA 273 specifies that the shear strength of reinforced concrete walls be calculated according to Section 21.6 of ACI 318-95. The applicable ACI equations are:

$$V_n = A_{cv}\,(2\sqrt{f'_{ce}} + \rho_n f_{ye})\ \text{for walls with a ratio of}$$
$$h_w / l_w \text{ greater than 2.0, and}$$

$$V_n = A_{cv}\,(3\sqrt{f'_{ce}} + \rho_n f_{ye})\ \text{for walls with a ratio of}$$
$$h_w / l_w \text{ less than 1.5}$$

FEMA 273 allows the use of these equations for walls with reinforcement ratios, ρ_n, as low as 0.0015 — below the 1995 ACI-specified minimum of 0.0025. For walls with reinforcement ratios below 0.0015, FEMA 273 specifies that the strength calculated at $\rho_n = 0.0015$ can still be used.

ATC-40 modifies the provisions of FEMA 273 and ACI 318-95 for wall shear strength. The principal modifications are that V_n need not be taken lower than $4\sqrt{f'_{ce}}\,A_{cv}$, and that $2\sqrt{f'_{ce}}$ is assumed for the concrete contribution to shear strength, regardless of the ratio of h_w / l_w. Reinforcement ratios less than 0.0025 are also addressed differently in the ATC-40 document, but in typical cases of light reinforcement, the $4\sqrt{f'_{ce}}\,A_{cv}$ lower limit governs the calculations.

The FEMA 273 and ATC-40 wall shear strength recommendations are design equations that do not explicitly consider:

- The effect of axial load on shear strength

- The distinction between shear strength at plastic-hinge zones versus that away from plastic-hinge zones

- The potential degradation of shear strength at plastic hinge zones

Equations for wall shear strength given in Paulay and Priestley (1992) recognize a significant increase in shear strength due to axial load level. The equations also recommend a much lower shear strength at plastic-hinge zones, accounting for potential degradation, than away from plastic-hinge zones.

If warranted by the specific conditions under evaluation, an approach similar to that used by Priestley et al. (1996) and Kowalsky et al. (1997) for columns can be used. The following shear strength equation:

$$V_n = V_c + V_s + V_p \qquad (5\text{-}1)$$

expresses the shear strength as the sum of three components: the contributions of the concrete, steel, and axial load. Each of these components is defined as follows:

$$V_c = \alpha \beta k_{rc} \sqrt{f'_{ce}}\, b_w (0.8 l_w) \qquad (5\text{-}2)$$

where k_{rc} is a function of ductility, as shown below:

$k_{rc} = 3.5$ for low ductility ($\mu_\Delta \le 2$) and away from plastic hinge regions.

$k_{rc} = 0.6$ for high ductility ($\mu_\Delta \ge 5$)

For values of ductility between the above limits, k_{rc} is calculated by linear interpolation.

The coefficient α accounts for wall aspect ratio, as considered in the ACI-318 equations:

$$\alpha = 3 - M/(0.8 l_w V) \qquad (5\text{-}3)$$

$$1.0 \le \alpha \le 1.5$$

The coefficient β accounts for longitudinal reinforcement ratio, as recognized by ASCE/ACI Task Committee 426 (1973):

$$\beta = 0.5 + 20 \rho_g \qquad (5\text{-}4)$$

$$\beta \le 1.0$$

where ρ_g is the ratio of total longitudinal reinforcement over gross cross-sectional area for the wall component.

$$V_s = \rho_n f_{ye}\, b_w h_d \qquad (5\text{-}5)$$

where h_d equals the height over which horizontal reinforcement contributes to shear strength, taken as ($l_w - c$)cot θ, where θ equals the angle, from the vertical, of the critical inclined shear crack. θ is taken as 35 degrees unless limited to larger angles by the potential corner-to-corner crack. Thus h_d does not exceed the clear height of a wall pier.

$$V_p = ((l_w - c)N_u)/(2M/V) \qquad (5\text{-}6)$$

M/V is taken as the larger of the values at the top and bottom of the wall pier. Thus $2M/V$ should not be less than the clear height of the wall pier.

These shear strength equations might also apply to coupling beams, for which l_w is the overall depth (measured vertically) of the coupling beam, and h_d is the horizontal length over which vertical stirrups contribute to shear strength.

c. Diagonal Compression (Web Crushing)

Walls and wall piers that have sufficient horizontal reinforcement to prevent a shear failure in diagonal tension may still suffer a shear failure associated with diagonal compression or web crushing. Web crushing behavior becomes more likely at higher levels of lateral deformation, and for walls with higher axial loads, N_u.

The web-crushing shear strength of a wall can be estimated according to the following equation (Oesterle et al., 1983):

$$V_{wc} = \frac{1.8 f'_{ce}}{1 + \left(600 - 2000\, \dfrac{N_u}{A_g f'_{ce}} \right)}\, b_w (0.8 l_w) \qquad (5\text{-}7)$$

where δ is the story drift ratio to which the wall component is subjected. The above equation applies to a typical range of axial loads for walls: $0 < N_u / A_g f'_{ce} < 0.09$. For walls with higher axial loads, V_{wc} is held constant at the value calculated for $N_u / A_g f'_{ce} = 0.09$. Thus, V_{wc} does not exceed:

$$V_{wc} = \frac{1.8 f'_{ce}}{1 + 420 \delta}\, b_w (0.8 l_w) \qquad (5\text{-}8)$$

The above expressions give a lower bound to the test data. Multiplying V_{wc} by 1.5 would give a reasonable upper bound to the web-crushing shear strength.

An alternative expression for the web-crushing shear strength is given in Section 5.44 of Paulay and Priestley (1992). This expression is based on displacement ductility rather than story drift and does not consider the effect of axial load.

The above procedures apply to the flexure/web-crushing behavior mode, and they indicate a degradation of web-crushing strength with increasing drift or ductility. Tests (Barda et al., 1976) have also shown preemptive web-crushing behavior; that is, web crushing that occurs at small displacement levels, before the wall has attained its flexural strength. The test results show that walls may suffer preemptive web crushing when shear stress levels exceed $12\sqrt{f'_{ce}}$ to $15\sqrt{f'_{ce}}$.

d. Sliding Shear

Sliding shear strength is assessed at construction joints and plastic hinge zones using the shear friction provisions of Section 11.7.4 of ACI 318-95. All reinforcement that crosses the potential sliding plane and is located within the wall section that resists shear is assumed to contribute to the sliding-shear strength.

Isolated Walls and Wall Piers. For isolated walls and wall piers, the potential sliding plane is a horizontal plane. Vertical reinforcement that crosses this plane and contributes to flexural strength also contributes to sliding-shear strength.

Shear transfer occurs primarily in the web of a wall section rather than in wall flanges. All vertical bars located in the web of the wall section, or within a distance b_w from the web, are considered effective as shear-friction reinforcement. For wall sections that have typical columns as boundary elements, the vertical bars in the wall web plus those in the boundary elements can be used for shear friction. For wall sections that have wide flanges as boundary elements, the vertical bars placed in the flanges, at a distance of more than b_w from the web, are not considered effective for shear friction (Paulay and Priestley, 1992).

It may be argued that only the reinforcement on the tension side of the neutral axis should be effective in contributing to shear friction strength, but such a recommendation has not been well established or tested.

Sliding-shear strength is investigated at construction joints and at plastic-hinge zones. The quality of the construction joint should be considered in establishing the appropriate coefficient of friction, μ, as specified in ACI 318. At plastic-hinge regions, increasing cyclic deformations cause horizontal flexural cracks at the potential sliding plane to open more widely, which results in a degradation of sliding-shear strength. In such a case, the effective coefficient of friction, μ, can be considered to be reduced.

A more detailed assessment of the sliding-shear strength of squat walls can be carried out according to the recommendations in Section 5.7 of Paulay and Priestley (1992).

Coupling Beams. If diagonal tension failures are prevented by sufficient stirrup reinforcement, and if diagonal bars are not used, sliding shear is likely to occur in short coupling beams at moderate-to-high ductilities. According to Paulay and Priestley (1992), there is a danger of sliding shear occurring in coupling beams whenever V_u exceeds $1.2\,(l_n/h)\sqrt{f'_{ce}}\,b_w d$, (assuming diagonal bars are not present and stirrups prevent a diagonal tension failure). The provisions of the 1997 UBC require diagonal bars in coupling beams when V_u exceeds $4\sqrt{f'_{ce}}\,b_w d$ and l_n/d is less than four.

For this document, in the absence of more detailed analyses, the sliding-shear strength of coupling beams may be assumed to be equal to $1.2\,(l_n/h)\sqrt{f'_{ce}}\,b_w$ d at high ductility levels and may be assumed equal to $3(l_n/h)\sqrt{f'_{ce}}\,b_w d$ at moderate ductility levels. Alternatively, a shear-friction approach could be considered for coupling beams.

5.3.7 Wall Boundary Confinement

For walls responding in flexure, boundary-tie reinforcement is usually needed in the plastic hinge regions to allow high ductility values to be achieved. Table 6-18 of FEMA 273 and Table 9-10 of ATC-40 reference the boundary confinement requirements of ACI 318-95, and both FEMA 273 and ATC-40 reference the 1994 Uniform Building Code (ICBO, 1994) and Wallace (1994, 1995). These references give substantially different recommendations for boundary tie requirements.

Paulay and Priestley (1992) and the New Zealand concrete code (SANZ, 1995) present more widely applicable recommendations for wall boundary ties. An adaptation of these recommendations is given below.

For walls to achieve *high ductility capacities*, boundary ties must meet the following criteria:

a. Walls with $c \le 0.15l_w$ and $\rho_l \le 400 / f_{ye}$:

    Boundary ties are not required.

b. Walls with $c \le 0.15l_w$ and $\rho_l > 400 / f_{ye}$:

    Boundary ties are necessary, as specified below, to prevent buckling of longitudinal bars:

    - Boundary ties extend over a length of the wall section at the compression boundary greater than or equal to c', taken as the larger of $c - 0.1l_w$ or $0.5c$, where c is the distance from the compression face to the neutral axis.

    - Boundary ties extend over a height of the wall at the plastic hinge region greater than or equal to $2l_p$.

    - Ties are spaced at no more than $6d_b$, where d_b is the diameter of the longitudinal bar being tied.

    - Each longitudinal bar is restrained against bar buckling by either a crosstie or a 90-degree bend of a hoop with d_{bt} greater than or equal to $0.25d_b$; or is restrained by a hoop leg parallel to the wall surface which spans not more than 14 in. between 90-degree bends of the hoop, with d_{bt} greater than or equal to $0.4d_b$. (d_{bt} is the diameter of the crosstie or hoop.)

c. Walls with $c > 0.15l_w$:

    Boundary ties are necessary to prevent buckling of longitudinal bars and to confine the concrete to achieve higher compressive strains. In addition to meeting the requirements of item (b) above, ties are provided so that:

$$A_{sh} \ge 0.2sh_c \left(\frac{f'_{ce}}{f_{yhe}} \right) \left(\frac{A_g}{A_{ch}} \right) \left(\frac{c}{l_w} - 0.10 \right) \qquad (5\text{-}9)$$

The term ρ_l is the local reinforcement ratio for flexural reinforcement, as defined below:

$$\rho_l = A_s/bs_l$$

where A_s is the area of vertical wall reinforcement in a layer spaced at s_l along the length of the wall, and where b is the width of the wall at the compression boundary.

Walls that do not meet the criteria for high ductility capacities, but which have some boundary ties in the plastic hinge region, spaced at no more that $10 d_b$, and that have dimensions $c \le 0.20l_w$, can be assumed to achieve *moderate ductility capacities* $(2 \le \mu_\Delta \le 5)$.

5.3.8 Lap Splice Strength

As specified in Section 6.4.5 of FEMA 273 and Section 9.5.4.5 of ATC-40, the strength of existing lap splices may be estimated according to the ratio of lap-length provided to the tension development length required by ACI 318-95.

Thus, the strength of lap splices can be taken as:

$$f_s = (l_b/l_d)f_{ye} \qquad (5\text{-}10)$$

where: f_s = stress capacity of the lap splice
l_b = provided lap-splice length
l_d = tension development length for straight bars, taken according to ACI 318, Chapter 12

Note that the tension development length, l_d, is used in the above equations without the 1.3 splice factor of ACI-318, because the specified lap-splice lengths prescribed for new design are conservative (ATC 1996).

For splices in plastic-hinge regions, the evaluation should consider that lap-splice slip may still be possible even if splice lengths are adequate according to the above criteria.

A method of assessing lap-splice strength and the ductility capacity of flexural plastic hinges that contain lap splices is given in Sections 5.5.4, 7.4.5, and 7.4.6 of Priestley et al. (1996). The method allows the calculation of strength based on a fundamental consideration of the mechanics of lap-splice slip.

When the lap-splice strength is less than that required to yield the reinforcement, the full moment strength of the section will not develop. Even when lap splices have sufficient capacity to yield the reinforcement, they may still slip when moderate ductility levels are reached. As developed for columns, the Priestley et al. (1996) method indicates that all lap splices may become prone to slipping when the concrete compressive strain reaches 0.002. The method gives an estimate of the degradation of lap splice strength with increasing ductility, which results in a loss of moment capacity down to a residual value based on axial force alone.

5.3.9 Wall Buckling

Thin wall sections responding in flexure may be prone to out-of-plane buckling, typically at higher ductility levels. The 1997 UBC prescribes a minimum wall thickness of 1/16 the clear story height for walls that require boundary confinement. Out-of-plane buckling is possible in plastic-hinge regions of walls even if they do not require confinement. Paulay and Priestley (1992, 1993) address the wall buckling phenomenon in detail, and the New Zealand concrete code (SANZ, 1995) provides design recommendations for minimum wall thickness based on the research.

Flanged or barbell-shaped wall sections are typically not vulnerable to buckling, unless the flange is unusually narrow, having a width, b, less than that specified below.

Based on the research, the following simplified criteria are recommended: Walls with width, b, equal to or greater than $l_u/16$ can be assumed to achieve high ductility capacity without buckling. Walls with b equal to $l_u/24$ can be assumed to be vulnerable to buckling at moderate-to-high ductility levels.

The length, l_u, is taken as the smaller of:

- The clear story height between floors bracing the wall in the out-of-plane direction, *and*

- $2.5l_p$ for single-curtain walls and walls with ρ_l greater than $200/f_{ye}$, *or* $2.0l_p$ for two-curtain walls with ρ_l less than or equal to $200/f_{ye}$.

The term b is the width of the wall at the compression boundary. The term ρ_l is the local reinforcement ratio for flexural reinforcement, as defined in Section 5.3.7.

FEMA 273 and ATC-40 do not address overall wall buckling.

5.4 Symbols for Reinforced Concrete

Symbols that are used in this chapter are defined below. Further information on some of the variables used (particularly those noted "per ACI") may be found by looking up the symbol in Appendix D of ACI 318-95.

A_{ch} = Cross sectional area of confined core of wall boundary region, measured out-to-out of confining reinforcement and contained within a length c' from the end of the wall, Section 5.3.7

A_{cv} = Net area of concrete section bounded by web thickness and length of section in the direction of shear force considered, in^2 (per ACI)

A_g = Gross cross sectional area of wall boundary region, taken over a length c' from the end of the wall, Section 5.3.7

A_{sh} = Total cross-sectional area of transverse reinforcement (including crossties) within spacing s and perpendicular to dimension h_c. (per ACI)

b = Width of compression face of member, in (per ACI)

b_w = Web width, in (per ACI)

c = Distance from extreme compressive fiber to neutral axis (per ACI)

c' = Length of wall section over which boundary ties are required, per Section 5.3.7

d_b = Bar diameter (per ACI)

d_{bt} = Bar diameter of tie or loop

f_c' = Specified compressive strength of concrete, psi (per ACI)

f_{ce}' = Expected compressive strength of concrete, psi

f_y = Specified yield strength of nonprestressed reinforcement, psi. (per ACI)

f_{ye} = Expected yield strength of nonprestressed reinforcement, psi.

f_{yh} = Specified yield strength of transverse reinforcement, psi (per ACI)

f_{yhe} = Expected yield strength of transverse reinforcement, psi

h_c = Cross sectional dimension of confined core of wall boundary region, measured out-to-out of confining reinforcement

h_d = Height over which horizontal reinforcement contributes to V_s per Section 5.3.6.b

h_w = Height of wall or segment of wall considered (per ACI)

k_{rc} = Coefficient accounting the effect of ductility demand on V_c per Section 5.3.6.b

l_p = Equivalent plastic hinge length, determined according to Section 5.3.3.

l_u = Unsupported length considered for wall buckling, determined according to 5.3.9

l_n = Beam clear span (per ACI)

l_w = Length of entire wall or segment of wall considered in direction of shear force (per ACI). (For isolated walls and wall piers equals horizontal length, for spandrels and coupling beams equals vertical dimension i.e., overall depth)

M_{cr} = Cracking moment (per ACI)

M_e = Expected moment strength at section, equal to nominal moment strength considering expected material strengths.

M_n = Nominal moment strength at section (per ACI)

M_u = Factored moment at section (per ACI)

M/V = Ratio of moment to shear at a section. When moment or shear results from gravity loads in addition to seismic forces, can be taken as M_u/V_u

N_u = Factored axial load normal to cross section occurring simultaneously with V_u; to be taken as positive for compression, negative for tension (per ACI)

s = Spacing of transverse reinforcement measured along the longitudinal axis of the structural member (per ACI)

s_l = spacing of vertical reinforcement in wall (per ACI)

V_c = Nominal shear strength provided by concrete (per ACI)

V_n = Nominal shear strength (per ACI)

V_p = Nominal shear strength related to axial load per Section 5.3.6

V_s = Nominal shear strength provided by shear reinforcement (per ACI)

V_u = Factored shear force at section (per ACI)

V_{wc} = Web crushing shear strength per Section 5.3.6.c

α = Coefficient accounting for wall aspect ratio effect on V_c per Section 5.3.6.b

β = Coefficient accounting for longitudinal reinforcement effect on V_c per Section 5.3.6.b

δ = Story drift ratio for a component, corresponding to the global target displacement, used in the computation of V_{wc}, Section 5.3.6.c

μ = Coefficient of friction (per ACI)

μ_Δ = Displacement ductility demand for a component, used in Section 5.3.4, as discussed in Section 6.4.2.4 of FEMA-273. Equal to the component deformation corresponding to the global target displacement, divided by the effective yield displacement of the component (which is defined in Section 6.4.1.2B of FEMA-273).

ρ_g = Ratio of total reinforcement area to cross-sectional area of wall.

ρ_l = Local reinforcement ratio in boundary region of wall according to Section 5.3.7

ρ_n = Ratio of distributed shear reinforcement on a plane perpendicular to plane of A_{cv} (per ACI). (For typical wall piers and isolated walls indicates amount of horizontal reinforcement.)

5.5   Reinforced Concrete Component Guides

The following Component Damage Classification
Guides contain details of the behavior modes for
reinforced concrete components. Included are the
distinguishing characteristics of the specific behavior
mode, the description of damage at various levels of
severity, and performance restoration measures.
Information may not be included in the Component
Damage Classification Guides for certain damage

severity levels; in these instances, for the behavior
mode under consideration, it is not possible to make
refined distinctions with regard to severity of damage.
See also Section 3.5 for general discussion of the use of
the Component Guides and Section 4.4.3 for
information on the modeling and acceptability criteria
for components.

RC1A	COMPONENT DAMAGE CLASSIFICATION GUIDE	System:	Reinforced Concrete
		Component Type:	Isolated Wall or Stronger Pier
		Behavior Mode:	Ductile Flexural

How to distinguish behavior mode:

By observation:

Wide flexural cracking and spalling should be concentrated in the plastic hinge zone, although minor flexural cracking (width not exceeding 1/8 in.) may extend beyond the plastic hinge zone. Shear cracks may occur but widths should not exceed 1/8 in. If cracks exceed this width, see RC1B. Vertical cracks and spalling may occur at the extreme fibers of the plastic hinge region (toe region). If there is spalling or crushing of concrete within the web or center area of the section, see RC1C. If reinforcing bars in the toe region buckle, see RC1E.

Ductile flexural behavior typically occurs in well-designed walls that have sufficient horizontal reinforcement and do not have heavy vertical (flexural) reinforcement.

Note: At low damage levels, damage observations will be similar to those for other behavior modes.

By analysis:

Strength in all other behavior modes, even after possible degradation, is sufficient to ensure that flexural behavior controls. Strength associated with shear, web crushing, sliding shear, and lap splices — taken for conditions of high ductility — exceeds moment strength. Foundation rocking strength exceeds moment strength. Boundary ties are sufficient to prevent bar buckling or loss of confinement, and wall thickness is sufficient to prevent overall buckling.

Refer to Evaluation Procedures for:

- Identifying plastic hinge locations and extent.

- Identifying flexural versus shear cracks.

- Calculation of moment, diagonal tension, web-crushing, sliding-shear, lap splice, and foundation rocking strength.

- Required boundary ties and wall thicknesses.

Severity	Description of Damage		Performance Restoration Measures
Insignificant	*Criteria:*	• No crack widths exceed 3/16 in., <u>and</u>	(Repairs may be necessary for restoration of nonstructural characteristics.)
		• No shear cracks exceed 1/8 in., <u>and</u>	
		• No significant spalling or vertical cracking	
$\lambda_K = 0.8$ $\lambda_Q = 1.0$ $\lambda_D = 1.0$	*Typical Appearance:* $2l_p$	Note: l_p *is length of plastic hinge. See Section 5.3.3*	

COMPONENT DAMAGE CLASSIFICATION GUIDE continued		RC1A
Severity	**Description of Damage**	**Performance Restoration Measures**
Slight $\lambda_K = 0.6$ $\lambda_Q = 1.0$ $\lambda_D = 1.0$	*Criteria:* • Crack widths do not exceed 1/4 in., <u>and</u> • No shear cracks exceed 1/8 in., <u>and</u> • No significant spalling or vertical cracking, <u>and</u> • No buckled or fractured reinforcement, <u>and</u> • No significant residual displacement. *Typical Appearance:* Similar to insignificant damage, except wider flexural cracks and typically more extensive cracking.	• Inject cracks $\lambda_K{}^* = 0.9$ $\lambda_Q{}^* = 1.0$ $\lambda_D{}^* = 1.0$
Moderate $\lambda_K = 0.5$ $\lambda_Q = 0.8$ $\lambda_D = 0.9$	*Criteria:* • Spalling or vertical cracking (or incipient spalling as identified by sounding) occurs at toe regions in plastic hinge zone, typically limited to the cover concrete, <u>and</u> • No buckled or fractured reinforcement, <u>and</u> • No significant residual displacement. *Typical Appearance:* Crack widths typically do not exceed 1/4 in. vertical cracking and/or spalling $2l_p$ Note: *l_p is length of plastic hinge. See Section 5.3.3*	• Remove and patch spalled and loose concrete. Inject cracks. $\lambda_K{}^* = 0.8$ $\lambda_Q{}^* = 1.0$ $\lambda_D{}^* = 1.0$
Heavy	Not Used	
Extreme	*Criteria:* • Reinforcement has fractured. *Typical Indications* • Wide flexural cracking typically concentrated in a single crack. • Large residual displacement.	• Replacement or enhancement required.

RC1B	COMPONENT DAMAGE CLASSIFICATION GUIDE	System: **Reinforced Concrete**
		Component Type: **Isolated Wall or Stronger Pier**
		Behavior Mode: **Flexure/Diagonal Tension**

How to distinguish behavior mode:

By observation:

For insignificant to moderate levels of damage, indications will be similar to those for RC1A, although shear cracking may begin at lower ductility levels. At higher levels of damage, one or more wide shear cracks begin to form.

Typically occurs in walls that have a low-to-moderate amount of horizontal reinforcement, and which may have heavy vertical (flexural) reinforcement. May be most prevalent in walls with intermediate aspect ratios, $M/Vl_w \approx 2$, but depending on the reinforcement, can occur over a wide range of aspect ratios.

By analysis:

Shear strength calculated for conditions of low ductility exceeds flexural capacity, but shear strength calculated for conditions of high ductility is less than the flexural capacity.

Foundation rocking strength exceeds moment strength. Boundary ties are sufficient to prevent buckling of longitudinal bars and loss of confinement prior to shear failure. Wall thickness is sufficient to prevent overall buckling prior to shear failure. Sliding shear strength is not exceeded.

Refer to Evaluation Procedures for:

- Identifying plastic hinge locations and extent.
- Identifying flexural versus shear cracks.

- Calculation of moment, diagonal tension, web-crushing, sliding-shear, lap splice, and foundation rocking strength.
- Required boundary ties and wall thickness.

Severity	Description of Damage		Performance Restoration Measures
Insignificant $\lambda_K = 0.8$ $\lambda_Q = 1.0$ $\lambda_D = 1.0$	*Criteria:*	• Shear crack widths do not exceed 1/16 in., <u>and</u> • Flexural crack widths do not exceed 3/16 in., <u>and</u> • No significant spalling or vertical cracking.	(Repairs may be necessary for restoration of nonstructural characteristics.)
	Typical Appearance: $2l_p$	Note: l_p is length of plastic hinge. See Section 5.3.3	

COMPONENT DAMAGE CLASSIFICATION GUIDE continued			RC1B
Severity	**Description of Damage**		**Performance Restoration Measures**
Slight	Not Used		
Moderate $\lambda_K = 0.5$ $\lambda_Q = 0.8$ $\lambda_D = 0.9$	*Criteria:*	• Shear crack widths do not exceed 1/8 in., <u>and</u> • Flexural crack widths do not exceed 1/4 in., <u>and</u> • Shear cracks exceed 1/16 in., <u>or</u> limited spalling (or incipient spalling as identified by sounding) occurs at web or toe regions, <u>and</u> • No buckled or fractured reinforcement, <u>and</u> • No significant residual displacement.	• Remove and patch spalled and loose concrete. Inject cracks.
	Typical Appearance:	Similar to insignificant damage except wider cracks, possible spalling, and typically more extensive cracking.	$\lambda_K{}^* = 0.8$ $\lambda_Q{}^* = 1.0$ $\lambda_D{}^* = 1.0$
Heavy $\lambda_K = 0.2$ $\lambda_Q = 0.3$ $\lambda_D = 0.7$ Note: λ_Q can be calculated based on shear strength at high ductility. See Section 5.3.6.	*Criteria:* *Typical Appearance:*	• Shear crack widths may exceed 1/8 in., but do not exceed 3/8 in. Higher cracking width is concentrated at one or more cracks. $2l_p$ Note: *l_p is length of plastic hinge. See Section 5.3.3*	• Replacement or enhancement is required for full restoration of seismic performance. • For <u>partial</u> restoration of performance, inject cracks $\lambda_K{}^* = 0.5$ $\lambda_Q{}^* = 0.8$ $\lambda_D{}^* = 0.8$
Extreme	*Criteria:* *Typical Indications*	• Reinforcement has fractured. • Wide shear cracking typically concentrated in a single crack.	• Replacement or enhancement required.

RC1C	COMPONENT DAMAGE CLASSIFICATION GUIDE	System: **Reinforced Concrete**
		Component Type: **Isolated Wall or Stronger Pier**
		Behavior Mode: **Flexure/Web Crushing**

How to distinguish behavior mode:

By observation:
For insignificant-to-moderate levels of damage, indications will be similar to those for RC1A and RC1B. At higher levels of damage, extensive diagonal cracking and spalling of web regions begins to occur.

Typically occurs in walls that have sufficient horizontal reinforcement, and that may have heavy vertical (flexural) reinforcement. May be more prevalent in low-rise walls, walls with higher axial loads, and in walls with flanges or heavy boundary elements.

By analysis:
Web crushing strength, calculated for high levels of story drift or ductility, is less than flexural strength.

Foundation rocking strength exceeds moment strength. Boundary ties are sufficient to prevent buckling of longitudinal bars and loss of confinement prior to web-crushing failure. Wall thickness is sufficient to prevent overall buckling prior to web crushing failure. Sliding shear strength is not exceeded.

Refer to Evaluation Procedures for:

- Identifying plastic hinge locations and extent.

- Identifying flexural versus shear cracks.

- Calculation of moment, diagonal tension, web-crushing, sliding-shear, lap splice, and foundation rocking strength.

- Required boundary ties and wall thickness.

Severity	Description of Damage		Performance Restoration Measures
Insignificant	$\mu_\Delta \le 3$	See **RC1B**	See **RC1B**
Slight	Not Used		
Moderate $\lambda_K = 0.5$ $\lambda_Q = 0.8$ $\lambda_D = 0.9$	*Criteria:*	• Shear crack widths do not exceed 1/8 in., <u>and</u> • Flexural crack widths do not exceed 1/4 in., <u>and</u> • Limited spalling (or incipient spalling as identified by sounding) occurs at web or toe regions, <u>or</u> shear cracks exceed 1/16 in., <u>and</u> • No buckled or fractured reinforcement, <u>and</u> • No significant residual displacement.	• Remove and patch spalled and loose concrete. Inject cracks. $\lambda_K{}^* = 0.8$ $\lambda_Q{}^* = 1.0$ $\lambda_D{}^* = 1.0$
Heavy $\lambda_K = 0.2$ $\lambda_Q = 0.3$ $\lambda_D = 0.7$	*Criteria:* *Typical Appearance:*	• Significant spalling of concrete in web, <u>and</u> • No fractured reinforcement. Note: *l_p is length of plastic hinge. See Section 5.3.3*	• Remove and patch all spalled and loose concrete. Inject cracks. $\lambda_K{}^* = 0.8$ $\lambda_Q{}^* = 1.0$ $\lambda_D{}^* = 1.0$
Extreme	*Criteria:*	• Heavy spalling and voids in web concrete, <u>or</u> significant residual displacement.	• Replacement or enhancement required.

RC1D	COMPONENT DAMAGE CLASSIFICATION GUIDE	System: **Reinforced Concrete**
		Component Type: **Isolated Wall or Stronger Pier**
		Behavior Mode: **Flexure/Sliding Shear**

How to distinguish behavior mode:

By observation:
For insignificant-to-moderate levels of damage, indications will be similar to those for RC1A. In the plastic hinge zone, flexural cracks join up across the section, which becomes a potential sliding plane. At higher levels of damage, degradation of the concrete and sliding along this crack begin to occur.

Typically occurs in low-rise walls that have sufficient horizontal reinforcement. Sliding may occur at horizontal construction joints. May be more prevalent in walls with lower axial loads, and in walls with flanges or heavy boundary elements. Unlikely to occur if diagonal reinforcement crosses the potential sliding plane.

By analysis:
Sliding shear strength is less than shear corresponding to moment strength.

Strength associated with diagonal tension, web crushing, and lap splices — taken for conditions of high ductility — exceeds moment strength. Foundation rocking strength exceeds moment strength. Boundary ties are sufficient to prevent buckling of longitudinal bars and loss of confinement prior to sliding. Wall thickness is sufficient to prevent overall buckling.

Boundary ties are insufficient to prevent bar buckling or provide adequate confinement.

Refer to Evaluation Procedures for:

• Identifying plastic hinge locations and extent.

• Identifying flexural versus shear cracks.

• Calculation of moment, diagonal tension, web-crushing, sliding-shear, lap splice, and foundation rocking strength.

• Required boundary ties and wall thickness.

Severity	Description of Damage		Performance Restoration Measures
Insignificant	See **RC1A**		See **RC1A**
Slight	See **RC1A**		See **RC1A**
Moderate	Not Used		
Heavy $\lambda_K = 0.4$ $\lambda_Q = 0.5$ $\lambda_D = 0.8$	*Criteria:*	• Development of a major horizontal flexural crack along the entire wall length, with some degradation of concrete along the crack, indicating that sliding has occurred. Possible small lateral offset at crack.	• Remove and patch all spalled or loose concrete. Inject cracks. $\lambda_K^* = 0.8$ $\lambda_Q^* = 1.0$ $\lambda_D^* = 1.0$
	Typical Appearance: Crack widths typically do not exceed 3/8 in. $2l_p$	Note: l_p is length of plastic hinge. See Section 5.3.3	
Extreme	*Criteria:*	• Significant lateral offset at sliding plane	• Replacement or enhancement required.

RC1E	COMPONENT DAMAGE CLASSIFICATION GUIDE	System:	**Reinforced Concrete**
		Component Type:	**Isolated Wall or Stronger Pier**
		Behavior Mode:	**Flexure/Boundary Compression**

How to distinguish behavior mode:

By observation:
For insignificant-to-moderate levels of damage, indications will be similar to those for RC1A (although spalling may occur at lower ductility levels). At higher levels of damage, boundary regions in plastic hinge zone begin to sustain spalling and crushing.

Flexure/boundary compression typically occurs in walls that have sufficient horizontal reinforcement and do not have well confined boundary regions. May be more prevalent in walls with a higher M/Vl_w ratio.

Caution: When vertical cracks or spalling at boundary regions is observed, boundary reinforcement should be exposed and inspected for buckling or cracking.

By analysis:
Strength in all other behavior modes, even after possible degradation, is sufficient to ensure that flexural behavior controls. Strength associated with shear, web crushing, sliding shear, and lap splices — taken for conditions of high ductility — exceeds moment strength. Foundation rocking strength exceeds moment strength. Wall thickness is sufficient to prevent overall buckling.

Boundary ties are insufficient to prevent bar buckling or provide adequate confinement.

Refer to Evaluation Procedures for:

- Identifying plastic hinge locations and extent.

- Identifying flexural versus shear cracks.

- Required boundary ties and wall thickness.

- Calculation of moment, diagonal tension, web-crushing, sliding-shear, lap splice, and foundation rocking strength.

Severity	Description of Damage		Performance Restoration Measures
Insignificant	See **RC1A**		See **RC1A**
Slight	See **RC1A**		See **RC1A**
Moderate	See **RC1A**		See **RC1A**
Heavy $\lambda_K = 0.4$ $\lambda_Q = 0.6$ $\lambda_D = 0.7$	*Criteria:*	• Spalling or vertical cracking occurs at toe regions in plastic hinge zone, <u>and</u> • Boundary longitudinal reinforcement is buckled <u>or</u> concrete within core of boundary regions (not just cover concrete) is heavily damaged. *Typical Appearance:* Crack widths typically do not exceed 3/8 in. buckled reinforcement and/or heavily damaged concrete Note: l_p is length of plastic hinge. See *Section 5.3.3*	• Remove spalled and loose concrete. Remove and replace buckled reinforcement. Provide additional ties around longitudinal bars of the critical boundary region, at the location of the replaced bars. Patch concrete. Inject cracks. $\lambda_K{}^* = 0.8$ $\lambda_Q{}^* = 1.0$ $\lambda_D{}^* = 1.0$
Extreme	See **RC1A**		See **RC1A**

RC2A	COMPONENT DAMAGE CLASSIFICATION GUIDE	System:	**Reinforced Concrete**
		Component Type:	**Weaker Pier**
		Behavior Mode:	**Ductile Flexural**

How to distinguish behavior mode:

By observation:
Wide flexural cracking and spalling should be concentrated in the plastic hinge zone, although minor flexural cracking (width not exceeding 1/8 in.) may extend beyond the plastic hinge zone. Shear cracks may occur but widths should not exceed 1/8 in. Vertical cracks and spalling may occur at the extreme fibers of the plastic hinge region.

Ductile flexural behavior typically occurs in well-designed, slender wall piers that have sufficient horizontal reinforcement and do not have heavy vertical (flexural) reinforcement.

Note: At low damage levels, damage observations will be similar to those for other behavior modes.

By analysis:
Strength in all other behavior modes, even after possible degradation, is sufficient to ensure that flexural behavior controls. Strength associated with shear, web crushing, sliding shear, and lap splices — taken for conditions of high ductility — exceeds moment strength. Foundation rocking strength exceeds moment strength. Boundary ties are sufficient to prevent bar buckling or loss of confinement, and wall thickness is sufficient to prevent overall buckling.

Refer to Evaluation Procedures for:

- Identifying plastic hinge locations and extent.
- Identifying flexural versus shear cracks.

- Calculation of moment, diagonal tension, web-crushing, sliding-shear, lap splice, and foundation rocking strength.
- Required boundary ties and wall thickness.

Severity	Description of Damage		Performance Restoration Measures
Insignificant	See **RC1A**		See **RC1A**
Slight	See **RC1A**		See **RC1A**
Moderate $\lambda_K = 0.5$ $\lambda_Q = 0.8$ $\lambda_D = 0.9$	*Criteria:*	• Spalling or vertical cracking (or incipient spalling as identified by sounding) occurs at toe regions in plastic hinge zone, typically limited to the cover concrete, <u>and</u> • No buckled or fractured reinforcement, <u>and</u> • No significant residual displacement. *Typical Appearance:* Crack widths typically do not exceed 1/4 in. Note: l_p *is length of plastic hinge. See Section 5.3.3* $2l_p$ $2l_p$	• Remove and patch spalled and loose concrete. Inject cracks. $\lambda_K^* = 0.8$ $\lambda_Q^* = 1.0$ $\lambda_D^* = 1.0$
Heavy	Not Used		
Extreme	See **RC1A**		See **RC1A**

RC2H	COMPONENT DAMAGE CLASSIFICATION GUIDE	System: **Reinforced Concrete**
		Component Type: **Weaker Pier**
		Behavior Mode: **Preemptive Diagonal Tension**

How to distinguish behavior mode:

By observation:
For lower levels of damage, indications will be similar to those for other behavior modes, although flexural cracks may not be apparent. Damage quickly becomes heavy when diagonal cracks open up. Because flexural reinforcement never yields, flexural cracks should not have a width greater than 1/8 in.

Preemptive diagonal shear typically occurs in wall piers that have inadequate (or no) horizontal reinforcement, and that may have heavy vertical reinforcement. May be more prevalent in wall piers with low M/Vl_w ratio.

By analysis:
Strength in shear at low ductility is less than the capacity corresponding to moment strength, foundation rocking strength, or lap-splice strength (at low ductility).

Refer to Evaluation Procedures for:

- Identifying flexural versus shear cracks.

- Calculation of moment, shear, lap-splice, and foundation rocking strength.

Severity	Description of Damage		Performance Restoration Measures
Insignificant $\lambda_K = 0.9$ $\lambda_Q = 1.0$ $\lambda_D = 1.0$	*Criteria:*	• No shear cracking <u>and</u> • Flexural crack widths do not exceed 1/8 in.	See **RC1A**
	Typical Appearance:	Similar to RC2A except no shear cracking and smaller crack widths.	
Slight	Not Used		
Moderate $\lambda_K = 0.5$ $\lambda_Q = 0.8$ $\lambda_D = 0.9$	*Criteria:*	• No crack widths exceed 1/8 in. <u>and</u> • No vertical cracking or spalling	• Inject cracks $\lambda_K^* = 0.8$ $\lambda_Q^* = 1.0$ $\lambda_D^* = 1.0$
	Typical Appearance:	Similar to insignificant damage except thin shear cracks may be present.	
Heavy $\lambda_K = 0.2$ $\lambda_Q = 0.3$ $\lambda_D = 0.7$ Note: λ_Q can be calculated based on shear strength at high ductility See Section 5.3.6	*Criteria:* *Typical Appearance:*	• Shear crack widths exceed 1/8 in., but do not exceed 3/8 in. Cracking becomes concentrated at one or more cracks. 	• Replacement or enhancement is required for full restoration of seismic performance. • For <u>partial</u> restoration of performance, Inject cracks. $\lambda_K^* = 0.5$ $\lambda_Q^* = 0.8$ $\lambda_D^* = 0.8$
Extreme	*Criteria:* *Typical Indications*	• Reinforcement has fractured. • Wide shear cracking typically concentrated in a single crack.	• Replacement or enhancement required

RC3B	COMPONENT DAMAGE CLASSIFICATION GUIDE	System:	**Reinforced Concrete**
		Component Type:	**Coupling Beam**
		Behavior Mode:	**Flexure/Diagonal Tension**

How to distinguish behavior mode:

By observation:

For insignificant-to-moderate levels of damage, indications will be similar to those for RC1A, although shear cracking may begin at lower ductility levels. At higher levels of damage, one or more wide shear cracks begin to form.

Flexure/Diagonal tension typically occurs in coupling beams that have inadequate stirrup reinforcement and that may have heavy horizontal (flexural) reinforcement. More prevalent in deeper beams than in shallower beams, but depending on the reinforcement, can occur over a wide range of aspect ratios.

By analysis:

Shear strength calculated for conditions of low ductility exceeds flexural capacity, but shear strength calculated for conditions of high ductility is less than the flexural capacity.

Web crushing strength and sliding shear strength are not exceeded.

Refer to Evaluation Procedures for:
- Identifying plastic hinge locations and extent.
- Identifying flexural versus shear cracks.

- Calculation of moment, diagonal tension, web-crushing, sliding-shear, and lap splice strength.
- Required boundary ties and wall thicknesses.

Severity	Description of Damage		Performance Restoration Measures
Insignificant	See **RC1B**		See **RC1B**
Slight	Not Used		
Moderate	See **RC1B**		See **RC1B**
Heavy $\lambda_K = 0.2$ $\lambda_Q = 0.3$ $\lambda_D = 0.7$ Note: λ_Q can be calculated based on shear strength at high ductility See Section 5.3.6	*Criteria:* • Shear crack widths may exceed 1/8 in., but do not exceed 3/8 in. Higher width cracking is concentrated at one or more cracks. *Typical Appearance:* 		• Replacement or enhancement is required for full restoration of seismic performance. • For <u>partial</u> restoration of performance, Inject cracks. $\lambda_K{}^* = 0.5$ $\lambda_Q{}^* = 0.8$ $\lambda_D{}^* = 0.8$
Extreme	See **RC1B**		See **RC1B**

RC3D	COMPONENT DAMAGE CLASSIFICATION GUIDE	System:	**Reinforced Concrete**
		Component Type:	**Coupling Beam**
		Behavior Mode:	**Flexure/Sliding Shear**

How to distinguish behavior mode:

By observation:

For insignificant-to-moderate levels of damage, indications will be similar to those for RC1A. Vertical flexural cracks join up across one or both ends of the section, which become a potential sliding plane. At higher levels of damage, degradation of the concrete and sliding along the critical crack begin to occur.

This behavior typically occurs in coupling beams that do not have diagonal reinforcement, but have sufficient stirrups to prevent diagonal tension failures.

Refer to Evaluation Procedures for:

● Identifying plastic hinge locations and extent.

● Identifying flexural versus shear cracks.

By analysis:

Sliding shear strength is less than shear corresponding to moment strength.

Strength associated with diagonal tension, web crushing, and lap splices for conditions of high ductility exceeds moment strength.

● Calculation of moment, diagonal tension, web-crushing, sliding-shear, and lap splice strength.

● Required boundary ties and wall thickness.

Severity	Description of Damage		Performance Restoration Measures
Insignificant	See **RC1D**		See **RC1D**
Slight	See **RC1D**		See **RC1D**
Moderate	Not Used		
Heavy $\lambda_K = 0.2$ $\lambda_Q = 0.3$ $\lambda_D = 0.7$	*Criteria:*	● Development of a major vertical flexural crack along the entire beam depth, with some degradation of concrete along the crack, indicating that sliding has occurred. Possible small lateral offset at crack. *Typical Appearance:* Crack widths typically do not exceed 3/8 in. 	● Remove and patch all spalled or loose concrete. Inject cracks. $\lambda_K^* = 0.8$ $\lambda_Q^* = 1.0$ $\lambda_D^* = 1.0$
Extreme	See **RC1D**		See **RC1D**

6: Reinforced Masonry

6.1 Introduction and Background

This section provides material relating to reinforced masonry (RM) construction and includes the Component Damage Classification Guides (Component Guides) in Section 6.5. Reinforced masonry component types and behavior modes are defined and discussed in Section 6.2. The overall damage evaluation procedure uses conventional material properties as a starting point. Section 6.3 provides supplemental information on strength and deformation properties for evaluating reinforced masonry components. Typical hysteretic behavior for reinforced masonry components and the interpretation of cracking are discussed in FEMA 307. The information presented on reinforced masonry components has been generated from a review of available empirical and theoretical data listed in the reference section and the annotated tabular bibliography in FEMA 307. These provide the user with further detailed resources on reinforced masonry component behavior.

Unreinforced masonry components (URM) are covered in Chapter 7 of this document. The distinction between reinforced and unreinforced masonry can sometimes be an issue. In those cases, masonry with less that 25 percent of the recommended minimum reinforcement specified in FEMA 273 should be considered unreinforced.

The most effective first step in identifying reinforced masonry components and their likely behavior modes is to place the structure in the context of the history of local construction practices, and to determine the type and amount, if any, of reinforcement used. There are examples of the use of iron to reinforce brick masonry construction in the 19th century; however, the widespread use of modern reinforced masonry did not begin until the 1930s. The use of reinforced masonry building systems was accelerated on the west coast following the 1933 Long Beach earthquake when the use of unreinforced masonry for new buildings in California was prohibited, so there is a distinct difference in building types in California before and after 1933. Reinforced masonry construction technology also developed in the east, although unreinforced masonry structures may still be built in some areas. The use of Portland cement mortars increased steadily from the beginning of the 20th century, as did the strength and quality of fired clay masonry. FEMA 274, Chapter 7, includes additional information on the history of masonry construction in the United States, as does the Brick Institute of America "Technical Notes on Brick Construction, No. 17." (BIA, 1988)

A wide variety of construction systems may be classified as reinforced masonry. The most common are:

- Fully-grouted hollow concrete block

- Partially-grouted hollow concrete block

- Fully-grouted hollow clay brick

- Partially-grouted hollow clay brick

- Grouted-cavity wall masonry (two wythes of clay brick or hollow units with a reinforced, grouted cavity)

Most of these are addressed in this section; however, the quantity and quality of experimental data available for each type varies considerably.

The last twenty-five years have seen a dramatic increase in masonry research over that in prior years, as evidenced by the proceedings of the International Brick/ Block Masonry Conferences (1969 - present), The North American Masonry Conferences (1976 - present), and the Canadian Masonry Symposia (1976 - present). Much of this work has been directed toward measuring strength and serviceability characteristics under gravity or wind loading or toward development of working-stress design methods. Since the early eighties, a growing number of studies have addressed the strength and deformation characteristics of reinforced masonry components under cyclic (simulated seismic) loading. Notable early studies include those at the University of California, San Diego (e.g., Hegemier et al., 1978), University of California, Berkeley (e.g., Hidalgo et al., 1978 and 1979), and the University of Canterbury at Christchurch, New Zealand (e.g., Priestley and Elder, 1982). In 1985, the Technical Coordinating Committee for Masonry Research (TCCMAR) organized the U.S.-Japan Coordinated Program for Masonry Building Research. The majority of experimental data available today for the complete load-displacement response of reinforced masonry under fully-reversed cyclic loads (static and dynamic) were generated in this program (Noland, 1990). The U.S.-Japan Coordinated Program for Masonry Building Research (often referred to as the

"TCCMAR program") included experimental and analytical studies on the seismic response of reinforced masonry materials, components, seismic structural elements, and complete building systems. Documentation of the data was thorough, and coordination of materials and methods between different research institutions was carefully controlled. Noland (1990) provides a complete list of experimental studies and associated publications.

Despite the variety of reinforced masonry systems in use, most of the TCCMAR research and earlier cyclic-loading studies were conducted with fully-grouted, hollow concrete block masonry. Most of the Component Damage Classification Guides for reinforced masonry in this document therefore apply most directly to fully-grouted concrete block masonry. A series of coordinated studies (Atkinson and Kingsley, 1985; Young and Brown, 1988; Hamid et al., 1989; Shing et al., 1991; Blondet and Mayes, 1991; Agbabian et al., 1989) have shown that the behavior character-istics of hollow concrete and hollow clay masonry in compression, in-plane flexure, and out-of-plane flexure are quite similar in terms of ductility and energy-dissipation characteristics, although clay masonry is generally of significantly higher strength. Clay masonry is also more likely to exhibit brittle characteristics and separation of faceshells from grout, whereas concrete masonry with well-designed grout can behave more homogeneously. For the purposes of this document, the behavior of fully-grouted hollow clay and hollow concrete masonry is assumed to be identical.

Relatively little work has been conducted on the seismic response of partially-grouted masonry. An extensive study of partially-grouted shear walls was conducted by NIST (Fattal, 1993), but the emphasis in reported results was on shear strength only. Schultz (1996) reports that in-plane response of partially-grouted walls with light horizontal reinforcement is characterized by vertical cracking at the junction of grouted and ungrouted vertical cells, propagating between horizontally grouted cells. Load degradation is associated with widening of the vertical cracks to 0.25" and greater. Masonry pier tests conducted at the University of California, Berkeley (Hidalgo et al., 1978; Chen et al., 1978; and Hidalgo et al., 1979) included several partially-grouted specimens. Damage patterns for these specimens were not so different from fully-grouted specimens, and strength was only mildly affected by partial grouting. However, deformation capacity was dramatically decreased relative to identical walls with full grouting.

Seismic response of grouted brick-cavity wall masonry has also received relatively little attention. The masonry pier tests conducted at UC Berkeley (Hidalgo et al., 1978; Chen et al., 1978; and Hidalgo et al., 1979) included 18 tests on two-wythe, grouted clay brick masonry. Failure modes were similar to those for hollow clay masonry, but tended to be more brittle, involving the development of vertical splitting cracks between the brick wythes and the grout. Horizontal reinforcement had little or no effect on the behavior of grouted brick-cavity walls failing in shear. This can be attributed to the rapid failure and delamination of the brick wythes, leaving a narrow and unstable grout-reinforced core that was incapable of developing a stable flexural compression zone.

Component Damage Classification guides for reinforced masonry reflect the availability of experimental data for each of the reinforced masonry systems. Reinforced masonry systems that are not well represented by experimental tests are not included in the guides.

6.2 Reinforced Masonry Component Types and Behavior Modes

6.2.1 Component Types

Component types for reinforced masonry are conceptually very similar to those for reinforced concrete (see Chapter 5). Table 6-1 lists four common reinforced masonry component types. Note that components are distinguished in terms of both geometric characteristics and behavior modes.

Each component defined in Table 6-1 may suffer from different types of damage, acting either in a pure behavior mode such as flexure, or, more likely, in a mixed mode such as flexure degrading to shear or sliding-shear failure. Table 6-2 outlines the likelihood of different behavior modes occurring in components RM1 through RM4, and references the relevant Component Guides in Section 6.5.

Table 6-3 outlines the manner in which the strength and deformation capacity of each behavior mode may be evaluated. A detailed description of each entry in Table 6-3 is given in Section 6.3. Additional example hysteresis curves are provided in FEMA 307, Section 3.

Table 6-1   Component Types for Reinforced Masonry

Component Type		Description
RM1	Stronger pier	Examples are cantilever walls that ultimately are controlled by capacity at their base (e.g., flexural plastic hinge, shear failure, rocking) and story-height wall piers that are stronger than spandrels that frame into them. Wall components may be rectangular (planar) or may include out-of-plane components (flanges) that can have a significant effect on the response.
RM2	Weaker pier	Wall piers controlled by shear failure (more likely) or flexural hinging at the top and bottom (less likely). Wall components may be rectangular (and planar) or may include out-of-plane components (flanges) that can have a significant effect on the response.
RM3	Weaker spandrel or coupling beam	Masonry beams that are weaker than the wall piers into which they frame. These are often controlled by shear capacity and less frequently by flexure.
RM4	Stronger spandrel or coupling beam	Masonry beams that are stronger than the wall piers into which they frame.

6.2.2   Behavior Modes with High Ductility

Reinforced masonry structural components with relatively high ductility exhibit some of the following common attributes:

- Wall piers with aspect ratios (height / length) of two or greater or spandrels with span to depth ratios of four or greater.

- Moderate levels of axial load ($P/A_g < 0.10f_{me}$). High axial loads decrease ductility by increasing the strain in the flexural compression zone, resulting in crushing at lower curvatures than in lightly-loaded walls. Walls with very low levels of axial load may be limited by sliding shear capacity.

- Relatively large flexural demand compared to corresponding shear. An example is a wall with flexible or weak spandrels. The lack of significant intermediate rotational restraint on the wall leads to cantilever behavior with relatively high M/V ratios as compared to frame behavior.

- Very small, or no, tension flange. Development of reinforcement in tension flanges can result in over-reinforced sections, dramatically limiting ductility.

- Uniformly distributed reinforcement.

- Sufficient shear reinforcement to ensure flexural response

The initial expected strength of a ductile reinforced masonry component in flexure is given by the in-plane moment strength as defined in Section 7.4.4 of FEMA 273. Flanges, particularly on the tension side, should be included as part of the critical section according to the limits set in FEMA 273. It is also important to consider

the effects of axial load, and to consider all reinforcement in the wall as effective. The theoretical basis for calculating flexural strength of reinforced masonry walls follows the well-established principles of ultimate strength design for reinforced concrete, and there is sufficient experimental data to support its use for masonry. For additional discussion, see Priestley and Elder (1982), Shing et al. (1991), Kingsley et al. (1994), and Seible et al. (1994b).

The displacement capacity of a ductile flexural wall can be determined with reasonable accuracy by idealizing it as a cantilever beam and calculating the flexural and shear deformations. Displacements following cracking, but prior to significant yielding, may be approximated using an effective cracked stiffness (Priestley and Hart, 1989). After yielding, the wall can be idealized as having an equivalent plastic-hinge zone at the base, and displacement can be calculated using the methods presented in Paulay and Priestley (1992).

With increasing distance from the plastic-hinge zone, the contribution of shear deformations to displacements is less significant, and a pure flexural model is sufficient. Seible et al. (1995) showed that at the maximum displacement in a five-story, full-scale reinforced masonry building, the shear deformation component of lateral displacement was as high as 50 percent at the first story, and less than 10 percent at the fifth floor. Priestley and Elder (1982), Leiva and Klingner (1991), Shing et al. (1990a, b), and Kingsley et al. (1994) provide additional experimental evidence to support the calculation of displacements in ductile flexural walls. The probable displacement capacity of ductile flexural walls should be at least four times the yield displacement, or one percent of building drift.

Table 6-2   *Likelihood of Earthquake Damage to Reinforced Masonry Components According to Component and Behavior Mode.*

Ductility		Behavior Mode	Wall Component Type			
			RM1 Stronger Pier	RM2 Weaker Pier	RM3 Weaker Spandrel	RM4 Stronger Spandrel
High ductility	A	Flexure	Common See Guide RM1A	Unlikely	Common See Guide RM3A	N/A
		Foundation rocking	May occur, but not considered See FEMA 273 or ATC-40	May occur, but not considered See FEMA 273 or ATC-40	N/A	N/A
Moderate ductility	B	Flexure / Diagonal shear	Common See Guide RM1B	Common See Guide RM2B	May occur Similar to Guide RM3A	N/A
	C	Flexure / Sliding shear	May occur See Guide RM1C	May occur Similar to Guide RM1C	Unlikely	N/A
	D	Flexure / Out-of-plane instability	May occur following large displacement cycles See Guide RM1D	Unlikely	Unlikely	N/A
	E	Flexure / Lap splice slip	May occur See Guide RM1E	Unlikely	May occur	N/A
	F	Pier rocking	May occur Similar to Guide RM1E	May occur Similar to Guide RM1E	N/A	N/A
Little or no ductility	G	Preemptive diagonal shear	Common Similar to Guide RM2G	Common See Guide RM2G	Common See Guide RM3G	N/A
	H	Preemptive sliding shear	May occur in poorly detailed wall Similar to Guide RM1C	May occur in poorly detailed wall Similar to Guide RM1C	N/A	N/A

Notes: • Shaded areas of the table with notation "*See Guide...*" indicate behavior modes for which a specific Component Guide is provided in Section 6.5. The notation "*Similar to Guide...*" indicates that the behavior mode can be assessed by using the guide for a different, but similar component type or behavior mode.
 • *Common* indicates that the behavior mode has been evident in postearthquake field observations and/or that experimental evidence supports a high likelihood of occurrence.
 • *May occur* indicates that a behavior mode has a theoretical or experimental basis, but that it has not been frequently reported in postearthquake field observations.
 • *Unlikely* indicates that the behavior mode has not been observed in either the field or the laboratory.
 • N/A indicates that the failure mode cannot occur for that component.

Table 6-3 Behavior Modes for Reinforced Masonry Components (Note: Hysteresis Curves from Shing et al., 1991)

Behavior Mode	Approach to calculate strength (use expected material values)	Approach to estimate displacement capacity	Ductility Category	Example hysteresis loop shape
A. Ductile Flexure	The expected strength under in-plane forces is limited by the development of the expected moment strength, M_e. This is calculated considering all distributed steel, axial loads, and the development of tension flanges, if present. Note that the maximum possible strength is greater than the expected strength, which may influence the governing mode. See Section 6.3.2a	The displacement capacity is limited by the maximum curvature attained within the effective plastic-hinge zone. Classical moment-curvature analysis may be used and related to displacement with some empirical calibration. See Sections 6.3.2b and 6.3.2c	High ductility capacity See Section 6.2.2	
B. Flexure / Shear	The initial expected strength is governed by M_e, as calculated for the ductile flexural mode. Strength degrades as the masonry component of the shear strength, V_m, degrades, with residual strength governed by the reinforcement component V_s. See Sections 6.3.2a and 6.3.3a	Displacement capacity at the initial expected strength may be estimated as the intersection of the flexural load-displacement curve with the degrading shear-strength envelope. See Section 6.3.3b	Moderate ductility capacity See Section 6.2.3	
C. Flexure / Sliding shear	The initial expected strength is governed by M_e, as calculated for the ductile flexural mode. Stiffness and strength degrade as sliding-shear mode develops, and hysteresis becomes pinched. The initial strength may be maintained, but only at large displacements. See Sections 6.3.2a and 6.3.4	Displacements due to sliding may be large. Displacement capacity for ductile flexural behavior may provide reasonable estimate. See Section 6.3.4b	Moderate-to-high ductility capacity See Section 6.2.2 and 6.2.3	

Table 6-3   Behavior Modes for Reinforced Masonry Components (Note: Hysteresis Curves from Shing et al., 1991) (continued)

D. Flexure / Out-of-plane stability	The initial expected strength is governed by M_e, as calculated for the ductile flexural mode. Following instability failure, strength drops rapidly. See Sections 6.3.2a and 6.3.5.	Displacement capacity is limited by the slenderness of the wall with respect to the height and/or the length. See Section 6.3.5.	Moderate-to-high ductility capacity See Section 6.2.2 and 6.2.3
E. Flexure / Lap splice slip	The initial expected strength is governed by M_e, as calculated for the ductile flexural mode. With failure of lap splices, strength degrades to rocking mode, limited by crushing of wall toes. See Sections 6.3.2a and 6.3.6	Displacement capacity at expected strength limited by lap-splice slip. See Section 6.3.6	Moderate ductility capacity See Section 6.2.3
G. Preemptive diagonal shear	The expected strength is reached before the development of the expected moment capacity and is governed by shear strength, V_e. See Section 6.3.3	No inelastic capacity. 	No ductility capacity See Section 6.2.4
H. Preemptive sliding shear	The expected strength is reached prior to the development of the expected moment capacity, and is governed by the sliding shear strength, V_{se}. See Section 6.3.4	Little displacement capacity, limited by crushing of bottom course of masonry and/or buckling of vertical reinforcement.	Little ductility capacity See Section 6.2.4

Damage in flexural walls is likely to include both horizontal and diagonal cracks of small size concentrated in the plastic-hinge region. Diagonal cracks typically propagate from horizontal, flexural cracks, and therefore have similar, regular spacing. At the large displacements, crushing may occur at the wall toes.

Another relatively ductile behavior mode is foundation rocking. This can occur if the rocking capacity of the foundation is less than the strength of the wall component it supports. Foundation components are covered in FEMA 273 and ATC-40.

6.2.3 Behavior Modes with Moderate Ductility

Moderately-ductile components initially behave similarly to highly-ductile components, but their ultimate displacement capacity is limited by the influence of less-ductile modes such as sliding or diagonal shear. The response of moderately-ductile components is difficult to predict analytically due to the complex interaction of moment, shear, and axial load, and less difficult to recognize in a damaged component. The majority of experimental data for reinforced masonry components falls into this moderately ductile category, For some examples, refer to Shing et al. (1991).

The initial strength is governed by the flexural capacity; however, the initial strength cannot be maintained at high ductility levels. Displacement capacity for moderate-ductility modes is difficult to calculate. Research is currently underway to improve the ability to predict displacements associated with diagonal shear modes of behavior, but there are currently no established guidelines, with the exception of the semi-empirical recommendations in FEMA 273.

At low levels of response, damage in moderately ductile components resembles that for ductile components, consisting primarily of horizontal flexural and diagonal shear cracks. The component response at larger displacements depends on the governing behavior mode, as described in the following paragraphs.

a. Flexure / Diagonal shear

Diagonal shear response is characterized by the growth of diagonal cracks accompanied by degrading strength. Eventually, cracks cross the entire length of the wall, and the residual strength of the wall is that provided by the horizontal reinforcement alone. Extensive experimental evidence is available to document this behavior mode, including Shing et al. (1991), Hidalgo et al. (1978), and Chen et al.(1978).

b. Flexure / Sliding shear

Walls may be susceptible to sliding-shear mechanisms when axial load levels are low, vertical reinforcement ratios are low, or when very large ductilities are achieved and the shear friction mechanism degrades. At low displacements, sliding may be observed as a simple lateral offset in a wall. At very large displacements, localized crushing of the bottom course of masonry can result, and vertical reinforcement can experience large lateral offsets.

c. Flexure / Out-of-plane instability

At high ductility levels, the flexural compression zone of slender walls may be susceptible to instability after the development of large tensile strains during previous cycles. This type of failure has been observed in laboratory tests of well-detailed, highly-ductile flexural walls, (see Paulay and Priestley, 1993) but it has not been noted in the field. Out-of-plane instability would not be expected in walls with flanges at the end of the wall, or in very thick walls.

d. Flexure / Lap splice slip

If starter bars with insufficient development length are located at the base of structural walls, overturning forces can result in bond degradation and eventual rocking of the wall on the foundation. Local damage may appear first as vertical cracks at the location of the lap splices, and eventually crushing at the wall toes. (Priestley et al., 1978).

6.2.4 Behavior Modes with Low Ductility

General characteristics of reinforced masonry components exhibiting low-ductility behavior include:

- Wall piers with aspect ratios (height / length) of less than 0.8 and spandrels with span-to-depth ratios of less than two

- High levels of axial load ($P_u/f_{me} A_g > 0.15$)

- Large tension flanges connected continuously to the component

- Large amounts of flexural reinforcement at component edges

- Light shear reinforcement relative to flexural reinforcement

Experimental research on non-ductile walls under cyclic load histories includes Shing et al. (1991), Hidalgo et al. (1978), Chen et al. (1978), and Hidalgo et al. (1979).

Flexural capacity does not govern the nonductile modes of failure. Strength is defined by the diagonal shear strength or the horizontal sliding strength. In the first case, diagonal shear failure causes the lateral capacity of the wall to be immediately reduced to the capacity of the horizontal reinforcement alone. In the latter case, preemptive sliding of the wall does not allow the development of the full flexural capacity, resulting in large displacements with little capacity to dissipate hysteretic energy. It should be noted that bed-joint sliding in URM components may be considered as relatively ductile behavior. Calculation of the shear strength of masonry structural walls is addressed by Leiva and Klingner (1991), Shing et al. (1991), and Anderson and Priestley (1992).

Components that experience preemptive, force-controlled failures cannot be considered to have dependable inelastic displacement capacity.

Diagonal shear failure can occur with little or no early indication of incipient failure. Damage is characterized by one or two dominant diagonal cracks of large width. Damage may ultimately include crushing and spalling in the central portion of the wall. Walls that fail in sliding shear may have very little cracking or damage outside the sliding joint. Ultimately, crushing and spalling of the base course of masonry units can occur.

6.3   Reinforced Masonry Evaluation Procedures

This section provides the basis for calculating the strength and deformation capacities of reinforced masonry components both before and after a damaging earthquake. Subsections are organized according to behavior modes.

6.3.1   Material Properties

The procedures for evaluating strength and deformation capacities presuppose the knowledge of component characteristics, including dimensions, amounts and location of reinforcement and the material properties. Methodologies for structural investigation and the evaluation of these parameters are given in Chapter 3. Additional guidelines for estimating masonry material properties are given in FEMA 273, Section 7.3.2.

If no information from testing is available, initial assumptions for expected material properties as given in FEMA 273 and summarized in Table 6-4 may be assumed.

a.   Masonry

Table 6-4   Initial Expected Clay or Concrete Masonry Properties

Condition	Expected Strength f_{me} (psi)	Elastic Modulus (psi)	Friction Coefficient
Good	900	$550\,f_{me}$	0.7
Fair	600	$550\,f_{me}$	0.7
Poor	300	$550\,f_{me}$	0.7

If observed failure modes are not consistent with the initial material strength values given above, actual expected values may be substantially greater (more than three times the table values).

b.   Reinforcing steel

Recommendations in FEMA 273 Section 6.4.2.2. are adopted here for yield strength of reinforcement. In the absence of applicable test data, the expected strength of yielding reinforcement, f_{ye}, is assumed to be equal to 1.25 times the nominal yield stress. A range of reinforcement strength values between 1.1 and 1.4 times the nominal yield strength can also be considered in the evaluation procedures.

6.3.2   Flexure

a.   Strength

The in-plane flexural capacity of a reinforced masonry wall with distributed reinforcement may be calculated based on the well-established principles of ultimate strength design, as stated in FEMA 273, Section

7.4.4.2.A. It is convenient to express this moment strength in terms of contributions from the masonry, the reinforcement, and the axial load independently, including all distributed reinforcement (Paulay and Priestley, 1992), as shown in Equation 6-1.

$$M_e = C_m\left(c - \frac{a}{2}\right) + \sum_{i=1}^{n}\left|f_{ye}A_{si}(c - x_i)\right| + P_u\left(\frac{l_w}{2} - c\right)$$ (6-1)

Where:

M_e = expected moment capacity of a masonry section
C_m = compression force in the masonry
f_{ye} = expected reinforcement yield strength
A_{si} = area of reinforcing bar i
x_i = location of reinforcing bar i
c = depth to the neutral axis
a = depth of the equivalent stress block
P_u = wall axial load
l_w = length of the wall

Note that all bars are considered to participate, and it is assumed for the purpose of calculating the moment that all bars are yielding. This expression arbitrarily sums moments about the neutral axis of the section. In some cases, it may be more convenient to use a different location. For example, the axial component often passes through the centroid of the section and can be eliminated from the summation of moments at the centroid; however, its effect on moment capacity must be included by proper calculation of C_m.

The expression for moment strength is valid for masonry walls with reinforcement concentrated in the wall boundaries. Walls with concentrated reinforcement, particularly when larger bar sizes are used, are vulnerable to grout flaws in the wall toes. These walls are more likely to develop lap-splice slip or sliding-shear behavior modes.

b. Deformation

A ductile flexural component can be idealized as a cantilever element with a zone of concentrated plastic rotation at the base (the equivalent plastic hinge). Paulay and Priestley (1992) provide a simple model for calculating displacements. At first yield, the displacement at the level of the horizontal force resultant (i.e., the effective height), is

$$\Delta_y = \frac{\phi_y h_e^2}{3}$$ (6-2)

Where:

ϕ_y = yield curvature of a masonry section
h_e = effective height of the wall

The maximum displacement capacity at the effective height is:

$$\Delta_p = \left(\phi_m - \phi_y\right)l_p\left(h_e - 0.5l_p\right)$$ (6-3)

Where:

ϕ_m = maximum plastic curvature of a masonry section
l_p = effective plastic-hinge length (see Section 6.3.2c)

The displacement ductility is:

$$\mu_\Delta = \frac{\Delta_y + \Delta_p}{\Delta_y} = 1 + \frac{\Delta_p}{\Delta_y}$$ (6-4)

c. Plastic-Hinge Length

The concept of a plastic hinge in masonry is adopted as a computational convenience to describe in simple terms the complex distribution of cracks and the localized inelastic deformations in reinforcement. While there is no plastic hinge at a point *per se*, there is a zone over which the curvature may be expected to exceed the yield curvature at large displacements. The following expression for plastic-hinge length has been shown to agree reasonably well with experimental results for reinforced masonry walls (Paulay and Priestley, 1993), and is given in FEMA 274 Section C7.4.4.3A:

$$l_p = 0.2l_w + 0.04h_e$$ (6-5)

Where:

l_w = length of the wall
h_e = height to the resultant of the lateral force
= M/V

See also Shing et al. (1990a, b).

For the purposes of this document, the plastic-hinge length is useful for calculating ultimate displacements in flexural walls, and for identifying the zone over which to expect degradation in shear strength with increasing ductility.

d. Flanges

When a flexural wall includes flanges, there is a potential to develop the flange reinforcement in flexural tension, thus increasing the flexural strength, and potentially decreasing the ductility capacity. In reinforced masonry, a flange can only be engaged when reinforcement and grout are continuous around the wall intersections. When calculating the flexural capacity of a flanged wall, FEMA 273, Section 7.4.4.2.C recommends that an effective flange width equal to 3/4 of the effective wall height ($h_e = M/V$) should be assumed, (He and Priestley, 1992; Seible et al., 1994b).

Damage patterns in fully-engaged flanges appear as horizontal cracks, and possibly as the continuation of diagonal cracks from the in-plane wall.

e. Coupling

Coupling between wall pier and spandrel components causes cyclic axial loads in the piers generated by shear in the spandrels. When the cyclic axial force is compressive, the pier strength is increased, and the ductility decreased. Similarly, when the axial force is tensile, the strength is decreased and the ductility is increased. For walls with relatively little gravity load, the tension force due to coupling can be sufficient to place the wall in a state of net tension. For the purposes of damage classification, the coupling-induced axial loads are to be considered when identifying the governing behavior mode.

Experimental data for reinforced masonry coupled walls is given in Seible et al. (1991), Paulay and Priestley (1992), and Merryman et al. (1990).

6.3.3 Shear

a. Strength

The in-plane shear capacity of an undamaged structural wall may be calculated using the recommended procedure in FEMA 273, Section 7.4.4.2.B, which may be expressed as the sum of three components corresponding to the contributions of masonry, reinforcement, and axial load, respectively:

$$V_e = V_m + V_s + V_p \qquad (6\text{-}6)$$

where:

$$V_m = \left[4.0 - 1.75 \left(\frac{M}{V l_w} \right) \right] A_g \sqrt{f_{me}} \qquad (6\text{-}7)$$

$$V_s = 0.5 \left(\frac{A_v}{s} \right) f_{ye} d_{vs} \qquad (6\text{-}8)$$

$$V_p = 0.25 P_u \qquad (6\text{-}9)$$

In FEMA 273, the V_p component is included as a part of the V_m component; they are expressed separately here to facilitate the following discussion of degrading shear strength. Note that in FEMA 273, the total shear strength V_e is limited to:

$$V_e = 6 A_g \sqrt{f_{me}} \text{ for walls with } M/V_d < 0.25 \qquad (6\text{-}10)$$

$$V_e = 4 A_g \sqrt{f_{me}} \text{ for walls with } M/V_d \geq 1.00 \qquad (6\text{-}11)$$

Equations 6-10 and 6-11 describe the maximum shear strength of an initially undamaged wall. As a flexural wall undergoes cyclic displacements, horizontal cracks initiate on the tension side of the wall and propagate towards the neutral axis, and diagonal cracks initiate near the center of the wall and propagate outward. As cracks open, horizontal reinforcement is engaged, and the mechanism of shear resistance in the masonry changes. The bulk of the masonry shear is transferred through the flexural compression zone – where the local shear strength is enhanced by the increasing compression stresses – and the remainder is transferred through aggregate interlock across the cracks. This mechanism degrades as both flexural and shear cracks open wider, until the capacity of the wall is reduced to nearly that of the horizontal reinforcement alone. Priestley, et al. (1994) have developed a model that captures this response for concrete columns, but such a relationship has not yet been developed and verified for masonry walls. Because it is nonconservative to ignore the degrading strength, however, the following relationship may serve for masonry in the area of the plastic-hinge zone until an improved model can be developed:

$$V_m = k\left[4.0 - 1.75\left(\frac{M}{Vl_w}\right)\right]A_g\sqrt{f_{me}} \qquad (6\text{-}12)$$

where $k=1$ for displacement ductility values less than 1.5, and reduces linearly to a value of 0.1 at a displacement ductility of 4, and further to 0.0 at a displacement ductility of 8.

b. Deformation

Deformation mechanisms of walls with a predominant shear mode behavior cannot be quantified as simply as those of walls with flexural behavior modes, particularly after significant diagonal cracking and after yielding of reinforcement. As an approximation, the flexural force-displacement relationship can be developed, as described in Section 6.3.2, and the degrading shear strength relationship in Equation 6-12 above may be used to identify the displacement at which shear modes of behavior begin to dominate the response.

6.3.4   Sliding

a. Strength

The ability of a structural masonry wall to resist sliding shear may be described in terms of shear friction across a crack or construction joint, as described in ACI 318-95, Chapter 11. This friction may be visualized as having two components, (Paulay and Priestley 1992), the first due to the friction associated with the axial load on the wall, and the second due to the friction associated with the clamping force provided by the vertical reinforcement across the sliding plane, thus:

$$V_{se} = \mu P_u + \mu A_{vf} f_{ye} \qquad (6\text{-}13)$$

Where:

P_u = wall axial load

A_{vf} = area of reinforcement crossing perpendicular to the sliding plane

f_{ye} = expected yield strength of reinforcement

μ = coefficient of friction at the sliding plane

Values for the coefficient of friction may be determined using the recommendations of ACI 318-95, Section 11.7.4.3. Atkinson et al. (1988) determined that for mortared brick masonry joints, a value of 0.7 represents an average expected value.

Uniformly distributed reinforcement is more effective in resisting sliding shear than is reinforcement concentrated at the ends of the wall. Distributed reinforcement leads to a larger flexural compression zone than does concentrated reinforcement, thus enhancing shear transfer across the plane. Distributed reinforcement is also located closer to the rough surfaces that generate the shear friction forces.

Wall components that are classified as RM1 may be particularly vulnerable to sliding-shear behavior, or, more specifically, flexural response that degrades to sliding-shear response. The reason for this vulnerability is that, at the large curvature ductilities developed in the plastic-hinge zones of flexural walls, horizontal cracks open wide and cause the reinforcement across the sliding plane to yield. As the cracks open, the potential to develop the shear friction mechanism degrades, leaving only the comparatively flexible dowel-action mechanism of the reinforcing bars (Paulay and Priestley, 1992). Under cyclic reversals at large curvature ductility, it is possible to open a horizontal crack across the entire length of a wall. Sliding behavior is well documented for reinforced masonry walls (Shing et al., 1991), particularly those with light axial loads and light vertical reinforcement (Seible et al., 1994a, b).

b. Deformation

The deformation limit for sliding-shear behavior modes may be governed by the fracture of bars (dowels) crossing the sliding plane, crushing of the base course of masonry, or degradation of the shear and flexure transfer mechanisms in the flexural compression zone of the wall as the wall slides beyond its support.

6.3.5   Wall Instability

Out-of-plane buckling in the compression zone of flexural walls has been observed in experiments (Paulay and Priestley, 1992), but has not been reported for actual masonry structures subjected to earthquakes. The phenomenon is associated with compression stresses in flexural reinforcement that has achieved large inelastic tensile strains in previous cycles. Until the reinforcement yields in compression and the flexural cracks close, the reinforcement must carry the entire flexural compression force alone, thus leaving the wall in the compression zone vulnerable to buckling. Masonry walls, where the reinforcement is typically centered, are particularly vulnerable. Paulay and Priestley (1993) have suggested a simplified design relationship to calculate the critical wall width for which instability may limit ductility. For building evaluation, it is useful to determine the maximum ductility that may be expected for a given wall thickness. The following relationships, where the thickness, length, and height of the wall are given by t, l, and h, may be used to identify walls for which stability may be a limiting factor:

For $\dfrac{t}{l_w} \leq \dfrac{1}{24}$ or $\dfrac{t}{h_e} \leq \dfrac{1}{18}$

the displacement ductility may
be no greater than $\mu_\Delta = 4$ (6-14)

For $\dfrac{t}{l_w} \geq \dfrac{1}{12}$ and $\dfrac{t}{h_e} \geq \dfrac{1}{8}$

the displacement ductility will not be
limited by stability   (6-15)

Experimental tests on slender masonry walls at large ductilities suggest that this relationship may be conservative (Seible et al., 1994a, b).

The lack of evidence for this type of failure in existing structures may be due to the large number of cycles at high ductility that must be achieved – most conventionally-designed masonry walls are likely to experience other behavior modes such as diagonal shear before instability becomes a problem.

6.3.6 Lap-Splice Slip

Relatively little research has been conducted specifically to investigate aspects of lap-splice slip that are unique to reinforced masonry as opposed to reinforced concrete (Hammons et al., 1994; Soric and Tulin, 1987). Experimental evidence of strength and/or deformation capacity of reinforced masonry components being limited by lap-splice slip failure has been noted in shear walls (Igarashi et al., 1993; Shing et al., 1991; Kubota and Murakami, 1988) and masonry beams (Okada and Kumazawa, 1987). Experimental studies in which lap splices were specifically avoided in plastic-hinge regions (Kingsley et al., 1994; Seible et al., 1994b; Shing et al., 1991) have shown superior performance over similar component tests including lap splices. In particular, the specimens without lap-splices in the plastic-hinge zones showed development of large curvature ductilities and well-distributed cracking in plastic hinge zones.

Research in Japan (for example, Seible et al., 1987; Okada and Kumazawa, 1987) has included the use of special spiral reinforcement for lap-splice confinement. Such reinforcement has been shown to limit successfully lap-splice slip, and to extend the effective plastic-hinge zone from one or two cracks to numerous cracks in the ends of masonry coupling beams or at the base of shear walls.

Studies to date indicate that lap splices in masonry are more susceptible to slip than are splices in concrete, because lap-splice regions in masonry are unlikely to include significant lateral confinement reinforcement. Hammons et al. (1994) found that splitting failure of masonry units in lap-splice regions was likely, regardless of lap-splice length, for bars #4 and greater in four-inch hollow units, #6 and greater in six-inch units, and #8 and greater in eight-inch units. While there are no experimental data on laps with more than two bars in a single grouted cell, it may be supposed that such bar configurations are susceptible to lap-splice failure.

For evaluation of lap-splice development length, l_d, refer to FEMA 222A, *NEHRP Recommended Provisions for Seismic Regulations for New Buildings*, Section 8.4.5. The lap-splice equation in FEMA 222A should be modified to include the expected yield strength rather than the characteristic yield strength, and to use a strength-reduction factor of 1.0.

Rocking in reinforced masonry walls may develop as a consequence of lap-splice failure at the base of shear walls. While the strength of the wall can be compromised dramatically, there is evidence to suggest that rocking can be a stable mechanism of energy dissipation (Priestley et al.,1978; Igarashi et al.,1993). The validity of such a mechanism depends on the deformation capacity of connected components relative to the increased displacement demand that results from rocking.

6.3.7 Masonry Beams

Because of the physical restrictions of typical hollow clay or concrete masonry units, it is difficult to provide satisfactory confinement reinforcement, and impossible to provide diagonal reinforcement of masonry spandrels or coupling beams. It is therefore difficult to avoid preemptive shear or, at best, flexure/shear behavior modes in masonry beams. Masonry beams which are detailed to allow ductile flexural response are likely to fall under the category of RM3 components.

Bending and shear capacity of reinforced masonry beams may be evaluated using the principles set forth in FEMA 222A (BSSC, 1994), incorporating expected material strengths rather than characteristic strengths, and setting strength reduction factors equal to 1.0.

Many tests have been conducted on masonry beams under gravity loading, but few have been conducted under reversed cyclic loading with boundary conditions representative of typical coupled wall systems (i.e. incorporating slabs). A number of studies have been conducted in Japan as a part of the JTCCMAR research program, including Matsuno et al. (1987), Okada and Kumazawa (1987), and Yamazaki et al. (1988a and 1988b).

6.4 Symbols for Reinforced Masonry

A_g = Gross crossectional area of wall

A_{si} = Area of reinforcing bar i

A_v = Area of shear reinforcing bar

A_{vf} = Area of reinforcement crossing perpendicular to the sliding plane

a = Depth of the equivalent stress block

c = Depth to the neutral axis

C_m = Compression force in the masonry

f_{me} = Expected compressive strength of masonry

f_{ye} = Expected yield strength of reinforcement

h_e = Effective height of the wall (height to the resultant of the lateral force) = M/V

l_d = Lap splice development length

l_p = Effective plastic hinge length

l_w = Length of the wall

M/V = Ratio of moment to shear (shear span) at a section

M_e = Expected moment capacity of a masonry section

P_u = Wall axial load

s = Spacing of reinforcement

t = Wall thickness

V_e = Expected shear strength of a reinforced masonry wall

V_m = Portion of the expected shear strength of a wall attributed to masonry

V_s = Portion of the expected shear strength of a wall attributed to steel

V_p = Portion of the expected shear strength of a wall attributed to axial compression effects

V_{se} = Expected sliding shear strength of a masonry wall

x_i = Location of reinforcing bar i

Δ_p = Maximum inelastic displacement capacity

Δ_y = Displacement at first yield

ϕ_m = Maximum inelastic curvature of a masonry section

ϕ_y = Yield curvature of a masonry section

μ_Δ = Displacement ductility

μ = Coefficient of friction at the sliding plane

6.5   Reinforced Masonry Component Guides

The following Component Damage Classification Guides contain details of the behavior modes for reinforced masonry components. Included are the distinguishing characteristics of the specific behavior mode, the description of damage at various levels of severity, and performance restoration measures. Information may not be included in the Component Damage Classification Guides for certain damage severity levels; in these instances, for the behavior mode under consideration, it is not possible to make refined distinctions with regard to severity of damage. See also Section 3.5 for general discussion of the use of the Component Guides and Section 4.4.3 for information on the modeling and acceptability criteria for components.

RM1A	COMPONENT DAMAGE CLASSIFICATION GUIDE	System:	**Reinforced Masonry**

Component Type:	**Stronger Pier**
Behavior Mode:	**Ductile Flexural**
Applicable Materials:	**Fully grouted hollow concrete or clay units**

How to distinguish behavior mode:

By observation:

Damage in an RM1 component with a flexural response is likely to be localized in a zone with a vertical extent equal to approximately twice the length of the wall. Both horizontal and diagonal cracks of small size (< 0.05 in.) and uniform distribution may be present. Diagonal cracks typically propagate from horizontal, flexural cracks, and therefore have similar, regular spacing. If shear deformations are localized to one or two diagonal cracks of large width, the behavior mode is likely to be Flexure/Shear or Preemptive Shear. If a permanent horizontal offset is visible, the behavior mode may be Flexure/Sliding Shear

Caution: At low damage levels, damage observations will be similar to those for other behavior modes.

By analysis:

A wall detailed to ensure ductile flexural response will have sufficient horizontal reinforcement to allow development of a flexural plastic hinge mechanism through stable and distributed yielding of the vertical bars at the base of the wall. The ultimate capacity of the horizontal reinforcement alone in the hinge zone should be greater than the shear developed at the moment capacity of the wall. Wall vertical loads are likely to be small.

Refer to Evaluation Procedures for:
- Evaluation of flexural response.
- Identifying flexural versus shear cracks.
- Crack evaluation.

Severity	Description of Damage	Performance Restoration Measures
Insignificant $\lambda_K = 0.8$ $\lambda_Q = 1.0$ $\lambda_D = 1.0$	*Criteria:* • No crack widths exceed 1/16", and • No significant spalling *Typical Appearance:*	Not necessary for restoration of structural performance. (Cosmetic measures may be necessary for restoration of nonstructural characteristics.)

COMPONENT DAMAGE
CLASSIFICATION GUIDE **continued**

$$\boxed{\text{RM1A}}$$

Severity	Description of Damage		Performance Restoration Measures
Slight $\lambda_K = 0.6$ $\lambda_Q = 1.0$ $\lambda_D = 1.0$	*Criteria:*	● No crack widths exceed 1/8" ● No significant spalling or vertical cracking	● Inject cracks $\lambda_K^* = 0.9$ $\lambda_Q^* = 1.0$ $\lambda_D^* = 1.0$
	Typical Appearance:	Similar to insignificant damage except cracks are wider and more extensive.	
Moderate $\lambda_K = 0.4$ $\lambda_Q = 0.9$ $\lambda_D = 1.0$	*Criteria:*	● Crack widths do not exceed 1/8" ● Moderate spalling of masonry unit faceshells or vertical cracking at toe regions ● No buckled or fractured reinforcement ● No significant residual displacement.	● Remove and patch spalled masonry and loose concrete. Inject cracks. $\lambda_K^* = 0.8$ $\lambda_Q^* = 1.0$ $\lambda_D^* = 1.0$
	Typical Appearance:	Similar to slight damage except cracks are wider and more extensive.	
Extreme	*Criteria:* *Typical Indications*	● Reinforcement has fractured. ● Wide flexural cracking (>1/8" residual) ● Large residual displacement ● Extensive crushing or spalling ● Visibly fractured or buckled reinforcing	● Replacement or enhancement required.

RM1B	COMPONENT DAMAGE CLASSIFICATION GUIDE	System:	**Reinforced Masonry**

Component Type:	**Stronger Pier**
Behavior Mode:	**Flexure / Shear**
Applicable Materials:	**Fully grouted hollow concrete or clay units**

How to distinguish behavior mode:

By observation:

Damage in an RM1 component with a flexural /shear response is typically localized to the base of the wall, within the plastic hinge region. Both horizontal and diagonal cracks will be present, with diagonal cracks predominant. Diagonal cracks may appear to be independent from horizontal, flexural cracks, and may propagate across the major diagonal dimensions. At heavy damage levels, shear deformations are likely to be localized to one or two diagonal cracks of large width. If a permanent horizontal offset is visible, the behavior mode may be Flexure/Sliding Shear

By analysis:

Analysis of a wall with a Flexure / Shear behavior mode may be difficult, with no clear distinction between the controlling mechanism of flexure (deformation-controlled) or shear (force-controlled). Calculated capacities should be in the same range. Wall axial loads may be moderate-to-high.

Refer to Evaluation Procedures for:
- Evaluation of flexural response.
- Evaluation of shear response
- Evaluation of plastic hinge length
- Identifying flexural versus shear cracks.
- Crack evaluation.

Severity	Description of Damage		Performance Restoration Measures
Insignificant $\lambda_K = 0.8$ $\lambda_Q = 1.0$ $\lambda_D = 1.0$	*Criteria:* • No crack widths exceed 1/16", <u>and</u> • No significant spalling *Typical Appearance:* 		Not necessary for restoration of structural performance. (Cosmetic measures may be necessary for restoration of nonstructural characteristics.)
Slight $\lambda_K = 0.6$ $\lambda_Q = 1.0$ $\lambda_D = 1.0$	*Criteria:* • No crack widths exceed 1/8", <u>and</u> • No significant spalling or vertical cracking *Typical Appearance:* Similar to insignificant damage except cracks are wider with more extensive cracking.		• Inject cracks $\lambda_K^* = 0.9$ $\lambda_Q^* = 1.0$ $\lambda_D^* = 1.0$

COMPONENT DAMAGE
CLASSIFICATION GUIDE **continued**

| RM1B |

Severity	Description of Damage		Performance Restoration Measures
Moderate $\lambda_K = 0.4$ $\lambda_Q = 0.8$ $\lambda_D = 0.9$	*Criteria:* • Crack widths do not exceed 3/16" • Moderate spalling of masonry unit faceshells or vertical cracking at toe regions • No buckled or fractured reinforcement • No significant residual displacement *Typical Appearance:*		• Remove and patch spalled masonry and loose concrete. • Inject cracks. $\lambda_K^* = 0.8$ $\lambda_Q^* = 1.0$ $\lambda_D^* = 1.0$
Extreme	*Criteria:* *Typical* *Indications*	• Reinforcement has fractured • Wide flexural cracking (> ¼"), typically concentrated in a single crack • Wide diagonal cracking, typically concentrated in one or two cracks • Crushing or spalling at wall toes of more than one-half unit height or width, delamination of faceshells from grout • Visibly fractured or buckled reinforcing	• Replacement or enhancement required.
	Typical Appearance		

RM1C	COMPONENT DAMAGE CLASSIFICATION GUIDE	System:	**Reinforced Masonry**

Component Type:	**Stronger Pier**
Behavior Mode:	**Flexure / Sliding Shear**
Applicable Materials:	**Fully or partially grouted hollow concrete or clay units**

How to distinguish behavior mode:

By observation:

Evidence of movement will appear first in the form of pulverized mortar across a bed joint or construction joint. If grout cores include shear keys into the slab below, short diagonal cracks initiating at the keys may be visible in the course above the sliding joint. After severe sliding, crushing of the bottom course of masonry may occur.

By analysis:

Walls with very light axial loads ($P/f'_m A_g \leq 0.05$) may be susceptible to sliding, as are walls with very light flexural reinforcement, or large ductility demands.

Refer to Evaluation Procedures for:

• Evaluation of sliding response.

Severity	Description of Damage		Performance Restoration Measures
Insignificant		See **RM1A**	See **RM1A**
Slight $\lambda_K = 0.5$ $\lambda_Q = 0.9$ $\lambda_D = 1.0$	*Typical Appearance*	As for **RM1A** or **RM1B** and:	Not necessary for restoration of structural performance. (Cosmetic measures may be necessary for restoration of nonstructural characteristics.)
Moderate $\lambda_K = 0.2$ $\lambda_Q = 0.8$ $\lambda_D = 0.9$	*Typical Appearance*	Similar to slight with more extensive cracking and movement	• Remove and patch spalled masonry and loose concrete. • Inject cracks. $\lambda_K^* = 0.8$ $\lambda_Q^* = 1.0$ $\lambda_D^* = 1.0$
Extreme	*Criteria*	• Permanent wall offset • Spalling and crushing at base	• Replacement or enhancement required.
	Typical Appearance As for **RM1A** or **RM1B** and:		

RM1D	COMPONENT DAMAGE CLASSIFICATION GUIDE	System:	**Reinforced Masonry**
		Component Type:	**Stronger Pier**
		Behavior Mode:	**Flexure / Out-of-Plane Instability**
		Applicable Materials:	**Fully or partially grouted hollow concrete or clay units**

How to distinguish behavior mode:

By observation:

As for any unstable behavior mode, there will be little evidence of impending failure. Instability of the compression toe is preceded by large horizontal flexural cracks with significant plastic strains in the reinforcement crossing the crack. Evidence of such cracks, particularly when distributed across the plastic hinge zone rather than localized, may indicate incipient failure. Following failure, the wall will have visible out-of-plane displacements and localized crushing.

Caution: At low damage levels, damage observations will be identical to those for the **RM1A** behavior mode.

By analysis:

Walls with a tendency for compression toe instability will have large flexural displacement capacity, and little possibility for shear failure, even at large ductilities. Wall thickness will be less than or equal to the critical wall thickness for instability.

Refer to Evaluation Procedures for:

● Evaluation of flexural response.

● Evaluation of wall instability

● Identification of the plastic hinge zone.

Severity	Description of Damage		Performance Restoration Measures
Insignificant	See **RM1A**		See **RM1A**
Slight	See **RM1A**		See **RM1A**
Moderate	See **RM1A**		See **RM1A**
Heavy $\lambda_K = 0.4$ $\lambda_Q = 0.5$ $\lambda_D = 0.5$	*Typical Appearance*		● Complete or partial replacement or enhancement required.
Extreme	*Criteria:*	● Compression toe of wall buckled. ● Reinforcement has fractured. ● Wide flexural cracking. ● Laterally displaced units. ● Localized crushing or spalling.	● Replacement or enhancement required.

RM1E	COMPONENT DAMAGE CLASSIFICATION GUIDE	System:	**Reinforced Masonry**

	Component Type:	**Stronger Pier**
	Behavior Mode:	**Flexure / lap splice slip**
	Applicable Materials:	**Fully or partially grouted hollow concrete or clay units**

How to distinguish behavior mode:

By observation:

Walls that are vulnerable to lap splice slip will exhibit flexural response, and possible flexure/shear response, until the lap splice capacity is exceeded. Observed damage will therefore be very similar to RM1A and RM1B until lap splice slip occurs. Lap splice slip failure is characterized by splitting of the masonry units parallel to the reinforcing bars.

Caution: At low damage levels, damage observations will be identical to those for the **RM1A** behavior modes.

By analysis:

Walls that are vulnerable to lap splice slip may have:

- Bar size greater than:
  #4 in 4 inch units
  #6 in 6 inch units
  #8 in 8 inch units

- Lap splice less than l_d

Refer to Evaluation Procedures for:

- Evaluation of flexural response.

- Evaluation of lap splice slip response.

- Evaluation of development length l_d.

Severity	Description of Damage		Performance Restoration Measures
Insignificant	See **RM1A** or **RM1B**		See **RM1A** or **RM1B**
Slight	See **RM1A** or **RM1B**		See **RM1A** or **RM1B**
Moderate $\lambda_K = 0.4$ $\lambda_Q = 0.5$ $\lambda_D = 0.8$	*Criteria: Typical Appearance*	• Vertical cracks at toe of wall, particularly in narrow dimension of wall.	See **RM1A** or **RM1B**

COMPONENT DAMAGE
CLASSIFICATION GUIDE **continued**

$\boxed{\text{RM1E}}$

Severity	Description of Damage		Performance Restoration Measures
Extreme	*Criteria:*	• Splitting of face shells at toe of wall • Crushing and delamination of faceshells from grout cores	• Replacement or enhancement required.
	Typical Appearance	• Wide flexural cracking and/or crushed units at base of wall • Pulverized mortar at base – evidence of rocking.	

RM2B	COMPONENT DAMAGE CLASSIFICATION GUIDE	System:	**Reinforced Masonry**

Component Type:	**Weaker Pier**
Behavior Mode:	**Flexure / Shear**
Applicable Materials:	**Fully grouted hollow concrete or clay units**

How to distinguish behavior mode:

By observation:

Damage in an RM2 component with a flexural/shear response may be localized to the first story, or it may be evident at a number of levels in story-height piers. Both horizontal and diagonal cracks may be present, with diagonal cracks predominant. Diagonal cracks may appear to be independent from horizontal flexural cracks, and propagate across the major diagonal dimensions. When severely damaged, shear deformations will be localized to one or two diagonal cracks of large width. If diagonal cracks are uniformly distributed and of small width, the behavior mode may be ductile flexure. If a permanent horizontal offset is visible, the behavior mode may include Flexure/Sliding Shear.

By analysis:

Analysis of a wall with a Flexure / Shear behavior mode may not indicate a clear distinction between the controlling mechanism of flexure (deformation controlled) or shear (force controlled). Calculated capacities should be in the same range. Wall axial loads may be moderate to high.

Refer to Evaluation Procedures for:

- Evaluation of flexural response.
- Evaluation of shear response.
- Identifying flexural versus shear cracks.
- Crack width discussion.

Severity	Description of Damage	Performance Restoration Measures
Insignificant $\lambda_K = 0.8$ $\lambda_Q = 1.0$ $\lambda_D = 1.0$	*Criteria:* • No crack widths exceed 1/16." • No significant spalling. *Typical Appearance:* May appear similar to flexure following small displacement cycles. Diagonal cracks often propagate from horizontal cracks.	Not necessary for restoration of structural performance. (Cosmetic measures may be necessary for restoration of nonstructural characteristics.)

COMPONENT DAMAGE
CLASSIFICATION GUIDE **continued**

$$\boxed{\text{RM2B}}$$

Severity	Description of Damage		Performance Restoration Measures
Slight $\lambda_K = 0.6$ $\lambda_Q = 1.0$ $\lambda_D = 1.0$	*Criteria:*	• No crack widths exceed 1/8". • No significant spalling or vertical cracking.	• Inject cracks. $\lambda_K* = 0.9$ $\lambda_Q* = 1.0$ $\lambda_D* = 1.0$
	Typical Appearance:	Similar to insignificant damage, except cracks are wider and cracking is more extensive.	
Moderate $\lambda_K = 0.4$ $\lambda_Q = 0.8$ $\lambda_D = 0.9$	*Criteria:*	• Crack widths do not exceed 3/16". • Moderate spalling of masonry unit faceshells or vertical cracking at toe regions. • No buckled or fractured reinforcement. • No significant residual displacement.	• Remove and patch spalled masonry and loose concrete. Inject cracks. • Consider horizontal fiber composite overlay.
	Typical Appearance:		$\lambda_K* = 0.8$ $\lambda_Q* = 1.0$ $\lambda_D* = 1.0$
Extreme	*Criteria:* *Typical Indications* *Typical Appearance*	• Reinforcement has fractured • Wide flexural cracking typically > ¼" concentrated in a single crack. • Wide diagonal cracking, typically concentrated in one or two cracks • Extensive crushing or spalling at wall toes, visible delamination of faceshells from grout 	• Replacement or extensive enhancement required.

RM2G	COMPONENT DAMAGE CLASSIFICATION GUIDE	System:	**Reinforced Masonry**

Component Type:	**Weaker Pier**
Behavior Mode:	**Preemptive Shear**
Applicable Materials:	**Fully grouted hollow concrete or clay units**

How to distinguish behavior mode:

By observation:

At low levels of damage, wall may appear similar to RM2B. Diagonal cracks may be visible before flexural cracks. Damage occurs quickly in the form of one or two dominant diagonal cracks. Subsequent cycles may cause crushing or face shell debonding at the center of the wall and/or at the wall toes.

By analysis:

Calculated shear load capacity, including both masonry and steel components, will be less than or equal to shear associated with flexural load capacity

Refer to Evaluation Procedures for:

- Evaluation of flexural response.
- Evaluation of shear response.
- Evaluation of crack patterns.
- Crack evaluation.

Severity	Description of Damage		Performance Restoration Measures
Insignificant $\lambda_K = 0.9$ $\lambda_Q = 1.0$ $\lambda_D = 1.0$	*Criteria:*	• No diagonal cracks. • Flexural crack <1/16". • No significant spalling.	Not necessary for restoration of structural performance. (Cosmetic measures may be necessary for restoration of nonstructural characteristics.)
	Typical Appearance:	No visible damage.	
Slight $\lambda_K = 0.8$ $\lambda_Q = 1.0$ $\lambda_D = 1.0$	*Criteria:*	• No crack widths exceed 1/16". • No significant spalling or vertical cracking.	• Inject cracks. $\lambda_K{}^* = 0.9$ $\lambda_Q{}^* = 1.0$ $\lambda_D{}^* = 1.0$
	Typical Appearance:	Similar to insignificant damage, except that small diagonal cracks may be present.	

COMPONENT DAMAGE
CLASSIFICATION GUIDE **continued**

$$\boxed{\text{RM2G}}$$

Severity	Description of Damage		Performance Restoration Measures
Moderate $\lambda_K = 0.5$ $\lambda_Q = 0.8$ $\lambda_D = 0.9$	*Criteria:*	• Crack widths do not exceed 1/16". • No spalling of masonry unit faceshells or vertical cracking at toe regions.	• Inject cracks. • Consider horizontally oriented fiber composite overlay. $\lambda_K^* = 0.8$ $\lambda_Q^* = 1.0$ $\lambda_D^* = 1.0$
	Typical Appearance:	May be several diagonal cracks, typically with one dominant crack.	
Heavy $\lambda_K = 0.3$ $\lambda_Q = 0.4$ $\lambda_D = 0.5$ See FEMA 307 for calculation of λ_Q	*Criteria:* *Typical Appearance:*	• Single dominant crack, may be > 3/8".	• Inject cracks. • Provide horizontally oriented fiber composite overlay. • Consider replacement.
Extreme	*Criteria:* *Typical Indications*	• Reinforcement has fractured. • Wide diagonal cracking, typically concentrated in one or two cracks. • Crushing or spalling at center of wall or at wall toes.	• Replacement or enhancement required.

RM3A	COMPONENT DAMAGE CLASSIFICATION GUIDE	System:	**Reinforced Masonry**

Component Type:	**Weaker Spandrel**
Behavior Mode:	**Flexure**
Applicable Materials:	**Fully grouted hollow concrete or clay units**

How to distinguish behavior mode:

By observation:

Masonry wall frames will develop numerous flexural cracks within the beam plastic hinge zones, with very little damage in the pier or joint regions. If significant damage develops in piers, component should be reclassified as RM2.

By analysis:

Wall frame dimensions and reinforcement satisfy the requirements of Section 2108.2.6 of the 1994 or 1997 *UBC*, or

the component can be shown by the principles of capacity design to develop flexural plastic hinges in the beams without developing the strength of the piers.

Refer to Evaluation Procedures for:

• Evaluation of flexural response.

• Identification of the plastic hinge zone.

Severity	Description of Damage		Performance Restoration Measures
Insignificant $\lambda_K = 0.9$ $\lambda_Q = 1.0$ $\lambda_D = 1.0$	*Criteria:*	• No crack widths exceed 1/16". • No significant spalling.	Not necessary for restoration of structural performance. (Cosmetic measures may be necessary for restoration of nonstructural characteristics.)
Slight $\lambda_K = 0.8$ $\lambda_Q = 0.9$ $\lambda_D = 1.0$	*Criteria:* *Typical Appearance:*	• No crack widths exceed 1/8". • No significant spalling. 	• Inject cracks. $\lambda_K{}^* = 0.8$ $\lambda_Q{}^* = 1.0$ $\lambda_D{}^* = 1.0$
Moderate $\lambda_K = 0.6$ $\lambda_Q = 0.8$ $\lambda_D = 1.0$	*Criteria:* *Typical Appearance:*	• No crack widths exceed ¼". • Minor spalling (less than one unit depth) in beam ends. 	• Replace spalled material. • Inject cracks. $\lambda_K{}^* = 0.8$ $\lambda_Q{}^* = 1.0$ $\lambda_D{}^* = 1.0$
Extreme	*Criteria:* *Typical Appearance*	• Reinforcement has fractured. • Wide flexural cracking (> 3/8"). • Significant crushing or spalling at junction of pier and beams.	• Replacement or enhancement required.

RM3G	COMPONENT DAMAGE CLASSIFICATION GUIDE	System:	**Reinforced Masonry**

	Component Type:	**Weaker Spandrel**
	Behavior Mode:	**Preemptive Shear**
	Applicable Materials:	**Fully or partially grouted hollow concrete or clay units**

How to distinguish behavior mode:

By observation:

Cracking in reinforced masonry beams may be concentrated at the beam ends or distributed over the beam. Development of a plastic hinge zone is unlikely because of the difficulty of providing sufficient confinement reinforcement. Visible cracking is often a continuation of cracks in slabs or other adjacent elements, so beam damage should be evaluated in the context of the system behavior. If damage patterns appear to be associated with ductile flexural response, component may be reclassified as RM3.

By analysis:

Expected shear strength will typically be less than shear associated with the development of a flexural yielding mechanism at each end of the beam. If a stable plastic hinge can develop, component may be reclassified as RM3.

Refer to Evaluation Procedures for:

• Evaluation of flexural response.

Severity	Description of Damage		Performance Restoration Measures
Insignificant $\lambda_K = 0.9$ $\lambda_Q = 1.0$ $\lambda_D = 1.0$	*Criteria*	• Hairline cracks only.	Not necessary for restoration of structural performance. (Cosmetic measures may be necessary for restoration of nonstructural characteristics.)
Moderate $\lambda_K = 0.8$ $\lambda_Q = 0.8$ $\lambda_D = 1.0$	*Criteria* *Typical Appearance*	• Cracks < 1/8".	• Inject cracks. $\lambda_K^* = 0.8$ $\lambda_Q^* = 1.0$ $\lambda_D^* = 1.0$
Heavy $\lambda_K = 0.3$ $\lambda_Q = 0.5$ $\lambda_D = 0.9$	*Criteria* *Typical Appearance*	• Cracks > 1/8".	• Inject cracks. • Repair spalled areas $\lambda_K^* = 0.8$ $\lambda_Q^* = 1.0$ $\lambda_D^* = 1.0$
Extreme	*Criteria:* *Typical Appearance*	• Reinforcement has fractured. • Crushing or spalling at beam ends. • Large diagonal cracks and / or spalling at center portion of beam.	• Replacement or enhancement required.

7: Unreinforced Masonry

7.1 Introduction and Background

7.1.1 Section Organization

This section summarizes evaluation methodologies and repair recommendations for earthquake-damaged unreinforced masonry (URM) bearing-wall buildings. Reinforced masonry is covered in Chapter 6. Masonry with less than 25 percent of the minimum reinforcement required by FEMA 273 (ATC, 1997a) should be considered unreinforced. This material supports and supplements the Component Damage Classification Guides (Component Guides) for URM components contained in Section 7.5. The section is organized as follows:

Section 7.1 discusses the various materials and structural systems used in URM buildings, evaluation of rehabilitated buildings, and the limitations of the URM guidelines.

Section 7.2 identifies typical URM elements, components, and behavior modes, as well as characteristics of the types of earthquake damage these components can experience. Behavior modes discussed include those resulting from in-plane and out-of-plane demands on walls and those occurring in other elements or due to the interrelationships between building elements. Information on the relative likelihood of occurrence of each damage type is included, where information is available. For in-plane behavior modes, the strength and displacement capacity of each mode is discussed along with uncertainties in capacity calculations.

Section 7.3 presents evaluation procedures for URM walls subjected to in-plane and out-of-plane demands. This section also identifies the testing that may be needed to provide information on required material properties. Symbols are listed in Section 7.4 and references are listed in the Reference section.

FEMA 307 provides a summary of the hysteretic behavior observed in experimental tests of URM specimens and commentary on the FEMA 273 force-displacement relationships. It describes the development of the λ-factors in the component guides of FEMA 306. FEMA 307 also provides a tabular summary of important experimental research and a list of other references on URM elements.

7.1.2 Material Types and Structural Framing

Unreinforced masonry is one of the oldest and most diverse building materials. Important material variables include masonry unit type, construction, and the material properties of various constituents.

Solid clay-brick unit masonry is the most common type of masonry unit, but there are a number of other common types, such as hollow clay brick, structural clay tile, concrete masonry, stone masonry, and adobe. There are additional subgroupings within each of these larger categories. For example, as shown in FEMA 274, structural clay tile has been classified into structural clay load-bearing wall tile, structural clay non-load-bearing tile (used for partitions, furring, and fireproofing), structural clay floor tile, structural clay facing tile, and structural glazed facing tile. Hollow clay tile (HCT) is a more common term for some types of structural clay tile. Concrete masonry units (CMU) can be ungrouted, partially grouted, or fully grouted. Stone masonry can be made from any type of stone, but sandstone, limestone, and granite are common. Other stones common in a local area are used as well. Sometimes materials are combined, such as brick facing over CMU backing, or stone facing over a brick backing.

Wall construction patterns also vary widely, with bond patterns ranging from common running bond in brick to random ashlar patterns in stone masonry to stacked bond in CMU buildings. The variety of solid brick bond patterns is extensive. Key differences include the extent of header courses, whether collar joints are filled, whether cavity-wall construction was used, and the nature of ties between facing and backing wythes. In the United States, for example, typical running-bond brick masonry includes header courses interspersed by about five to six stretcher courses. Header courses help tie the wall together and allow it to behave in a more monolithic fashion for both in-plane and out-of-plane demands. The UCBC (ICBO, 1994) has specific prescriptive requirements on the percentage, spacing, and depth of headers. Facing wythes not meeting these requirements must be considered as veneer and are therefore not used to determine the effective thickness of the wall. Veneer wythes must be tied to the backing to help prevent out-of-plane separation and falling hazards. Although bed and head joints are routinely filled with mortar, the extent of collar-joint fill varies widely. Completely filled collar joints with metal ties between wythes help the wall to behave in a more monolithic fashion for out-of-plane demands. One form of construction where interior vertical joints are deliberately not filled is cavity-wall construction. Used in many northeastern United States buildings, the cavity helps provide an insulating layer and a means of dissipating moisture. The cavity also reduces the out-of-plane capacity of the wall.

Material properties—such as compressive, tensile, and shear strengths and compressive and shear moduli—vary widely among masonry units, brick, and mortar. An important issue for in-plane capacity is the relative strength of masonry and mortar. Older mortars typically used a lime/sand mix and are usually weaker than the masonry units. With time, cement was added to the mix and mortars became stronger. When mortars are stronger than the masonry, strength may be enhanced, but brittle cracking through the masonry units may be more likely to occur, resulting in lower deformation capacity.

Given the wide range of masonry units, construction, and material properties, developing a comprehensive methodology for the evaluation of earthquake damage is difficult. The methodology in this document is most directly relevant to solid clay masonry laid in running bond with a typical spacing of header courses. Additional issues that should be considered for different conditions are identified in some cases.

For additional general background on URM materials, see ABK (1981a), FEMA 274, and Rutherford and Chekene (1997).

There is significant diversity in the characteristics of the structural systems used in URM bearing wall buildings. A primary issue is the rigidity of floor and roof diaphragms. While the 1994 UCBC includes provisions for both flexible and rigid diaphragms, the original ABK research, upon which it was based, primarily addressed flexible wood diaphragms (ABK, 1984). While such wood-diaphragm buildings are the most common, there are a substantial number of buildings with more rigid diaphragms, particularly in areas outside California. These include concrete slabs spanning between steel beams, hollow concrete planks, and brick and HCT arches spanning between steel beams.

Buildings with rigid floor and roof diaphragms will respond to earthquake shaking in a substantially different manner from those with flexible diaphragms. When rigid diaphragms are used, the traditional model of building behavior used in the building code is that of a lumped-mass system. In this model, the diaphragms represent the mass, and the vertical elements (such as walls) are flexible and are the primary source of the dynamic response experienced by the building. In contrast, in URM buildings with flexible diaphragms, the ABK model assumption is that the ground motion is applied to the ends of the flexible diaphragms without significant amplification. Any amplification that occurs is caused by the dynamic response of the diaphragm and the coupled out-of-plane walls. In some cases, the diaphragm may yield, limiting the forces that can be transmitted to the in-plane walls. In rigid-diaphragm URM buildings, diaphragm yielding is unlikely, and the frequency of response of the wall and diaphragms is likely to be much closer.

The methodology in this document is most directly relevant to URM bearing-wall buildings with flexible diaphragms. While FEMA 273 generally separates diaphragm issues from wall issues, there can be interrelationships between the two. Such issues are pointed out where appropriate.

7.1.3 Seismically Rehabilitated URM Buildings

When evaluating earthquake damage to unreinforced masonry buildings, it is important to determine the extent of seismic rehabilitation work that may have been performed, because this can affect the interpretation and significance of the damage. For example, if the building has not been seismically rehabilitated, horizontal cracking near the floor lines may be related to buckling of a slender wall. However, if wall-to-diaphragm ties have been installed in a rehabilitation effort, then the cracking may be related to an out-of-plane bed-joint sliding-shear failure in the mortar below the wall ties. If the wall is backed by shotcrete as is commonly done in a seismic rehabilitation, then the masonry wall cracking may be less significant than if there were no concrete present.

URM buildings have been the focus of seismic-hazard mitigation policies, evaluation and rehabilitation standards, and seismic-strengthening efforts. On the west coast, in California in particular, a substantial number of URM buildings have been rehabilitated. Seismic strengthening practices and standards have evolved over the years, and the scope of work and expected performance of the rehabilitated buildings varies significantly. It is important to appreciate the variety of potential rehabilitation work that may be encountered in the field. Using FEMA 273 terminology, performance objectives used in URM rehabilitation include:

- Limited Partial Rehabilitation efforts, such as those that address only certain specific elements such as parapets (e.g., San Francisco Parapet Safety Program).

- More substantial Partial Rehabilitation efforts such as the San Francisco "Bolts-plus" provisions that address parapets, wall-diaphragm ties, and wall bracing.

- Reduced risk rehabilitation programs such as the City of Los Angeles Division 88 and RGA requirements (Division 88, 1985 and SEAOSC, 1986), the *Uniform Code for Building Conservation*, and FEMA 178 (BSSC, 1992)—all of which address the complete lateral-load-resisting system, but which require only a single-level check of the life safety performance level instead of the two-level check required by FEMA 273. Retrofit approaches using these methodologies may be capable of meeting the Basic Safety Objective if it can be shown that the building can meet the collapse prevention performance level for the BSE-2 earthquake as defined in FEMA 273.

- Basic Safety Objectives, such as the Field Act strengthening requirements for California elementary and secondary public schools.

- Enhanced Rehabilitation Objectives to limit damage and increase functionality have been implemented for a few select buildings.

While many voluntary and mandatory strengthening programs are in place in the west, and the number of rehabilitated URM buildings continues to grow nationwide, the vast majority of the rehabilitated URM buildings are those in Southern California strengthened to Division 88 (City of Los Angeles, 1985), RGA (SEAOSC, 1986) and similar standards and those in San Francisco that have complied with an earlier parapet safety program, but have not yet completed more stringent current requirements.

In general, this document is intended to be used with unrehabilitated buildings. The guidelines for in-plane and out-of-plane wall behavior assume, for example, that the wall has not been strengthened with other materials such as concrete, adhered fabric, or ferrocement overlays. If the URM wall has been strengthened with shotcrete or cast-in-place concrete, then the evaluating engineer will have to exercise judgment about the significance of the damage using both the provisions of Chapter 5 for reinforced concrete and those of Section 7. Generally, greater attention should be given to the damage in the concrete elements because they are usually the primary or intended lateral-force-resisting element. Fabrics and overlays have been the subject of experimental testing, but very limited rehabilitation has actually been performed using these techniques; as a result, they are not addressed in these provisions.

Even though the focus is on unrehabilitated buildings, the guidelines contain descriptions of damage and commentary on its interpretation for certain selected elements in rehabilitated buildings. Such commentary is based on observations from the 1987 Whittier, 1989 Loma Prieta, and 1994 Northridge, California, earthquakes.

7.2 Unreinforced Masonry Component Types and Behavior Modes

7.2.1 Non-Wall Components

For the procedures in this document, structural systems are subdivided into elements, which are further subdivided into components that can be related to specific modes of behavior during seismic shaking. Components within URM bearing wall buildings include parapets, appendages, wall-diaphragm ties, diaphragms, and walls. The focus of this document is in-plane wall behavior modes. For other building types, such as concrete wall buildings, earthquake damage is primarily related to the in-plane behavior of the wall or displacement incompatibility between the walls and other elements. For URM bearing wall buildings, however, other elements and components figure prominently in observations of actual damage. In many cases, non-wall component damage may occur before in-plane damage to the wall becomes significant. These other features and their common behavior modes are discussed briefly below. The remainder of the document focuses on wall elements and components.

Figure 7-1   *Diagram of Parapet Failure (from Rutherford and Chekene, 1990)*

Parapets: These short extensions of walls above the roof typically occur at the perimeter of the buildings and are primarily present for fire safety or aesthetic reasons. As originally constructed, they are not braced back to the roof and are thus susceptible to brittle flexural out-of-plane failure (see Figures 7-1 and 7-2). Braced parapets typically fail at the connections between the parapet and the brace (see wall-diaphragm tie failures for similar examples).

Appendages: This category includes veneer, cornices, friezes, pediments, dentils, brackets, statuary, and finials—in short, any minor masonry feature that is susceptible to falling. Damage may result from excessive accelerations of appendages and deformations that cause connection failures between the appendage and the structure; delamination of veneer

can result from missing or inadequate ties; deformation incompatibility between the appendage and structure can cause cracking and spalling; and pounding against adjacent buildings can lead to localized falling hazards (see Figures 7-3 and 7-4).

Wall-Diaphragm Ties: In the United States, wall-to-diaphragm ties in existing URM buildings are generally limited to low-strength tension connections called "government anchors" or "dog anchors," in which one end of a steel bar is embedded one wythe in from the outer face of the wall and the other end is hammered into the side of a wood joist. These ties typically occur where joists bear on the walls, not where they are parallel to the walls. Wall-diaphragm separation due to inadequate or missing tension ties can lead to out-of-plane failures of walls; missing shear ties can lead to the

Figure 7-2   Photo of Parapet Failure (from Rutherford and Chekene, 1990).

diaphragm sliding along the in-plane walls and then pushing against the walls perpendicular to the movement, resulting in corner damage to the walls (see Figures 7-5, 7-6 and 7-7). In rehabilitated buildings, a variety of tie failures have been observed, including bond failures between the masonry and the cementitious grout used in older drilled dowel connections, cone failures due to shallow embedment and/or weak masonry, pullthrough of through-plated anchors, and bed-joint sliding near ties. Wall-diaphragm failures are often associated with thin walls (such as two-wythe walls at upper stories), poor mortar conditions, and lack of sufficient overburden pressure.

Floor and Roof Diaphragms: Three categories of diaphragms can be identified: rigid concrete slab diaphragms, flexible wood and metal diaphragms, and intermediate systems such as hollow concrete planks and brick and HCT arches spanning between steel beams. In flexible diaphragms, excessive deflections can lead to out-of-plane wall damage. Hollow concrete plank systems may lack adequate interconnections to function as a continuous load path. Brick and HCT arch systems may be susceptible to vertical failure if beams separate locally. See Figure 7-8 for diagrams of some of the diaphragm types. Rigid concrete diaphragms are

generally not significantly damaged or the source of damage to other elements in URM bearing wall buildings, but they do affect the dynamic behavior of the building.

Table 7-1 summarizes behavior modes for non-wall URM elements, providing the source of the deficiency, type of damage, and intensity of ground shaking usually required to produce the damage. The intensity of ground shaking is a qualitative judgment based on actual earthquake reconnaissance and damage collection efforts. Individual buildings may respond in a different manner. Even though some types of damage do not generally lead to collapse, they can nonetheless endanger life safety either within the building or around the perimeter due to localized falling hazards. A simple example is an exterior parapet failure, which rarely leads to building collapse, but still poses a risk to pedestrians adjacent to the wall. Damage to non-URM wall elements is typically brittle or force-controlled, and it often does not affect the overall force-displacement relationship for the building. The component guides in this volume focus on wall damage. For non-URM wall elements that can affect the overall force-displacement relationship, such as wall-

Table 7-1   Behavior Modes for Non-Wall URM Elements

Element	Source of Deficiency	Behavior Mode	Intensity of Ground Shaking Usually Required to Produce Behavior Mode
Parapet	Out-of-plane flexural tension	Parapet falls	Low-to-moderate
Appendages	Connection failure	Falling hazard	Low-to-moderate
	Missing headers or veneer ties	Veneer delamination	Moderate
	Stiffness incompatibility	Cracking and spalling	Low-to-moderate
	Pounding	Local spalling	Moderate
Wall-Diaphragm Ties			
Tension Ties	Inadequate or missing ties	Wall-diaphragm separation	Moderate
Shear Ties	Diaphragm slides	Perpendicular walls punched out	Moderate
Diaphragms			
Rigid Concrete Slab	Not typically damaged	NA	NA
Hollow Concrete Plank	Lack of interconnection	Incomplete load path, excessive deflection	Moderate-to-high
HCT/Brick Arch on Steel Beams	Lack of tie rods between steel beams	Localized falling hazard as beams separate	Moderate-to-high
Flexible	Inadequate strength/stiffness	Excessive deflection can cause wall damage	Moderate-to-high

diaphragm ties, it is important to include their behavior and characteristics in modeling efforts.

7.2.2 Wall Components

URM wall elements can be subdivided into five Component Types as shown in Figure 7-9, based on the mode of inelastic behavior. Figure 7-9 also shows some of the common behavior modes. The majority of modes relate to in-plane damage, but out-of-plane damage can occur as well in each of the systems, often in combination with in-plane damage. The five component types are described below.

URM1: Solid cantilever walls. Such walls are typically found adjacent to other buildings or on alleys, and they act as cantilevers up from the foundation.

URM2: This component is a weak pier in a perforated wall. In this system, inelastic deformation occurs in the piers.

URM3: This component is a weak spandrel in a perforated wall. Inelastic deformation occurs first in the spandrels, which may create multistory piers similar to URM1 or URM2 and then lead to inelastic deformation and damage in the piers.

URM4: This component is a strong spandrel in a weak pier-strong spandrel mechanism. By definition, it should not suffer damage, and it is not discussed further in the report.

URM5: Perforated wall with panel zone weak joints. Inelastic deformation occurs in the region where the pier and spandrel intersect. Such damage is not observed generally in experimental tests, nor is it seen in actual earthquakes, except at outer piers of upper stories. In this document, such damage is considered a case of corner damage and, when caused by in-plane demands, is addressed as part of the URM3 spandrel provisions.

Table 7-2 summarizes behavior types for URM wall components identified above. The table provides an

Table 7-2   *Behavior Modes for URM Walls*

Ductility Category	Behavior Mode	Likelihood of Occurrence and Damage Guide Reference		
		Solid Wall (URM1)	Weak Piers (URM2)	Weak Spandrels (URM3)
Higher Ductility	Foundation Rocking	Common in field; no experiments; see text	NA	NA
	Wall-Pier Rocking	Possible; similar to URM2A Guide	Common in field; experiments done; see URM2A Guide	NA
	Bed Joint Sliding	Common in field; has experiments; similar to URM2B	Common in field; experiments done; see URM2B Guide	Unlikely; no guide
	Bed Joint Sliding at Wall Base	Possible; similar to URM2B Guide	NA	NA
	Spandrel Joint Sliding	NA	NA	Common in field; no experiments; see URM3D Guide
Moderate Ductility	Rocking/Toe Crushing	Seen in experiments; similar to URM2A Guide	Possible; similar to URM2A Guide	NA
	Flexural Cracking/Toe Crushing/Bed Joint Sliding	Seen in experiments; see URM1F Guide	Possible; similar to URM1F Guide	Unlikely; no guide
	Flexural Cracking/Diagonal Tension	Possible	Seen in experiments; similar to URM2K Guide	Unlikely
	Flexural Cracking/Toe Crushing	Seen in experiments; see URM1H Guide	Possible; similar to URM1H guide	Possible; no guide
	Spandrel Unit Cracking	NA	NA	Common in field; see URM3I guide
Little or No Ductility	Corner Damage	Common in field; no experiments; no specific guide; see text	NA	Common in outer pier of upper stories; no specific guide; see text
	Preemptive Diagonal Tension	Possible; similar to URM2K guide	May be common in field; seen in experiments; see URM2K Guide	May be common in field; no experiments; similar to URM2K Guide
	Preemptive Toe Crushing	Theoretical; similar to URM1H Guide	Theoretical; similar to URM1H Guide	Unlikely; no guide
	Out-of-Plane Flexural Response	Common in field; see URM1M Guide	Possible; similar to URM1M Guide	Unlikely; no guide

Notes: • Shaded areas of the table with notation "*See ...Guide*" indicate behavior modes for which a specific Component Guide is provided in Section 7.5. The notation "*Similar to ...Guide*" indicates that the behavior mode can be assessed by using the guide for a different, but similar component type or behavior mode.
   • *Common in field:* Frequently observed in earthquakes, even those of moderate size.
   • *Possible:* May not be explicitly documented but is assumed to have occurred or to occur in future.
   • *Seen in experiments:* Reported in experiments and may have occurred in earthquakes.
   • *Theoretical:* Theoretically possible but not widely reported, if at all.
   • *Unlikely:* Although reported for another component, it is unlikely for this one.
   • NA indicates that the failure mode cannot occur for this component.

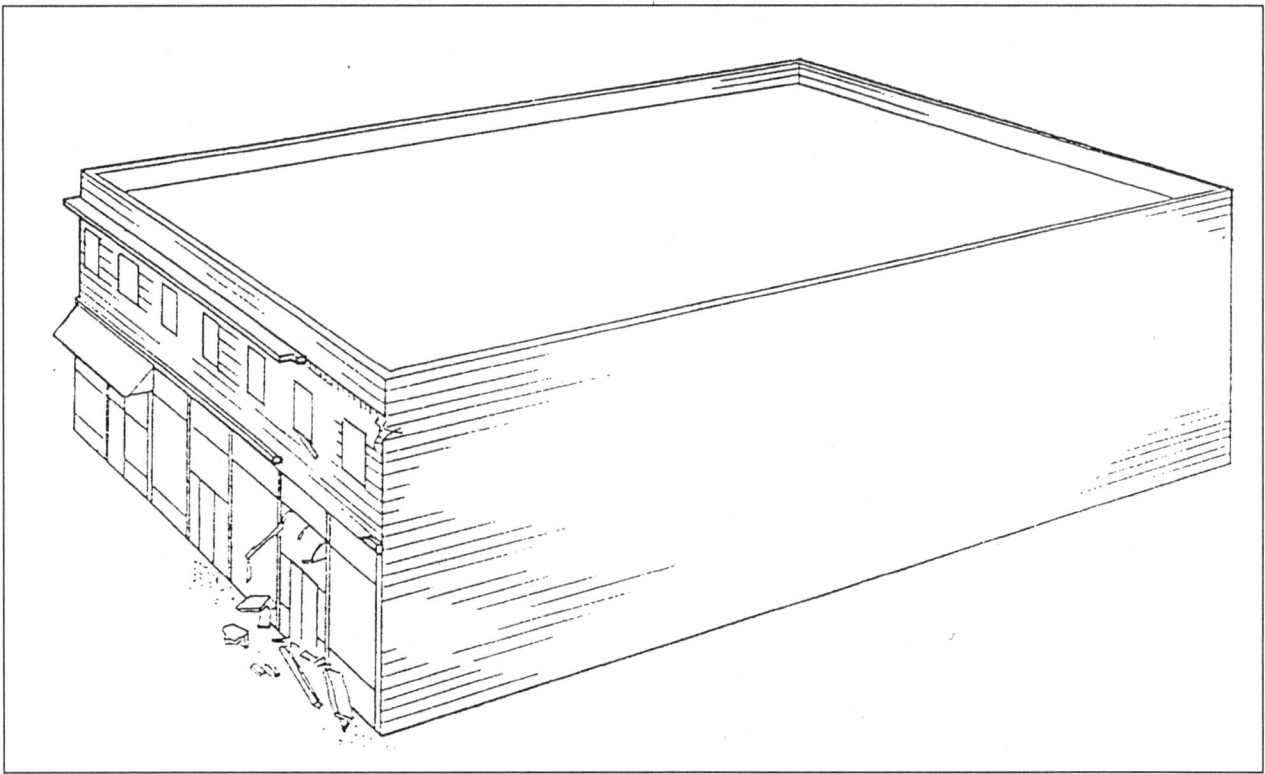

Figure 7-3   Diagram of an Appendage Failure (from Rutherford and Chekene, 1990)

indication of the ductility associated with a behavior mode, and the frequency with which such modes are observed. There are several categories for frequency of occurrence: "common in field" means that the mode has been frequently observed in earthquakes, even those of moderate size; "possible" means that the mode may not be explicitly documented in the literature, but it seems reasonable to assume it has occurred or will occur in earthquakes; "seen in experiments" means that such modes have been reported in experiments and may have occurred in some earthquakes; "theoretical" means that the mode, while theoretically possible, has not been widely reported in the literature, if at all. "Unlikely" means that, even though a mode has been reported for another component, it is unlikely to occur in the component under consideration. The modes listed in Table 7-2 are described in greater detail in the following sections. Component Guides have been developed for the damage resulting from the common modes and are shaded in Table 7-2; the designations such as "URM2A" refer to a specific guide.

7.2.3   Foundation Rocking

Rocking of a wall and its foundation on the supporting soil has been observed in the field. Though recognized as a potentially favorable mode of nonlinear response and a source of damping rather than significant damage, excessive rocking could theoretically lead to some instability and nonstructural damage in the superstructure, particularly if various walls rock out-of-phase or elements attached to the walls cannot tolerate the wall drift. See FEMA 273 and ATC-40 for an evaluation methodology for foundation rocking.

7.2.4   Wall-Pier Rocking

In the wall-pier rocking behavior mode, after flexural cracking develops at the heel, the wall or pier acts as a rigid body rotating about the toe. The rocking mode typically occurs when material shear capacity is high, piers are slender, and compressive stress is low. Post-cracking deformations can be large and relatively stable for many cycles. The ultimate limit state and the deflection at which it occurs are not well defined by the research, but three possible damage types can occur: in-

**1994 Northridge
Earthquake:
Santa Monica.**

**1989 Loma Prieta
Earthquake:
Watsonville.**

Figure 7-4   Photos of Appendage Failures (from Rutherford &Chekene, 1990)

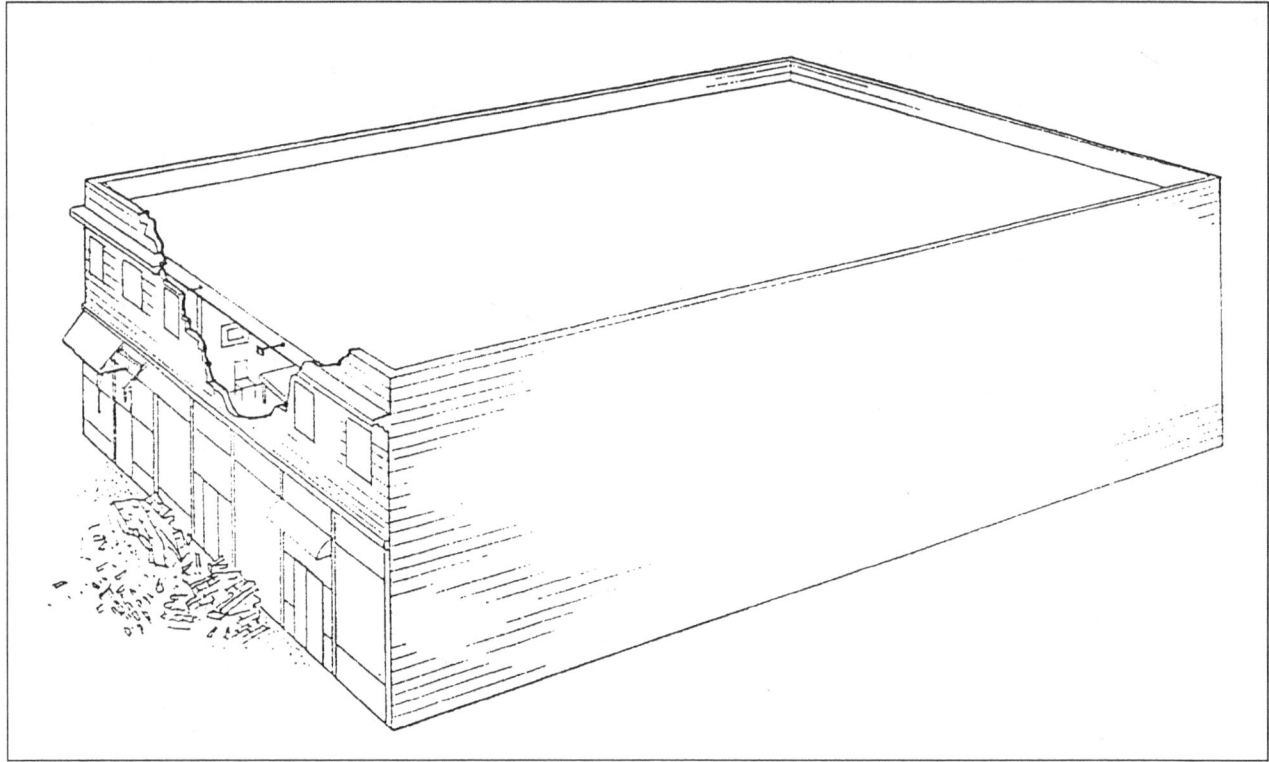

Figure 7-5 *Diagram of Wall-Diaphragm Tension Tie Failure (from Rutherford and Chekene, 1990)*

plane overturning, gradual degradation and softening of the pier, and excessive out-of-plane residual displacements ("walking") of the pier, leading to instability.

The strength and displacement capacities of an element in rocking are based on FEMA 273, where the "d" drift value of $0.4h_e/L$ was established. Currently available experimental results are insufficient to determine the relative influence of number of cycles, drift, and ductility on rocking degradation. In the field, it is often difficult to find evidence of rocking, because the cracks close at the completion of shaking. However, horizontal cracks at the top and bottom of piers have been observed, particularly as pointing mortar spalls.

7.2.5   Bed-Joint Sliding

In this type of behavior, sliding occurs on bed joints. Commonly observed both in the field and in experimental tests, there are two basic forms: sliding on a horizontal plane, and a stair-stepped diagonal crack where the head joints open and close to allow for movement on the bed joint. See Figure 7-10 for an

example of typical stair-stepped bed-joint sliding observed in the field. Pure bed-joint sliding is a ductile mode with significant hysteretic energy dissipation capability. If sliding continues in the absence of one of the less-ductile modes noted in the sections that follow, then gradual degradation of the cracking region occurs until instability is reached. Theoretically possible, but not widely reported, is the case of stair-stepped cracking in which sliding goes so far that an upper brick slides off a lower brick.

The strength and displacement capacities (in shear) for bed-joint sliding are based on FEMA 273, which uses a Mohr-Coulomb model originally developed as part of the ABK research. A Mohr-Coulomb model includes a bond and friction component. There are many uncertainties in this model, in relating the *in-situ* testing to the model, and in the *in-situ* testing itself. The FEMA 273 equation for shear capacity is:

$$V_{bjs} = v_{me} A_n = A_n[0.75(0.75v_{te} + P_{CE}/A_n)]/1.5 \quad (7\text{-}1)$$

Figure 7-6   Photo of Wall-Diaphragm Tension Tie Failure (from Rutherford and Chekene, 1990)

where:

v_{te} = the average test value from in-place testing

P_{CE} = the expected gravity compressive force

A_n = the area of net mortared/grouted section.

The model was calibrated with limited empirical tests using brick units, which resulted in the first 0.75 coefficient. Calibrations for other types of masonry such as ungrouted CMU or HCT have not been done. The model is most appropriate for estimating strength before cracking; after cracking the bond capacity will be eroded, and the strength is likely to be based on only the friction portion of the equation. Significant strength degradation has been observed in experiments at drifts of 0.3-0.4% which are likely to correspond to complete erosion of bond capacity. See Sections 7.3 and FEMA 307 for the implications of strength degradation due to sliding.

The v_{te} value representing bond strength is derived from the average of the individual push test values, v_{to}, adjusted for dead load by the equation:

$$v_{to} = V_{test}/A_b - p_{D+L}, \tag{7-2}$$

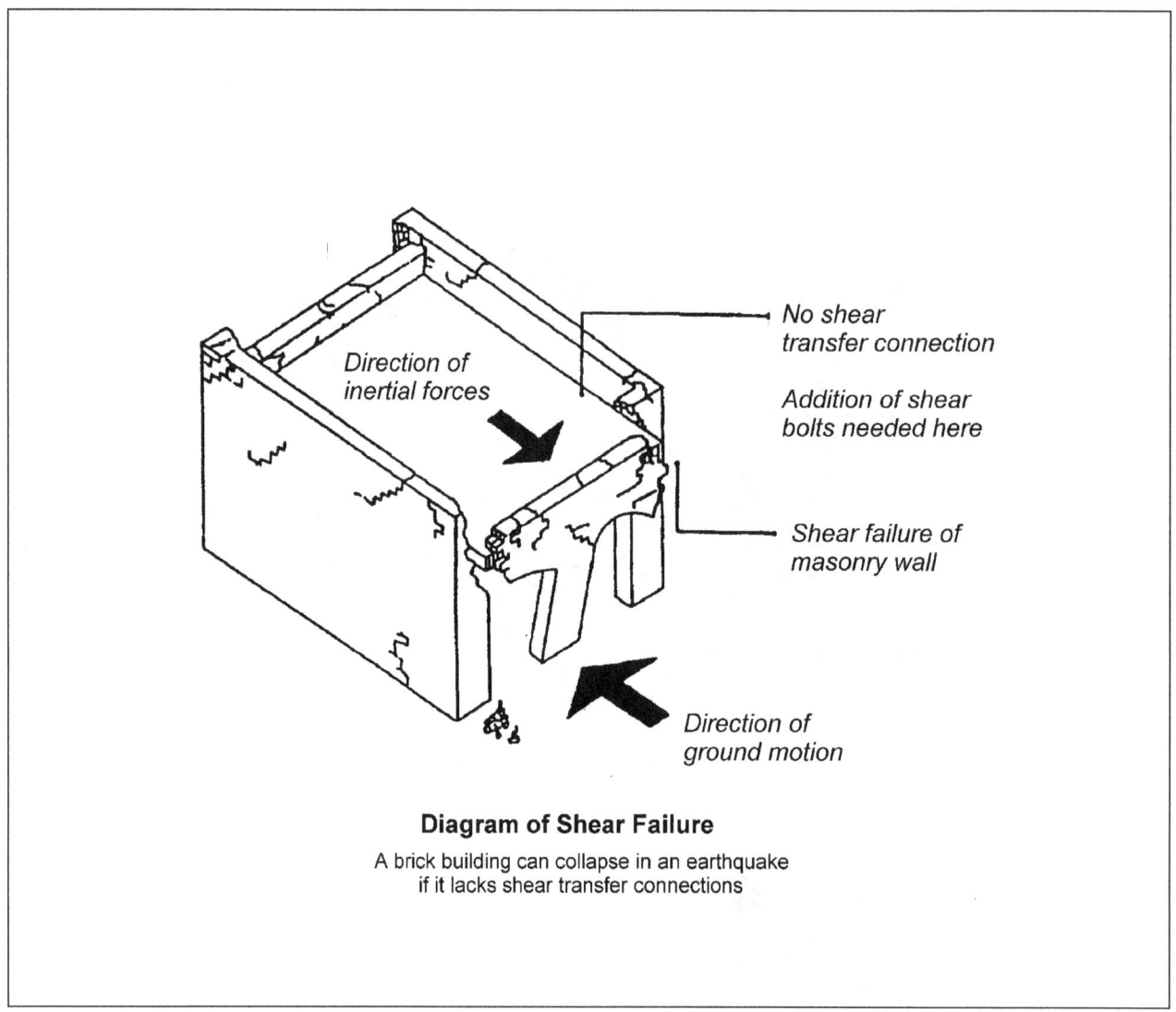

Diagram of Shear Failure

A brick building can collapse in an earthquake
if it lacks shear transfer connections

Figure 7-7   Diagram of Wall-Diaphragm Shear Tie Failure (from City of Los Angeles, 1991)

where:

V_{test} =   the test value

A_b =   the net mortared area of the bed joints above and below the test

p_{D+L} =   the estimated gravity stress at the test location.

A_b does not include the potential resistance of the collar joint; the influence of collar-joint fill is applied later to the v_{me} equation with the second 0.75 factor. This factor is waived if collar-joint fill is not present. While this simplifies the data reduction, it is less accurate than addressing the effect of fill in the collar joint at the test location itself. In many instances, it is difficult to determine the extent of the collar-joint fill. It can also be difficult to determine the actual gravity stress at a test location in walls with irregular openings. In some cases, flat-jack testing can be used to estimate gravity stresses. In FEMA 273, a 100 psi limit is set on v_{te}; although such a limit may be appropriate for design purposes, it is inappropriate for evaluating actual damage. The 1.5 factor in the v_{te} equation is to relate the average shear of V/A_n to the critical shear value of 1.5 V/A_n, as derived from a parabolic distribution of shear

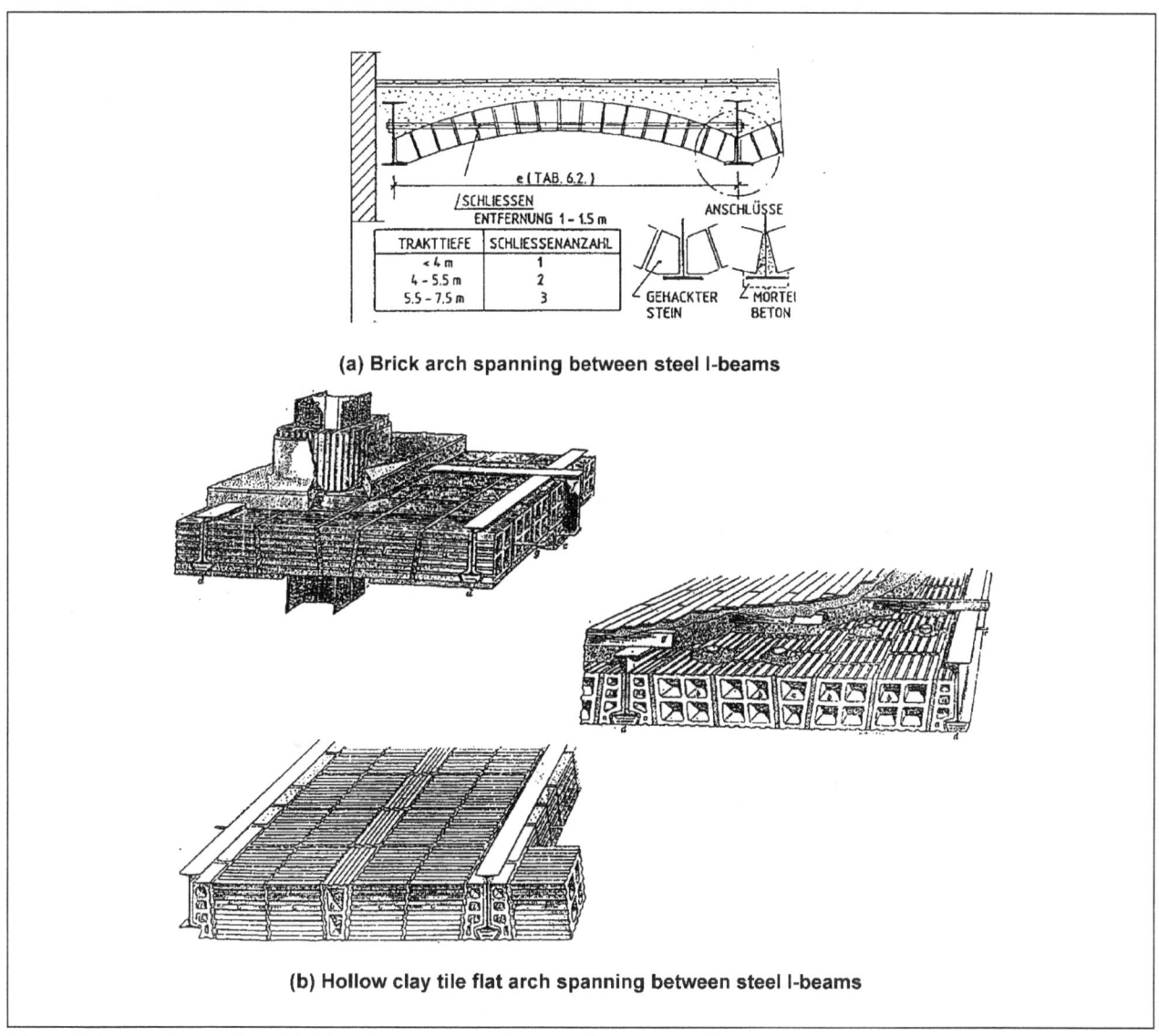

(a) Brick arch spanning between steel I-beams

(b) Hollow clay tile flat arch spanning between steel I-beams

Figure 7-8   *Examples of Various Masonry Diaphragms (from Rutherford and Chekene, 1997)*

in a rectangular section. ABK (1984) indicates that the 1.5 factor may overestimate the critical shear in long walls without openings.

Finally, the in-place push test has a number of uncertainties. Experience has shown that test results can vary substantially within the same masonry class, with coefficients of variation of 0.30 or more when the required number of tests are performed. This may be due to actual material variations, but it is also probably due in part to the uncertainty in determining when "either a crack can be seen or slip occurs"—the

governing criteria in *UBC* Standard 21-6 (ICBO, 1994) for determining the test load, V_{test}. Alternatives have been proposed to define V_{test} as the load occurring when the load-deflection curve stiffness is reduced to a certain percentage of the initial stiffness, or more simply, when a certain threshold deflection is reached.

7.2.6 Bed-Joint Sliding at Wall Base

Observed in experiments, this mode is a variation of bed-joint sliding in which the sliding occurs on the

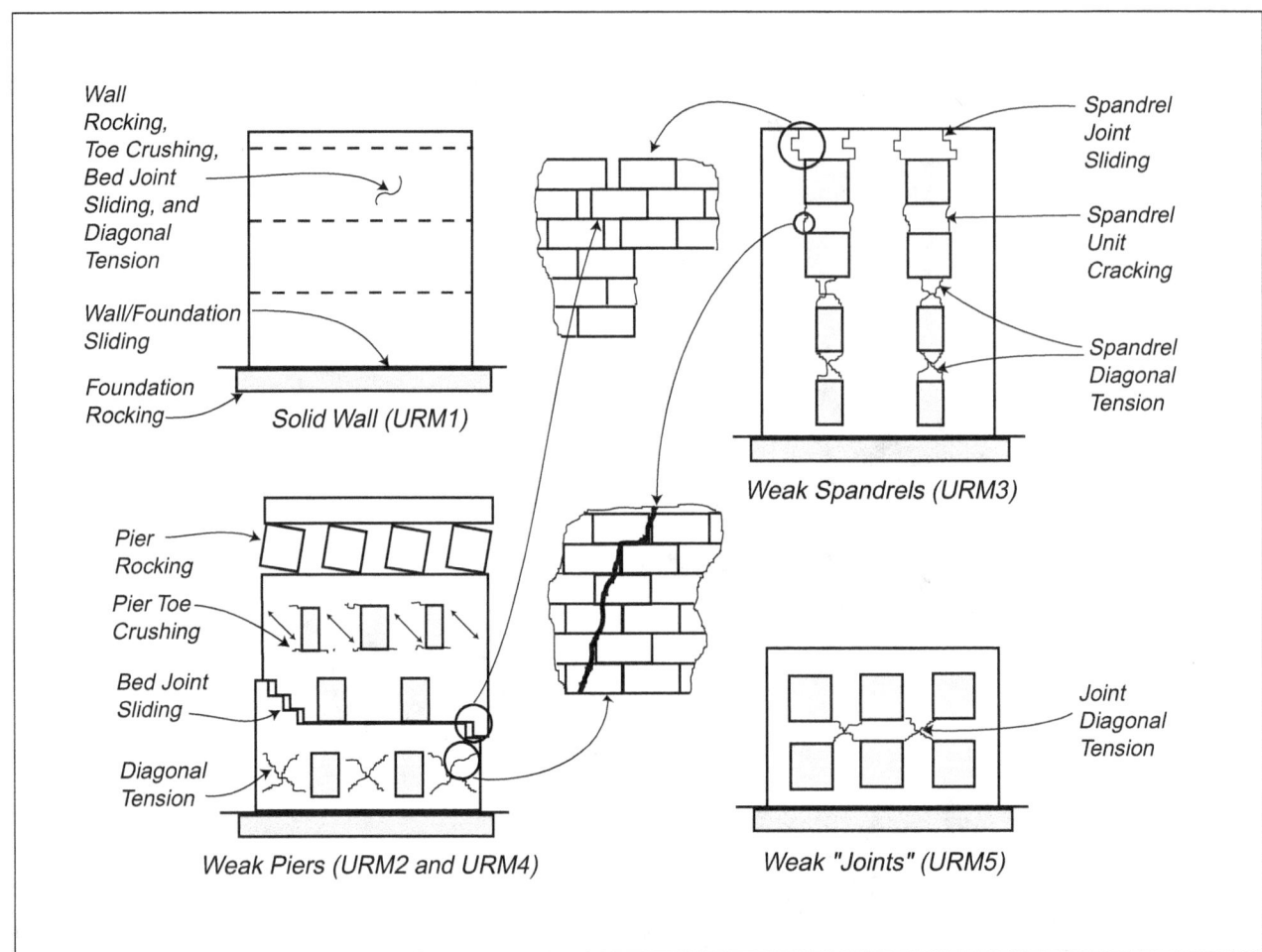

Figure 7-9 **URM Wall Components**

surface where the URM wall meets the foundation. Strength and displacement capacities are assumed to be similar to bed-joint sliding as described in Section 7.2.5.

7.2.7 Spandrel-Joint Sliding

Commonly observed in the field in running bond masonry, this form of bed-joint sliding in the ends of the spandrel resembles interlocked fingers pulling apart, and it occurs when the in-plane moment capacity of the spandrel is reached, but before the shear capacity of the spandrel is reached. This mode can be relatively ductile and can allow for significant drift, provided a reliable lintel is present. As the spandrel displaces, the nonlinear mechanism of response may move to other portions of the wall such as the piers.

The strength and displacement capacity are based on a modified version of the bed-joint sliding equation in FEMA 273. See Section 7.3 for details.

7.2.8 Rocking/Toe Crushing

This sequence of damage occurs when rocking (see Section 7.2.4) continues for several cycles, followed by an abrupt loss of capacity occurs as the toe crushes. Specimen W3 of Abrams and Shah (1992) is an example of such a phenomenon.

There is insufficient testing to determine parameters that would allow for an analytical determination of when rocking will degenerate into toe crushing. Intuitively, piers with higher axial stress and those subjected to higher drift levels and repeated cycles would be more likely to experience toe crushing. Because of the lack of data, this behavior is combined

Figure 7-10   Photo of Bed Joint Sliding

in this document with typical rocking behavior, and rocking capacity is set conservatively in FEMA 273.

7.2.9   Flexural Cracking/Toe Crushing/Bed Joint Sliding

In this sequence of behavior, flexural cracking occurs at the heel, but rocking does not begin. Instead, shear is redistributed to the toe, seismic forces increase, and a compression failure occurs in the toe. Diagonal cracks form, oriented toward the corners. Initial toe crushing is followed immediately by the ultimate limit state of bed-joint sliding.

This sequence was observed in Specimens W1, W2, and W3 of Manzouri et al. (1995). Initial capacity appears to be close to the FEMA 273 equation for toe crushing with a final capacity close to the frictional strength of the mortar. Under repeated cyclic loading, the toes may eventually deteriorate to the point of vertical instability. See FEMA 307 for commentary on this mode.

7.2.10   Flexural Cracking/Diagonal Tension

In this mode of behavior, flexural racking occurs at the heel, but not rocking. Shear is redistributed to the toe, seismic forces increase, a diagonal tension crack

develops, and capacity can be rapidly lost. Cracking typically occurs in the units as well as the joints. See Section 7.2.14 for more details.

Strength capacity is assumed to be the same as the FEMA 273 equation for diagonal tension capacity; displacements of approximately 0.5% have been observed in tests.

7.2.11   Flexural Cracking/Toe Crushing

In this sequence of behavior, flexural cracking occurs at the heel, but not rocking. Shear is redistributed to the toe, the seismic load increases, and a compression failure occurs in the toe. This type of behavior typically occurs in stockier walls with $L/h_{eff} > 1.25$. Based on laboratory testing of cantilever specimens, four steps can usually be identified. First, flexural cracking happens at the base of the wall, but it does not propagate all the way across the wall. This can also cause a series of horizontal cracks to form above the heel. Second, sliding occurs on bed joints in the central portion of the pier. Third, diagonal cracks form at the toe of the wall. Finally, large cracks form at the upper corners of the wall. Failure occurs when the triangular portion of wall above the crack rotates off the crack or the toe crushes so significantly that vertical load carrying capability is compromised.

Testing is limited to five monotonic specimens in Epperson and Abrams (1989), which all exhibited similar behavior, and Specimen W2 in Abrams and Shah (1992), which was tested with quasistatic reversed-cyclic loading. The strength is well-predicted by the FEMA 273 toe crushing equation. Toe crushing is considered in FEMA 273 to be a force-controlled mode, but moderate ductility was observed in Epperson and Abrams (1989), with drift values at conclusion of the test equal to 0.2-0.4%. In the Abrams and Shah (1992) test, even higher drift appears to have been achieved. Thus, some moderate degree of nonlinear capacity is possible. As noted above, behavior is similar to the Manzouri et al. (1995) tests, except that bed-joint sliding at the base did not occur at higher drifts.

7.2.12 Spandrel-Unit Cracking

In this type of damage, the moment at the end of the spandrel is not relieved by sliding, but instead causes brittle vertical cracking though the masonry units. Depending on the lintel construction, this can lead to a local falling hazard. It also can alter the height of the piers.

The cracked portion of the spandrel is assumed to lack both shear and tensile capacity. As a result, only the uncracked section is assumed to contribute to the strength and displacement capacity. See Section 7.3 for details.

7.2.13 Corner Damage

This form of damage is commonly observed at the intersection of the roof and walls subjected to in-plane and out-of-plane demands in moderate earthquakes. See Figures 7-11 and 7-12 for a diagram and photo of this damage type. Although not studied in experiments, it is likely to result from a combination of any of three possible causes:

- When a roof diaphragm without shear anchorage moves parallel to the walls subjected to in-plane demands, the walls subjected to out-of-plane demands may be punched outward. The tensile capacity of the wall (from the strength of the bed joints) is exceeded locally, and the wall corner falls.

- Damage may be exacerbated by cracking resulting from the horizontal spanning of the wall. In a

horizontal span, there is bending restraint at the ends of the wall due to the returns around the corner. If the resulting moment at the wall ends exceeds the capacity, a vertical crack occurs at the corner.

- For walls with openings near the corner, in-plane demands force moments into the joint between the outer pier at the top story and the adjacent spandrel. The moment places tensile demands on the head and bed joints at the pier/spandrel intersection, causing a diagonal crack to form.

Capacities are difficult to identify for the first two causes; the third is covered in a methodology developed in Section 7.3.

7.2.14 Preemptive Diagonal Tension

In this behavior mode, a diagonal tension crack forms without significant ductile response. Typical diagonal tension cracking—resulting from strong mortar, weak units, and high compressive stress—can be identified by diagonal cracks ("X" cracks) that propagate through the units. In many cases, the cracking is sudden and brittle, and vertical load capacity drops quickly. The cracks may then extend to the toe, and the triangles above and below the crack separate. In a few cases, the load drop may be more gradual with cracks increasing in size and extent with each cycle.

A second form of diagonal tension cracking exists with weak mortar, strong units, and low compressive stress, when the cracks propagate in a stair-stepped manner in head and bed joints. In Specimen MA of Calvi and Magenes (1994), this behavior was observed (Magenes, 1997).

Capacity is based on the FEMA 273 equation for diagonal tension. This equation requires calculation of masonry tensile strength, but there is no direct test for this value. As a substitute, the bed-joint mortar strength is used. This strength value only applies to the mortar, not the masonry units. Thus, there is a great deal of uncertainty in diagonal tension-strength calculations.

7.2.15 Preemptive Toe Crushing

In this form of damage, compression at the toe causes crushing without significant ductile response, such as rocking. There are no reported experimental observations of such behavior.

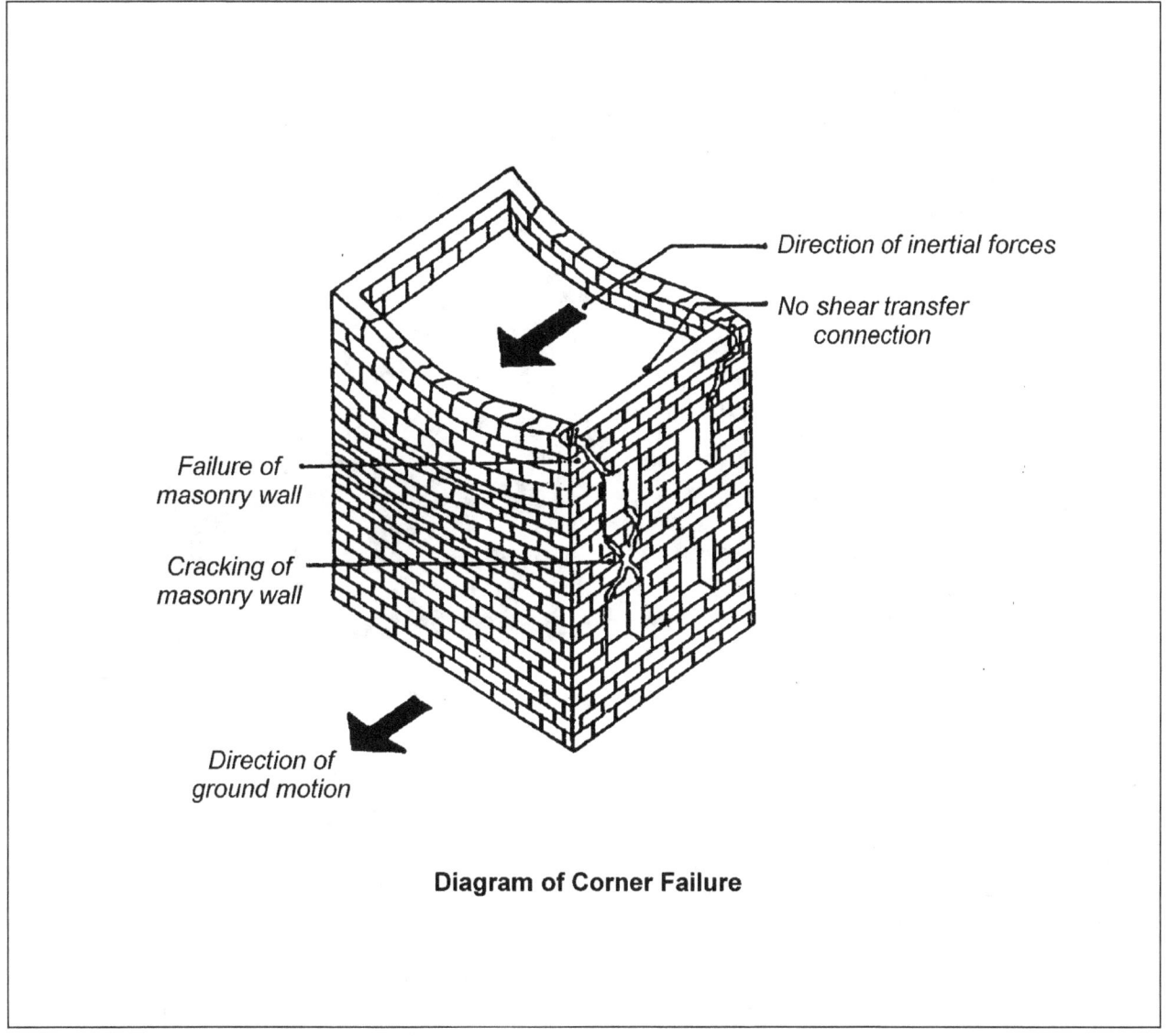

Direction of inertial forces

No shear transfer connection

Failure of masonry wall

Cracking of masonry wall

Direction of ground motion

Diagram of Corner Failure

Figure 7-11   Diagram of Corner Damage (from City of Los Angeles, 1991)

The FEMA 273 equation for toe crushing is used. Post-crack displacement capacity is assumed to be negligible.

7.2.16 Out-of-Plane Flexural Response

Out-of-plane failures are common in URM buildings. Usually they occur due to the lack of adequate wall ties as discussed in Table 7-1. When floor and roof ties are adequate, the wall may fail due to out-of-plane bending between floor levels. One mode of failure observed in experiments is rigid-body out-of-plane rocking occurring on three cracks: one at the top of the wall, one at the bottom, and one at midheight. The ultimate limit

state is that the walls rock too far and overturn. Important variables identified by ABK (1984) and Adham (1985) were the vertical stress on the wall, the height-to-thickness ratio of the wall, and the input velocity provided to the wall by the diaphragms. As rocking increases, the mortar and masonry units at the crack locations can degrade, and residual offsets can occur at the crack planes.

ABK (1984) and Adham (1985) provide a graph, based on ABK (1981c), showing the relationship between the velocity at the top and bottom of the piers, the overburden ratio of superimposed load over wall load,

Figure 7-12   Photo of Corner Damage (from Rutherford and Chekene, 1990)

and the height-to-thickness ratio. The graph indicates that walls meeting the requirements have a "98% probability of survival." The authors did not provide relationships between damage and spectral acceleration or peak ground acceleration because input velocities were found to be a better predictor of wall performance. Therefore, it is difficult to relate damage to the spectral accelerations and nonlinear static procedures used in documents such as FEMA 273 for in-plane motions. As a result, the prescriptive h/t ratio concept is retained in this document.

7.2.17 Other Modes

A review of the literature provides a substantial number of specimens that cannot be easily placed in the above set of categories. Most common are tests that report "diagonal cracking" but do not specify stair-stepping bed-joint sliding, diagonal tension cracking, or diagonally-oriented compressive splitting cracks.

In addition, given the geometric complexity and material variation inherent in unreinforced masonry walls, localized stress concentrations can develop that are difficult to predict.

7.3 Unreinforced Masonry Evaluation procedures

7.3.1 Overview

This section contains evaluation procedures for analytical determination of expected behavior mode strengths and capacities. It is to be used in concert with the component guides contained in Section 7.5. Only in-plane and out-of-plane behavior of walls are addressed; for information on non-URM wall element behavior modes, see Section 7.2.

The analytical procedures described below help establish or confirm the expected inelastic mechanism of response so that component types and behavior modes are correctly identified. See previous portions of the document for details on how these capacity curves are to be used. The sophistication of the force-deformation relationship of a multi-story URM wall can vary widely, primarily depending on whether spandrel effects and global overturning effects are taken into consideration. Existing standards--such as ABK (1984), Division 88 (City of Los Angeles, 1985), RGA (SEAOSC, 1986), UCBC (ICBO, 1994) and FEMA 178

(BSSC, 1992)--do not have specific provisions for modeling of spandrels, and, in provisions for weak pier systems, these standards also do not provide provisions for the impact of global overturning on individual piers. FEMA 273 (ATC, 1997a) also does not provide explicit guidance for these issues.

For this document, it is considered acceptable to ignore potential spandrel and global overturning effects in a perforated wall element if spandrel damage is not observed in the field. In such a case, the spandrels are assumed to have sufficient capacity and the inelastic mechanism of response is assumed to be a weak pier system. Consequently, the nonlinear static analysis need only consider the force-displacement relationships of the piers in the wall element. See Section 7.3.2 for specific evaluation procedures.

See Section 7.3.3 for specific evaluation procedures for solid wall components.

If spandrel damage is observed, then the model of the wall should include spandrel components. In many cases, inelastic behavior in spandrels will transform an initial system of strong piers and weak spandrels into a system of weak piers and strong spandrels, as the strength of the spandrel diminishes. See Section 7.3.4 for an example and specific evaluation procedures.

In this document, out-of-plane wall and pier behavior are separated from in-plane behavior. Out-of-plane capacity and its potential reduction due to observed damage is to be evaluated and reported separately. See Section 7.3.5 for specific evaluation procedures. When significant out-of-plane damage is observed, it may have an effect on the wall for the force-deformation curve oriented parallel or in-plane to the wall. For such cases, the component guides in Section 7.5 contain λ-factors to apply for in-plane loading.

The analytical procedures require the determination of material properties which can be obtained through testing or by assumption and verification. FEMA 273 (ATC, 1997a) provides the scope and details of testing for determining v_{me} (in-place push tests), f'_{me} (extracted or mockup prism tests or in-situ flatjack tests), and modulus of elasticity E (extracted prisms or flatjack testing). Conservative default values are also given to be used in lieu of testing. To determine the value of average flat-wise compressive strength of the brick, use ASTM C-67.

7.3.2 Evaluation Procedures for In-Plane Behavior of Piers in Walls with Weak Pier - Strong Spandrel Mechanisms

Evaluation of pier capacity is a three-step process:

a. **Step 1: Calculate Capacities for Individual Behavior Modes**

Determine capacities for each of the following five values:

• Rocking (V_r):

$$V_r = 0.9 \alpha P_{CE}(L/h_{eff}) \qquad (7\text{-}3)$$

where:

α = factor equal to 0.5 for fixed-free cantilever wall, or equal to 1.0 for a fixed-fixed pier
P_{CE} = expected vertical axial compressive force per load combinations in FEMA 273 (ATC, 1997a)
L = length of the wall
h_{eff} = height to resultant of lateral force. For piers with regular opening, h_{eff} is the clear height of pier; for irregular openings, see Kingsley (1995). The parameter h_{eff} may be varied to reflect observed crack patterns. See Figure 7-14 for an example.

• Bed joint sliding with bond plus friction (V_{bjs1}) and with friction only (V_{bjs2}):

$$V_{bjs1} = v_{me}A_n \qquad (7\text{-}4)$$

where:

v_{me} = bond plus friction strength of mortar, as defined in FEMA 273 (1997a)
A_n = area of net mortared/grouted section

and:

$$\begin{aligned} V_{bjs2} &= v_{friction}A_n \\ &= [0.75(P_{CE}/A_n)/1.5][A_n] = 0.5P_{CE} \qquad (7\text{-}5) \end{aligned}$$

- Diagonal tension (V_{dt}):

$$V_{dt} = f'_{dt} A_n(\beta)(1 + f_{ae}/f'_{dt})^{1/2} \qquad (7\text{-}6)$$

where:

f'_{dt} = diagonal tension strength, assumed as v_{me}, per FEMA 273 (1997a)

β = 0.67 for $L/h_{eff} < 0.67$, L/h_{eff} when $0.67 \geq L/h_{eff} \leq 1.0$, and 1.0 when $L/h_{eff} > 1$

- Toe crushing (V_{tc}):

$$V_{tc} = \alpha P_{CE}(L/h_{eff})(1 - f_{ae}/0.7f'_{me}) \qquad (7\text{-}7)$$

where:

f_{ae} = expected vertical axial compressive stress as defined in FEMA 273 (ATC, 1997a)

f'_{me} = expected masonry compressive strength

b. Step 2: Determine Predicted Behavior Mode and Capacity:

Differentiate piers by aspect ratio and applied vertical stress to determine which behavior mode is predicted as follows. Unless otherwise noted, force-displacement relationships are per FEMA 273 (ATC, 1997a).

Piers with aspect ratios of $L/h_{eff} > 1.25$

- If V_{bjs1} is the lowest value and less than 0.75 of V_{tc} or V_r, then URM2B (bed joint sliding) is the predicted mode, with an initial capacity of V_{bjs1}.

- If V_{tc} or V_r are the lowest values and are less than 0.75 of V_{bjs1}, then URM1H (flexural cracking/toe crushing) is the predicted mode, and V_{tc} is the predicted capacity. Assume this mode is force-controlled.

- If V_{tc}, V_r, and V_{bjs1} are lower than V_{dt}, $0.75\,V_{bjs1} \leq V_{tc} \leq V_{bjs1}$ and $0.75\,V_{bjs1} \leq V_r \leq V_{bjs1}$, then a sequence of URM1F (flexural cracking/toe crushing/bed joint sliding) is the predicted mode. Use V_{tc} for the initial capacity up to a "d" drift of

0.4% and V_{bjs2} for the final capacity from "d" to an "e" of 0.8%.

- If one of the categories above is not met, then the predicted capacity is the lowest of V_r, V_{bjs1}, V_{dt}, and V_{tc} and the associated behavior mode is as follows:

  - V_r : Mode URM2A (wall-pier rocking)

  - V_{bjs1}: Mode URM2B (bed joint sliding)

  - V_{dt} : Mode URM2K (preemptive diagonal tension)

  - V_{tc} : Mode URM2L (preemptive toe crushing)

Piers with aspect ratios of $L/h_{eff} \leq 1.25$

- If V_r or V_{tc} are the lowest values and $f_{ae} < 100$ psi, then URM2A is the predicted mode with V_r as the initial capacity.

- If V_r or V_{tc} are the lowest values and $f_{ae} \geq 150$ psi, then diagonal cracking with limited ductility, such as URM2G (flexural cracking/diagonal tension) is the predicted mode with V_{tc} as the capacity.

- If one of the categories above is not met, then the predicted capacity is the lowest of V_r, V_{bjs1}, V_{dt} and V_{tc} and the associated behavior mode is as follows:

  - V_r : Mode URM2A (wall-pier rocking)

  - V_{bjs1}: Mode URM2B (bed joint sliding)

  - V_{dt} : Mode URM2K (preemptive diagonal tension)

  - V_{tc} : Mode URM2L (preemptive toe crushing)

c. Step 3: Compare Predicted Mode with Observed Field Damage:

If field damage is consistent with predicted damage shown in the damage guide, then assume the component and damage classification and the capacity are correct.

If field damage is inconsistent with predicted damage shown in the damage guide, return to analysis at Step 1 and vary assumptions as to material properties and possible alternative modes. Consider, for example, for f'_{dt}, using 1/30 of the value of average flat-wise compressive strength of the brick in lieu of v_{me}. This test is standardized in ASTM C-67.

7.3.3 Evaluation Procedures for In-Plane Behavior of Solid Wall Components

Evaluation procedures for solid walls are similar to those for piers in walls with weaker pier-stronger spandrel mechanisms. Equations in Section 7.3.2 may be used, with the appropriate use of α=0.5. For the rocking equation in Section 7.3.2, the weight of the pier is ignored for simplicity, since it is assumed to be only a small fraction of the superimposed vertical load. When the weight of the wall represents a significant fraction of the vertical load, then the rocking equation may be modified as follows.

$$V_r = 0.9\alpha(P_{CE} + W_W)L/h_{eff} \qquad (7\text{-}8)$$

where:

α = factor equal to 0.5 for fixed-free cantilever wall

P_{CE} = expected vertical axial compressive force per load combinations in FEMA 273 (ATC, 1997a). This superimposed load is assumed to act at the center of the wall coincident with the location of the weight of the wall.

W_W = expected weight of the wall

L = length of the wall

h_{eff} = height to resultant of lateral force

7.3.4 Evaluation Procedures for In-Plane Behavior of Perforated Walls with Spandrel Damage

There is no methodology to analyze spandrels in the literature. As a placeholder until research is carried out, the following procedures have been developed. Procedures are given for estimating the moment and shear capacity of an uncracked spandrel, and for damaged walls that have experienced spandrel joint sliding or spandrel unit cracking. Examples are given of

how to address the implications of spandrel cracking on in-plane behavior of perforated walls.

a. Capacities of an Uncracked Spandrel

Moment Capacity. The moment capacity of the uncracked spandrel is assumed to be derived from the interlock between the bed joints and collar joint at the interface between the pier and the spandrel. See Figure 7-13. An elastic stress distribution is assumed across the end of the spandrel with the neutral axis located at the centerline of the spandrel height. It is assumed that the bed joint and collar joint capacities can be linearly superimposed to produce a resultant force. Both tension and compressive resultants are assumed to be derived from the mortar shear strength. (Note that alternative formulations are possible that use the compressive strength of the masonry to develop the compressive force.) Irregularities due to header courses are ignored. The uncracked moment capacity M_{spun} is then the product of the resultant force and the effective distance between the resultant.

- Uncracked bed joint shear stress, (v_{bjun}):

$$v_{bjun} = 0.75(0.75\,v_{te} + \gamma P_{CE}/A_n)/1.5 \qquad (7\text{-}9)$$

where:

v_{te} = the average test value from in-place testing

P_{CE} = the expected vertical axial compressive force per load combinations in FEMA 273 (ATC, 1997a) at the adjacent pier

A_n = the area of net mortared/grouted section of the adjacent pier

γ = 0.5. This arbitrary value indicates that the vertical axial stress on the spandrel bed joints at the end of the spandrel is assumed to be approximately half of the axial stress within the pier above the pier/spandrel joint.

- Uncracked collar joint shear stress, (v_{cun}):

$$\begin{aligned} v_{cun} &= 0.75(0.75\,v_{te} + \gamma P_{CE}/A_n)/1.5 \\ &= 0.375v_{te} \end{aligned} \qquad (7\text{-}10)$$

where:

v_{te} = the average test value from in-place testing

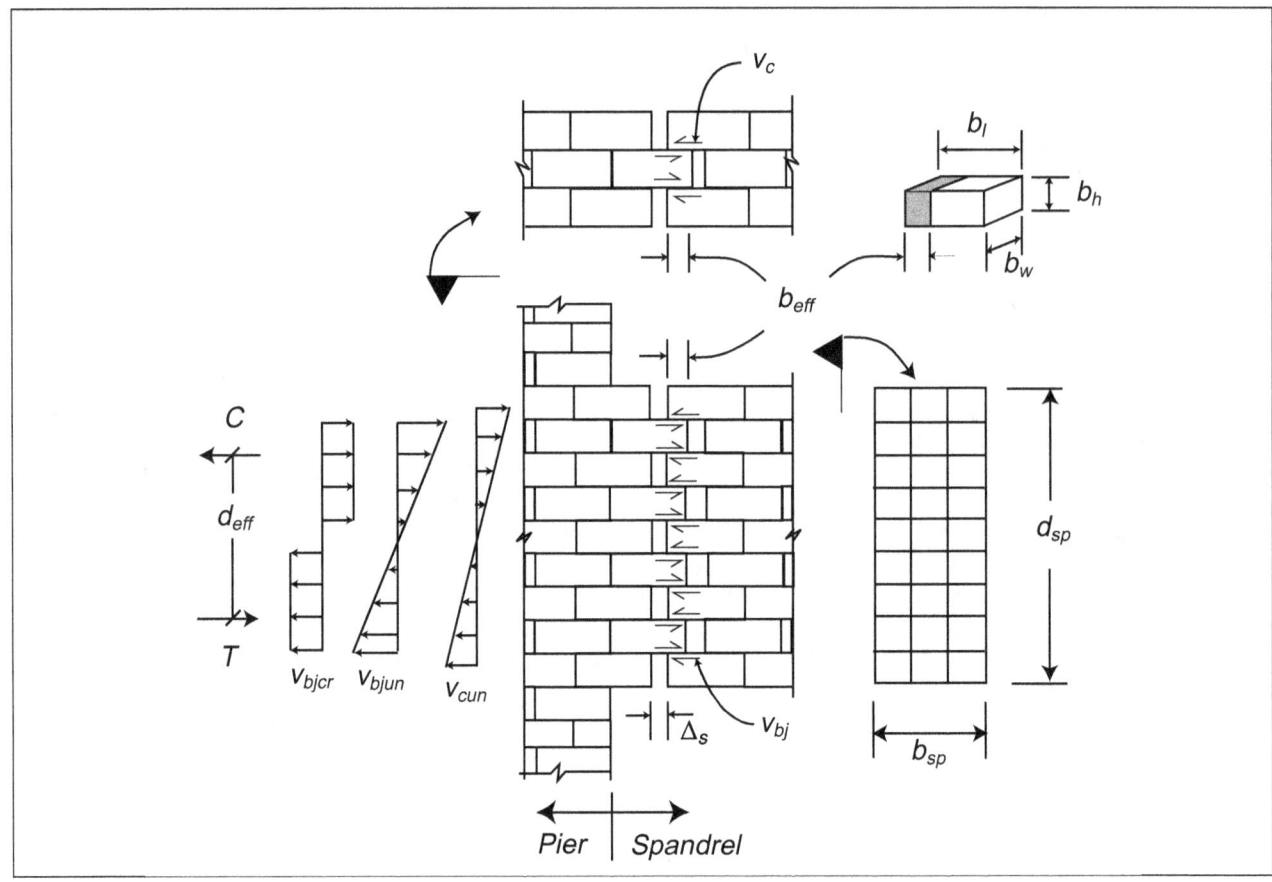

Figure 7-13   Spandrel Joint Sliding

$P_{CE}=$ the expected vertical axial compressive force per load combinations in FEMA 273 (ATC, 1997a) at the adjacent pier

A_n =   the area of net mortared/grouted section of the adjacent pier

γ =   0. This arbitrary value indicates that the axial stress on the spandrel collar joints at the end of the spandrel is assumed to be negligible.

- Effective length of interface for an uncracked spandrel, (b_{effun}):

$$b_{effun}= b_l/2 \qquad (7\text{-}11)$$

where:

$b_l/2=$ half the length of the masonry unit

- Number of rows of bed joints, (NR):

$$NR= 0.5(d_{sp}/b_h) \qquad (7\text{-}12)$$

Round NR down to the nearest whole number

where:

b_h =   height of the brick unit plus the bed joint thickness

d_{sp} =   depth of the spandrel

- Resultant tensile and compressive result forces, (T=C):

$$T= [(v_{bjun}) (b_w) (b_{effun}) + (v_{cun}) (b_h) (b_{effun}) (NB\text{-}1)] (\eta) \qquad (7\text{-}13)$$

where:

b_w =   width of the brick unit

b_h = height of the brick unit
NB = number of brick wythes
η = NR/2 or for more sophistication:
= $\Sigma_{i=1,NR} [(d_{sp}/2 - b_h(i)) / (d_{sp}/2 - b_h)]$

- Uncracked moment of spandrel, (M_{spun}):

$$M_{spun} = (d_{effun})(T) \qquad (7\text{-}14)$$

where:
d_{effun} = distance between T and C
= $(2/3)(d_{sp})$

Shear Capacity. The shear capacity of the spandrel is derived here from the equation for diagonal tensile capacity for the pier as follows.

- Diagonal tension (V_{spun}):

$$V_{spun} = f'_{dt}\, d_{sp}\, b_{sp}(\beta)\,(1 + f_{ae}/f'_{dt})^{1/2} \qquad (7\text{-}15)$$

where:
f'_{dt} = diagonal tension strength, assumed as v_{me}, per FEMA 273 (ATC, 1997a)
β = 0.67 for $L_{sp}/d_{sp} < 0.67$, L_{sp}/d_{sp} when $0.67 \geq L_{sp}/d_{sp} \leq 1.0$, and 1.0 when $L_{sp}/d_{sp} > 1$
L_{sp} = length of spandrel
f_{ae} = expected horizontal axial stress in the pier
= 0, unless known.
b_{sp} = width of spandrel

With $f_{ae}=0$, this equation then reduces to:
$$V_{spun} = f'_{dt}\, d_{sp}\, b_{sp}(\beta) \qquad (7\text{-}16)$$

b. Capacities of a Cracked Spandrel with Spandrel Joint Sliding

Moment Capacity. The moment capacity of the cracked spandrel with spandrel joint sliding is derived similar to the procedure given for an uncracked spandrel. Again see Figure 7-13.

- Cracked bed joint shear stress, (v_{bjcr}):

$$v_{bjcr} = 0.75(\varepsilon v_{te} + \gamma P_{CE}/A_n)/1.5$$
$$= 0.25\, P_{CE}/A_n \qquad (7\text{-}17)$$

where:

v_{te} = the average test value from in-place testing
ε = 0. The bond strength of the mortar is assumed to be lost.
P_{CE} = the expected vertical axial compressive force per load combinations in FEMA 273 (ATC, 1997a) at the adjacent pier
A_n = the area of net mortared/grouted section of the adjacent pier
γ = 0.5. This arbitrary value indicates that the vertical axial stress on the spandrel bed joints at the end of the spandrel is assumed to be approximately half of the axial stress within the pier above the pier/spandrel joint.

- Cracked collar joint shear stress, (v_{ccr}):

$$v_{ccr} = 0.75\,(\varepsilon v_{te} + \gamma P_{CE}/A_n)/1.5 = 0 \qquad (7\text{-}18)$$

where:
ε = 0. The bond strength of the mortar is assumed to be lost.
γ = 0. This arbitrary value indicates that the axial stress on the spandrel collar joints at the end of the spandrel is assumed to be negligible.

- Effective length of interface for a cracked spandrel, (b_{effcr}):

$$b_{effcr} = b_l/2 - \Delta_s \qquad (7\text{-}19)$$

where:
$b_l/2$ = half the length of the masonry unit
Δ_s = average slip (can be estimated as average opening width of open head joint)

- Number of rows of bed joints, (NR):

$$NR = 0.5(d_{sp}/b_h) \qquad (7\text{-}20)$$

Round NR down to the nearest whole number

where:
b_h = height of the brick unit plus the bed joint thickness
d_{sp} = depth of the spandrel

- Resultant tensile and compressive result forces, (T=C):

$$T= (v_{bjcr}) (b_w) (b_{effcr}) (NR) \qquad (7\text{-}21)$$

where:
b_w = width of the brick unit

- Cracked moment of spandrel, (M_{spcr}):

$$M_{spcr}= (d_{effcr}) (T) \qquad (7\text{-}22)$$

where:
d_{effcr} = distance between T and C
= $(1/2)(d_{sp})$

Shear Capacity. The shear capacity of a cracked spandrel with spandrel joint sliding is assumed to be the same as that of an uncracked spandrel provided keying action between bricks remains present at the end of the spandrel. Shear is resisted by bearing on the bed joints of the interlocked units.

c. Capacities of a Cracked Spandrel with Spandrel Unit Cracking

The moment and shear capacity of an cracked spandrel with spandrel unit cracking is derived similar to the procedure given above for an uncracked spandrel. The only modification is that the effective depth of the spandrel is reduced to only the amount of uncracked masonry remaining.

d. Examples of the Implications of Spandrel Cracking

Figure 7-14 shows a wall line with some cracking at the ends of spandrels. This section qualitatively discusses corner damage and gives quantitative procedures for assessing the impact of spandrel cracking on adjacent piers.

Corner Damage. One of the potential causes of corner damage is shown at the top of Figure 7-14 where the moment and shear at the end of the spandrel are resisted only by the weight of the masonry near the joint, direct tension on the head joints and bed joints, and shear in the collar joints. When these fairly weak capacities are exceeded, a diagonally-oriented crack propagates from the upper corner of the opening across the last pier. Since the crack is inclined, the effective height of the last pier is increased. For loading to the right, it may be appropriate to move the superimposed dead load closer

to the center of the pier in the evaluation of the pier rocking capacity of the last pier.

Multistory Pier Rocking. As noted above, it is assumed in this document that if there is no spandrel damage, then a weak pier-strong spandrel model should be used. On the other hand, if the spandrels are fully cracked, then there will be no bending rigidity provided by the spandrel, and the pier rocking should be assessed using a multistory pier. When the spandrels have a reduced capacity, it is necessary to determine the capacity of both the typical single story pier rocking and the multistory case. Figure 7-14 shows an example. To assess the multistory rocking capacity of Mechanism 1 in the figure the following procedure may be used:

- Assume that the shear imparted by the spandrel on the pier (V_{sp}) is

$$V_{sp}=2M_{sp}/L_{sp} \qquad (7\text{-}23)$$

where:
M_{sp} = the bending capacity of the spandrel as determined from previous sections
L_{sp} = the length of the spandrel

- Assume a distribution of acceleration within the wall line. Using ABK (1984) assumptions, the acceleration is uniform up the wall. Further assume for this example that loads tributary to each level are the same so that $V_{rR} = V_{r2}$

- Sum the moments around the pier toe at the first story so that:

$$\Sigma M=0$$
$$=(V_{rR})(h_1 +h_2+d_{sp}) +(V_{r2}) (h_1+d_{sp}/2) -$$
$$(P_{DLR}) (0.9L) - (V_{sp} + (1/2) (P_{DL2}))(0.9L)$$
$$- 2M_{sp} \qquad (7\text{-}24)$$

Substituting for V_{r2} and V_{sp} gives:
$$V_{rR}=[0.9 (P_{DLR}+ P_{DL2}/2 + 2M_{sp}/ L_{sp}) + 2M_{sp}] /$$
$$[2h_1 + h_1+ 3d_{sp}/2] \qquad (7\text{-}25)$$

where:
V_{rR} = the shear at the second story
h_1 = height of the first story pier
h_2 = height of the second story pier
d_{sp} = depth of the spandrel

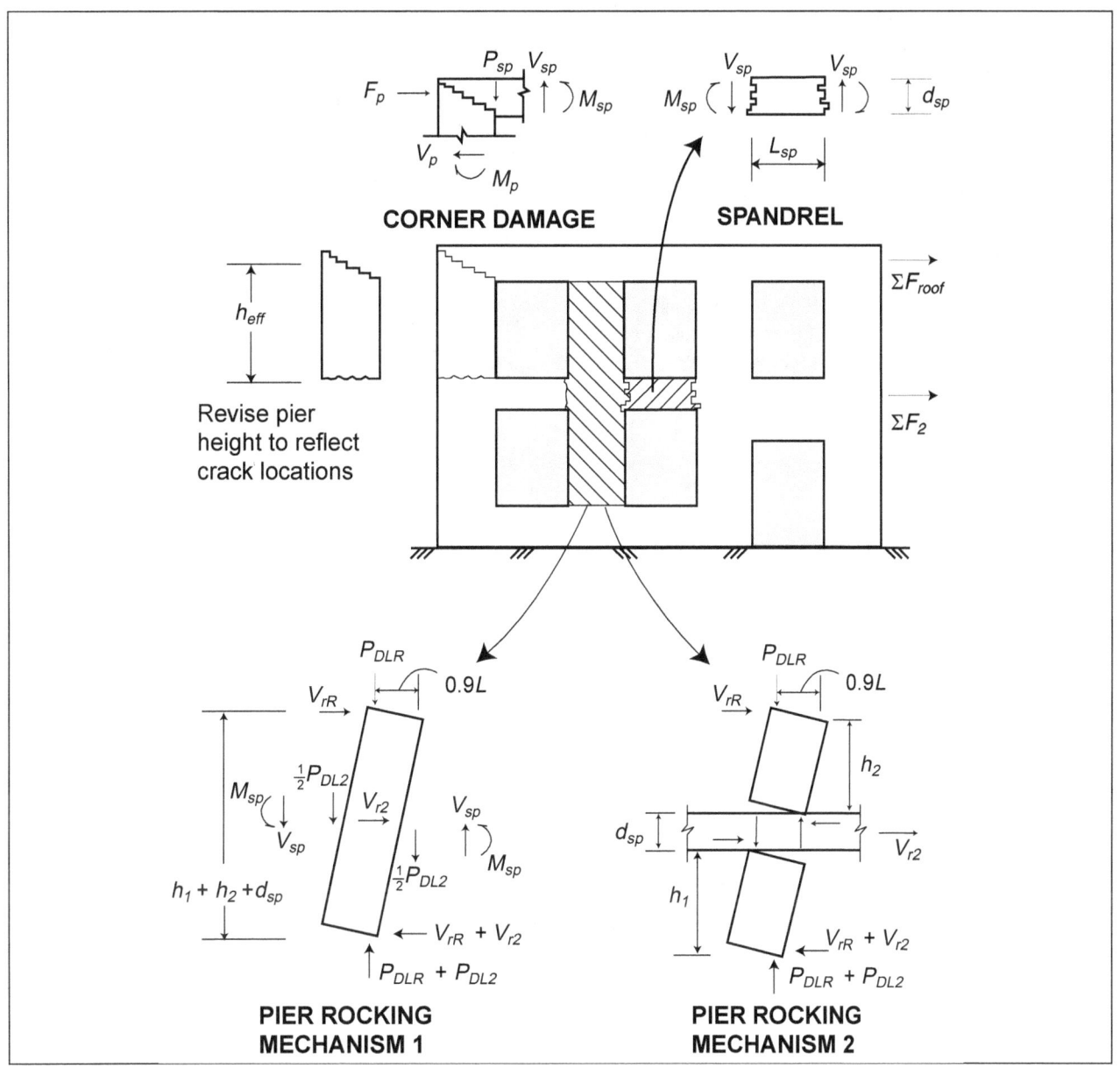

Figure 7-14   *Implications of Spandrel Cracking*

P_{DLR} = the expected gravity load at the roof

P_{DL2} = the expected gravity load at the second story, assumed here to be split equally on both sides of pier. Note that the pier weight is ignored for simplicity.

Single Story Pier Rocking. The rocking capacity of the piers in Mechanism 2 is given here. For the second story:

$$V_{rR} = 0.9L \, (P_{DLR}) / h_2 \qquad (7\text{-}26)$$

For the first story:

$$(V_{rR} + V_{r2})(h_1) = (P_{DLR} + P_{DL2}) (0.9L) \qquad (7\text{-}27)$$

Substituting for V_{r2} gives:

$$V_{rR} = 0.9L \, [(P_{DLR} + P_{DL2}) / 2h_1 \qquad (7\text{-}28)$$

Governing Mode. To determine the governing mode of behavior, compare the values for V_{rR} for multi-story and single-story rocking. The lowest rocking capacity will govern for determining the pier rocking strength.

Shear capacities for bed joint sliding, toe crushing and diagonal tension should then be compared. Use the equations in Section 7.3.2.

7.3.5 Evaluation Procedures for Out-of-Plane Behavior of Wall and Pier Components

Prescriptive strength and deformation acceptance criteria for out-of-plane wall demands are contained in FEMA 273 (ATC, 1997a). For the immediate occupancy performance level, flexural stresses should not exceed the tensile capacity of the wall. Thus, for any level of damage above Insignificant, where by definition some flexural cracking has occurred, the damaged wall should be considered as not meeting the immediate occupancy performance level.

For the collapse prevention and life safety performance levels, Table 7-4 of FEMA 273 tabulates permissible h/t ratios for walls without prior damage. For damaged walls, the Component Guides in this volume specify $\lambda_{h/t}$-factors which, when multiplied by the ratios in Table 7-4 (FEMA 273), give permissible h/t ratios for damaged walls.

The Component Guide also gives λ-factors for use when the damaged wall is a component in a force deformation curve oriented parallel to the length of the wall.

7.4 Symbols for Unreinforced Masonry

Symbols used in the unreinforced masonry sections of FEMA 306 and 307 are the same as those given in Section 7.9 of FEMA 273, except for the following additions and modifications.

C Resultant compressive force in a spandrel, lb

L_{sp} Length of spandrel, in.

M_{spcr} Expected moment capacity of a cracked spandrel, lb-in.

M_{spun} Expected moment capacity of an uncracked spandrel, lb-in.

V_{spcr} Expected diagonal tension capacity of a cracked spandrel, lb

V_{spun} Expected diagonal tension capacity of an uncracked spandrel, lb

NB Number of brick wythes in a spandrel

NR Number of rows of bed joints in a spandrel

T Resultant tensile force in a spandrel, lb

V_{bjs1} Expected shear strength of wall or pier based on bed joint shear stress, including both the bond and friction components, lb

V_{bjs2} Expected shear strength of wall or pier based on bed joint shear stress, including only the friction component, lb

V_{sp} Shear imparted on the spandrel by the pier, lb

V_{dt} Expected shear strength of wall or pier based on diagonal tension using v_{me} for f'_{dt}, lb

V_{tc} Expected shear strength of wall or pier based on toe crushing using v_{me} for f'_{dt}, lb

W_w Expected weight of a wall, lb

b_{effcr} Effective length of interface for a cracked spandrel, in.

b_{effun} Effective length of interface for an uncracked spandrel, in.

b_h Height of masonry unit plus bed joint thickness, in.

b_l Length of masonry unit, in.

b_w Width of brick unit, in.

d_{sp} Depth of spandrel, in.

d_{effcr} Distance between resultant tensile and compressive forces in a cracked spandrel, in.

d_{effun} Distance between resultant tensile and compressive forces in an uncracked spandrel, in.

f'_{dt} Masonry diagonal tension strength, psi

v_{bjcr} Cracked bed joint shear stress, psi

v_{bjun} Uncracked bed joint shear stress in a spandrel, psi

v_{ccr} Cracked collar joint shear stress in a spandrel, psi

v_{cun} Uncracked collar joint shear stress in a spandrel, psi

β =0.67 when $L/h_{eff}<0.67$, =L/h_{eff} when $0.67 \leq L/h_{eff} \leq 1.0$, and =1.0 when $L/h_{eff}>1$

Δ_s Average slip at cracked spandrel (can be estimated as average opening width of open head joint), in.

ε Factor for estimating the bond strength of the mortar in spandrels

γ Factor for coefficient of friction in bed joint sliding equation for spandrels

η Factor to estimate average stress in uncracked spandrel. Equal to NR/2 or, for more sophistication, use $\Sigma_{i=1,NR} [(d_{sp}/2 - b_h(i))/(d_{sp}/2 - b_h)]$

$\lambda_{h/t}$ Factor used to estimate the loss of out-of-plane wall capacity to damaged URM walls

μ_Δ Displacement ductility demand for a component, used in Section 5.3.4, and discussed in Section 6.4.2.4 of FEMA 273. Equal to the component deformation corresponding to the global target displacement, divided by the effective yield displacement of the component (which is defined in Section 6.4.1.2B of FEMA 273).

7.5 Unreinforced Masonry Component Guides

The following Component Damage Classification Guides contain details of the behavior modes for unreinforced masonry components. Included are the distinguishing characteristics of the specific behavior mode, the description of damage at various levels of severity, and performance restoration measures. Information may not be included in the Component Damage Classification Guides for certain damage severity levels; in these instances, for the behavior mode under consideration, it is not possible to make refined distinctions with regard to severity of damage. See also Section 3.5 for general discussion of the use of the Component Guides and Section 4.4.3 for information on the modeling and acceptability criteria for components.

URM2A	COMPONENT DAMAGE CLASSIFICATION GUIDE	System: **URM**
		Component Type: **Weaker Pier**
		Behavior Mode: **Wall-Pier Rocking**

How to distinguish behavior mode:

By observation:

Rocking-critical piers form horizontal flexural cracks at the top and bottom of piers. Because the cracks typically close as the pier comes back to rest at the end of ground shaking, these cracks can be quite subtle when only a few cycles of rocking have occurred and when pier drift ratios during shaking were small. As damage increases, softening of the pier can occur due to cracking, and the pier may begin to "walk" out-of-plane at the top and bottom. At the highest damage levels, crushing of units at the corners can occur.

By analysis:

As damage increases to the Moderate level and beyond, some small cracking within the pier may occur. Confirm by analysis that rocking governs over diagonal tension and bed joint sliding.

Caution: If horizontal cracks are located directly below wall-diaphragm ties, damage may be due to bed joint sliding associated with tie damage. If a horizontal crack is observed at midheight of the pier, see URM1M.

Refer to Evaluation Procedures for:

• In-plane wall behavior: See Section 7.3.2.

Level	Description of Damage		Typical Performance Restoration Measures
Insignificant $\lambda_K = 0.8$ $\lambda_Q = 1.0$ $\lambda_D = 1.0$ $\mu_\Delta \leq 1.5$	*Criteria:*	• Hairline cracks/spalled mortar in bed joints at top and bottom of pier.	Not necessary for restoration of structural performance. (Measures may be necessary for restoration of nonstructural characteristics.)
	Typical Appearance: 		

COMPONENT DAMAGE
CLASSIFICATION GUIDE **continued**

 URM2A

Level	Description of Damage		Typical Performance Restoration Measures
Slight		Not used.	
Moderate $\lambda_K = 0.6$ $\lambda_Q = 0.9$ $\lambda_D = 1.0$ $\Delta/h_{eff} \leq$ $h_{eff}/L_{eff}*$ 0.4%	*Criteria:*	1. Hairline cracks/spalled mortar in bed joints at top and bottom of pier. 2. Possible hairline cracking/spalled mortar in bed joints within piers. *Typical Appearance:*	Replacement or enhancement is required for full restoration of seismic performance. For partial restoration of performance: ● Repoint spalled mortar. $\lambda_K* = 0.8$ $\lambda_Q* = 0.9$ $\lambda_D* = 1.0$
Heavy $\lambda_K = 0.4$ $\lambda_Q = 0.8$ $\lambda_D = 0.7$ $\Delta/h_{eff} \leq$ $h_{eff}/L_{eff}*$ 0.8%	*Criteria:* *Typical Appearance:*	1. Hairline cracks/spalled mortar in bed joints at top and bottom of pier, plus one or more of: 2. Hairline cracking/spalled mortar in bed joints within piers, but bed joints typically do not open. 3. Possible out-of-plane or in-plane movement at top and bottom of piers ("walking"). 4. Crushed/spalled bricks at corners of piers.	Replacement or enhancement is required for full restoration of seismic performance. For partial restoration of performance: ● Replace/drypack damaged units ● Repoint spalled mortar ● Inject cracks $\lambda_K* = 0.8$ $\lambda_Q* = 0.9$ $\lambda_D* = 1.0$
Extreme	*Criteria:* *Typical Indications*	● Vertical load-carrying ability is threatened. ● Significant out-of-plane or in-plane movement at top and bottom of piers ("walking"). ● Significant crushing/spalling of bricks at corners of piers.	● Replacement or enhancement required.

URM2B	COMPONENT DAMAGE CLASSIFICATION GUIDE	System:	**URM**
		Component Type:	**Weaker Pier**
		Behavior Mode:	**Bed Joint Sliding**

How to distinguish behavior mode:

By observation:

In this type of behavior, sliding occurs on bed joints. Commonly observed both in the field and in experimental tests, there are two basic forms: sliding on a horizontal plane, and a stair-stepped diagonal crack where the head joints open and close to allow for movement on the bed joint. Note that, for simplicity, the figures below only show a single crack, but under cyclic loading, multiple cracks stepping in each direction are possible. Pure bed joint sliding is a ductile mode with significant hysteretic energy absorption capability. If sliding continues without leading to a more brittle mode such as toe crushing, then gradual degradation of the cracking region occurs until instability is reached. Theoretically possible, but not widely reported, is the case of stair-stepped cracking when sliding goes so far that an upper brick slides off a lower unit.

By analysis:

Stair-stepped cracking may resemble a form of diagonal tension cracking; confirm by analysis that bed joint sliding governs over diagonal tension.

Refer to Evaluation Procedures for:

• In-plane wall behavior: See Section 7.3.2.

Level	Description of Damage	Typical Performance Restoration Measures
Insignificant $\lambda_K = 0.9$ $\lambda_Q = 0.9$ $\lambda_D = 1.0$ $\mu_\Delta \le 1.5$	*Criteria:* 1. Hairline cracks/spalled mortar in head and bed joints either on a horizontal plane or in a stair-stepped fashion have been initiated, but no off-set along the crack has occurred and the crack plane or stair-stepping is not continuous across the pier. 2. No cracks in masonry units. *Typical Appearance:* 	Not necessary for restoration of structural performance. (Measures may be necessary for restoration of nonstructural characteristics.)

COMPONENT DAMAGE
CLASSIFICATION GUIDE **continued**

Level	Description of Damage		Typical Performance Restoration Measures
Slight	Not used.		
Moderate $\lambda_K = 0.8$ $\lambda_Q = 0.6^*$ $\lambda_D = 1.0$ *As an alternative, calculate as V_{bjs2}/V_{bjs1} $\Delta/h_{eff} \leq 0.4\%$	*Criteria:*	1. Horizontal cracks/spalled mortar at bed joints indicating that in-plane offset along the crack has occurred and/or opening of the head joints up to approximately 1/4", creating a stair-stepped crack pattern. 2. 5% of courses or fewer have cracks in masonry units. *Typical Appearance:* 	Replacement or enhancement is required for full restoration of seismic performance. For partial restoration of performance: ● Repoint spalled mortar and open head joints. ● Inject cracks and open head joints 　$\lambda_K^* = 0.8$ 　$\lambda_Q^* = 0.8^*$ 　$\lambda_D^* = 1.0^*$ *In some cases, grout injection may actually increase strength, but decrease deformation capacity, by changing behavior from bed joint sliding to a less ductile behavior mode (see FEMA 307, Section 4.1.3).
Heavy $\lambda_K = 0.6$ $\lambda_Q = 0.6^*$ $\lambda_D = 0.9$ *As an alternative, calculate as V_{bjs2}/V_{bjs1} $\Delta/h_{eff} \leq 0.8\%$	*Criteria:*	1. Horizontal cracks/spalled mortar on bed joints indicating that in-plane offset along the crack has occurred and/or opening of the head joints up to approximately 1/2", creating a stair-stepped crack pattern. 2. 5% of courses or fewer have cracks in masonry units. *Typical Appearance:* 	Replacement or enhancement is required for full restoration of seismic performance. For partial restoration of performance: ● Repoint spalled mortar and open head joints. ● Inject cracks and open head joints. 　$\lambda_K^* = 0.8$ 　$\lambda_Q^* = 0.8^*$ 　$\lambda_D^* = 1.0^*$ *In some cases, grout injection may actually increase strength, but decrease deformation capacity, by changing behavior from bed joint sliding to a less ductile behavior mode (see FEMA 307, Section 4.1.3).
Extreme	*Criteria:* *Typical Indications*	● Vertical load-carrying ability is threatened. ● Stair-stepped movement is so significant that upper bricks have slid off their supporting brick. ● Cracks have propagated into a significant number of courses of units. ● Residual set is so significant that portions of masonry at the edges of the pier have begun or are about to fall.	● Replacement or enhancement required.

URM3D	COMPONENT DAMAGE CLASSIFICATION GUIDE	System:	**URM**
		Component Type:	**Weaker Spandrel**
		Behavior Mode:	**Spandrel Joint Sliding**

How to distinguish behavior mode:

By observation:

Commonly observed in the field in running bond masonry, this form of bed joint sliding is characterized by predominantly vertical cracks at the ends of the spandrel, which look like interlocked fingers being pulling apart. This mode can be relatively ductile and allow for significant drift, provided a reliable lintel is present. As the spandrel displaces, the nonlinear mechanism of response may move to other portions of the wall such as the piers.

By analysis:

No analysis is typically necessary to distinguish this mode. Analytical procedures are provided to estimate the reduction in capacity due to damage.

Refer to Evaluation Procedures for:

• In-plane wall behavior: See Section 7.3.4.

Level	Description of Damage		Typical Performance Restoration Measures
Insignificant $\lambda_K = 0.9$ $\lambda_Q = 0.9$ $\lambda_D = 1.0$ $\mu_\Delta \leq 1.5$	*Criteria:*	1. Staggered hairline cracks/spalled mortar in head and bed joints in up to 3 courses at the ends of the spandrel. No cracks in units.	Not necessary for restoration of structural performance. (Measures may be necessary for restoration of nonstructural characteristics.)
	Typical Appearance: 		
Slight	Not used		

COMPONENT DAMAGE
CLASSIFICATION GUIDE **continued**

$\boxed{\text{URM3D}}$

Level	Description of Damage	Typical Performance Restoration Measures
Moderate $\lambda_K = 0.8$ $\lambda_Q = 0.4^1$ $\lambda_D = 0.9$ 1. As an alternative, calculate per Section 7.3.4	*Criteria:* 1. Staggered hairline cracks/spalled mortar at the ends of the spandrel in head and bed joints indicating that in-plane offset along the crack has occurred and opening of the head joints up to approximately 1/4". No cracks in units. 2. No vertical slip of the spandrel. *Typical Appearance:* 	Replacement or enhancement is required for full restoration of seismic performance. For partial restoration of performance: • Repoint spalled mortar and open head joints. • Inject cracks and open head joints. $\lambda_K^* = 0.8$ $\lambda_Q^* = 0.8^1$ $\lambda_D^* = 1.0^1$ 1. In some cases, grout injection may actually increase strength, but decrease deformation capacity, by changing behavior from bed joint sliding to a less ductile behavior mode (see FEMA 307, Section 4.1.3).
Heavy $\lambda_K = 0.6$ $\lambda_Q = 0.4^1$ $\lambda_D = 0.9$ 1. As an alternative, calculate per Section 7.3.4	*Criteria:* 1. Staggered hairline cracks/spalled mortar at the ends of the spandrel in head and bed joints, indicating that in-plane offset along the crack has occurred and opening of the head joints up to approximately 1/2". No cracks in units. 2. Possibly some deterioration of units at bottom ends of spandrel, but no vertical slip of the spandrel. 3. Possibly spandrel rotation with respect to the pier. *Typical Appearance:* 	Replacement or enhancement is required for full restoration of seismic performance. For partial restoration of performance: • Repoint spalled mortar and open head joints • Inject cracks and open head joints $\lambda_K^* = 0.8$ $\lambda_Q^* = 0.8^1$ $\lambda_D^* = 1.0^1$ 1. In some cases, grout injection may actually increase strength, but decrease deformation capacity, by changing behavior from bed joint sliding to a less ductile behavior mode (see FEMA 307, Section 4.1.3).
Extreme	*Criteria:* • Vertical load-carrying ability is threatened. *Typical Indications* • Sliding and/or deterioration of the units is so significant that keying action between bricks is lost. • Lintel support has separated from the pier.	• Replacement or enhancement required.

URM1F	COMPONENT DAMAGE CLASSIFICATION GUIDE	System:	**URM**
		Component Type:	**Solid Wall**
		Behavior Mode:	**Flexural Cracking/ Toe Crushing/Bed Joint Sliding**

How to distinguish behavior mode:

By observation:

This type of moderately ductile behavior has been experimentally observed in walls with $L/h_{eff} \approx 1.7$ in which bed joint sliding and toe crushing strength capacities are similar. Damage occurs in the following sequence. First, flexural cracking occurs at the heel of the wall. Then diagonally-oriented cracks appear at the toe of the wall, typically accompanied by spalling and crushing of the units. In some cases, toe crushing is immediately followed by a steep inclined crack propagating upward from the toe. Next, sliding occurs along a horizontal bed joint near the base of the wall, accompanied in some cases by stair-stepped bed joint sliding at upper portions of the wall. With repeated cycles of loading, diagonal cracks increase. Eventually, crushing of the toes or excessive sliding leads to failure.

By analysis:

At higher damage levels, cracking may be similar to URM1H; however, in URM1F, the bed joint sliding will occur at the base of the wall, in addition to the center of the wall. Confirm by analysis that bed joint sliding capacities are sufficiently low to trigger URM1F.

Caution: At low damage levels, flexural cracking may be similar to cracking that occurs in other modes.

Refer to Evaluation Procedures for:

• In-plane wall behavior: See Section 7.3.2

Level	Description of Damage	Typical Performance Restoration Measures
Insignificant $\lambda_K = 1.0$ $\lambda_Q = 1.0$ $\lambda_D = 1.0$ $\mu_\Delta \leq 1.5$	_Criteria:_ 1. Horizontal hairline cracks in bed joints at the heel of the wall. 2. Possibly diagonally-oriented cracks and minor spalling at the toe of the wall. _Typical Appearance:_	Not necessary for restoration of structural performance. (Measures may be necessary for restoration of nonstructural characteristics.)
Slight	Not used	

COMPONENT DAMAGE
CLASSIFICATION GUIDE **continued**

$$\boxed{\text{URM1F}}$$

Level	Description of Damage	Typical Performance Restoration Measures

Moderate
 	
$\lambda_K = 0.9$
$\lambda_Q = 0.6^1$
$\lambda_D = 0.9$
 	
1. As an alternative, calculate as V_{bjs2}/V_{tc}
 	
$\Delta/h_{eff} \le 0.8\%$
	Criteria: 1. Horizontal cracks/spalled mortar at bed joints at or near the base of the wall indicating that in-plane offset along the crack has occurred up to approximately 1/4".
	     2. Possibly diagonally-oriented cracks and spalling at the toe of the wall. Cracks extend upward several courses.
	     3. Possibly diagonally-oriented cracks at upper portions of the wall which may be in the units.
	
	Typical Appearance:
		• Replace/drypack damaged units.
		• Repoint spalled mortar and open head joints.
		• Inject cracks and open head joints.
		• Install pins and drilled dowels in toe regions.
		
		$\lambda_K{}^* = 1.0^1$
		$\lambda_Q{}^* = 1.0^1$
		$\lambda_D{}^* = 1.0^1$
		
		1. In some cases, grout injection may actually increase strength, but decrease deformation capacity, by changing behavior from bed joint sliding to a less ductile behavior mode (see FEMA 307, Section 4.1.3).

Heavy
 	
$\lambda_K = 0.8$
$\lambda_Q = 0.6^1$
$\lambda_D = 0.9$
 	
1. As an alternative, calculate as V_{bjs2}/V_{tc}
 	
$\Delta/h_{eff} \le 1.2\%$
	Criteria: 1. Horizontal bed joint cracks near the base of the wall similar to Moderate, except width is up to approximately 1/2".
	     2 Possibly extensive diagonally-oriented cracks and spalling at the toe of the wall. Cracks extend upward several courses.
	     3 Possibly diagonally-oriented cracks up to 1/2" at upper portions of the wall.
	
	Typical Appearance:
		• Replace/drypack damaged units.
		• Repoint spalled mortar and open head joints.
		• Inject cracks and open head joints.
		• Install pins and drilled dowels in toe regions.
		
		$\lambda_K{}^* = 1.0^1$
		$\lambda_Q{}^* = 1.0^1$
		$\lambda_D{}^* = 1.0^1$
		
		1. In some cases, grout injection may actually increase strength, but decrease deformation capacity, by changing behavior from bed joint sliding to a less ductile behavior mode (see FEMA 307, Section 4.1.3).

Extreme	*Criteria:* • Vertical load-carrying ability is threatened
	
	Typical Indications • Stair-stepped movement is so significant that upper bricks have slid off their supporting brick.
	     • Toes have begun to disintegrate.
	     • Residual set is so significant that portions of masonry at the edges of the pier have begun or are about to fall.
		• Replacement or enhancement required.

URM2K	COMPONENT DAMAGE CLASSIFICATION GUIDE	System: **URM**
		Component Type: **Weaker Pier**
		Behavior Mode: **Diagonal Tension**

How to distinguish behavior mode:

By observation:

Typical diagonal tension cracking—resulting from strong mortar, weak units, and high compressive stress—can be identified by diagonal cracks ("X" cracks) that propagate through the units. In many cases, the cracking is sudden, brittle, and vertical load capacity drops quickly. The cracks may then extend to the toe and the triangles above and below the crack separate. In a few cases, the load drop may be more gradual with cracks increasing in size and extent with each cycle. A second form of diagonal tension cracking also has been experimentally observed with weak mortar, strong units and low compressive stress where the cracks propagate in a stair-stepped manner in head and bed joints. The first (typical) case is shown below.

By analysis:

Since the stair-stepping form of cracking would appear similar to the early levels of stair-stepped bed joint sliding, confirm by analysis that diagonal tension governs over bed joint sliding. Since deterioration at the corners in the Heavy damage level may resemble toe crushing, also confirm that diagonal tension governs over toe crushing.

Refer to Evaluation Procedures for:

• In-plane wall behavior: See Section 7.3.2.

Level	Description of Damage	Typical Performance Restoration Measures
Insignificant $\lambda_K = 1.0$ $\lambda_Q = 1.0$ $\lambda_D = 1.0$ $\mu_\Delta \leq 1$	*Criteria:* 1. Hairline diagonal cracks in masonry units in fewer than 5% of courses. *Typical Appearance:* 	Not necessary for restoration of structural performance. (Measures may be necessary for restoration of nonstructural characteristics.)

COMPONENT DAMAGE
CLASSIFICATION GUIDE **continued**

Level	Description of Damage		Typical Performance Restoration Measures
Slight	Not used.		
Moderate $\lambda_K = 0.8$ $\lambda_Q = 0.9$ $\lambda_D = 1.0$ $\mu_\Delta \approx 1\text{-}1.5$	*Criteria:*	1. Diagonal cracks in pier, many of which go through masonry units, with crack widths below 1/4". 2. Diagonal cracks reach or nearly reach corners. 3. No crushing/spalling of pier corners. *Typical Appearance:*	• Repoint spalled mortar. • Inject cracks. $\lambda_K{}^* = 0.8$ $\lambda_Q{}^* = 1.0$ $\lambda_D{}^* = 1.0$
Heavy $\lambda_K = 0.4$ $\lambda_Q = 0.8$ $\lambda_D = 0.7$ $\mu_\Delta > 1.5$	*Criteria:*	1. Diagonal cracks in pier, many of which go through masonry units, with crack widths over 1/4". Damage may also include: 2. Some minor crushing/spalling of pier corners and/or 3. Minor movement along or across crack plane. *Typical Appearance:*	Replacement or enhancement is required for full restoration of seismic performance. For <u>partial</u> restoration of performance: • Replace/drypack damaged units. • Repoint spalled mortar. • Inject cracks. $\lambda_K{}^* = 0.8$ $\lambda_Q{}^* = 0.8$ $\lambda_D{}^* = 1.0$
Extreme	*Criteria:* *Typical Indications*	• Vertical load-carrying ability is threatened • Significant movement or rotation along crack plane. • Residual set is so significant that portions of masonry at the edges of the pier have begun or are about to fall.	• Replacement or enhancement required.

URM1H	DAMAGE CLASSIFICATION AND REPAIR GUIDE	System:	**URM**
		Component Type:	**Solid Wall**
		Behavior Mode:	**Flexural Cracking/ Toe Crushing**

How to distinguish behavior mode:

By observation:

This type of behavior typically occurs in stockier walls with $L/h_{eff} > 1.25$. Based on laboratory testing, four steps can usually be identified. First, flexural cracking happens at the base of the wall, but it does not propagate all the way across the wall. This can also cause a series of horizontal cracks to form above the heel. Second, sliding occurs on bed joints in the central portion of the pier. Third, diagonal cracks form at the toe of the wall. Finally, large cracks form at the upper corners of the wall. Failure occurs when the triangular portion of wall above the crack rotates off the crack or the toe crushes so significantly that vertical load is compromised. Note that, for simplicity, the figures below only show a single crack, but under cyclic loading, multiple cracks stepping in each direction are possible.

By analysis:

Stair-stepped cracking may resemble a form of bed joint sliding; confirm by analysis that toe crushing governs over bed joint sliding.

Refer to Evaluation Procedures for:

• In-plane wall behavior: See Section 7.3.2

Level	Description of Damage		Typical Performance Restoration Measures
Insignificant $\lambda_K = 0.9$ $\lambda_Q = 1.0$ $\lambda_D = 1.0$ $\mu_\Delta \leq 1.5$	*Criteria:*	1. Horizontal hairline cracks in bed joints at the heel of the wall. 2. Horizontal cracking on 1-3 cracks in the central portion of the wall. No offset along the crack has occurred and the crack plane is not continuous across the pier. 3 No cracks in masonry units.	Not necessary for restoration of structural performance. (Measures may be necessary for restoration of nonstructural characteristics.)
	Typical Appearance:		
Slight	Not used		
Moderate	Not used		

COMPONENT DAMAGE
CLASSIFICATION GUIDE **continued**

Level	Description of Damage	Typical Performance Restoration Measures
Heavy $\lambda_K = 0.8$ $\lambda_Q = 0.8$ $\lambda_D = 1.0$ $\Delta/h_{eff} \leq 0.3\%$	*Criteria:* 1. Horizontal hairline cracks in bed joints at the heel of the wall. 2. Horizontal cracking on 1-3 cracks in the central portion of the wall. Some offset along the crack may have occurred. 3. Diagonal cracking at the toe of the wall, likely to be through the units, and some of units may be spalled. *Typical Appearance:* 	Replacement or enhancement is required for full restoration of seismic performance. For <u>partial</u> restoration of performance: • Repoint spalled mortar. • Inject cracks. $\lambda_K{}^* = 0.9$ $\lambda_Q{}^* = 0.9$ $\lambda_D{}^* = 1.0$
Extreme $\lambda_K = 0.6$ $\lambda_Q = 0.6$ $\lambda_D = 0.9$ $\Delta/h_{eff} \leq 0.9\%$	*Criteria:* 1. Horizontal hairline cracks in bed joints at the heel of the wall. 2. Horizontal cracking on 1 or more cracks in the central portion of the wall. Offset along the crack will have occurred. 3. Diagonal cracking at the toe of the wall, likely to be through the units, and some of units may be spalled. 4. Large cracks have formed at upper portions of the wall. In walls with aspect ratios of $L/h_{eff} > 1.5$, these cracks will be diagonally oriented; for more slender piers, cracks will be more vertical and will go through units. *Typical Appearance:* 	Replacement or enhancement is required for full restoration of seismic performance. For <u>partial</u> restoration of performance: • Replace/drypack damaged units. • Repoint spalled mortar. • Inject cracks. • Install pins and drilled dowels in toe regions. $\lambda_K{}^* = 0.9$ $\lambda_Q{}^* = 0.8$ $\lambda_D{}^* = 1.0$

URM3I	COMPONENT DAMAGE CLASSIFICATION GUIDE	System:	**URM**
		Component Type:	**Weak Spandrel**
		Behavior Mode:	**Spandrel Unit Cracking**

How to distinguish behavior mode:

By observation:

In this type of behavior, the moment at the end of the spandrel is not relieved by sliding, but instead causes brittle vertical cracking though the masonry units. Cracking propagates rapidly as displacement increases and cycles continue. Depending on the lintel construction, this can lead to a local falling hazard. It also increases the effective height of the piers. As the spandrel displaces, the nonlinear mechanism of response may move to other portions of the wall such as the piers.

By analysis:

No analysis is typically necessary to distinguish this mode.

Refer to Evaluation Procedures for:

• In-plane wall behavior: See Section 7.3.4.

Level	Description of Damage	Typical Performance Restoration Measures
Insignificant $\lambda_K = 0.9$ $\lambda_Q = 0.9$ $\lambda_D = 1.0$ $\mu_\Delta \leq 1.5$	*Criteria:* 1. Predominantly vertical cracks/spalled mortar through no more than one unit at the ends of the spandrel. *Typical Appearance:* 	Not necessary for restoration of structural performance. (Measures may be necessary for restoration of nonstructural characteristics.)
Slight	Not used.	

COMPONENT DAMAGE
CLASSIFICATION GUIDE **continued**

$$\boxed{\text{URM3I}}$$

Level	Description of Damage	Typical Performance Restoration Measures
Moderate	Not Used	

Heavy
 	
 $\lambda_K = 0.2$
 $\lambda_Q = 0.4^1$
 $\lambda_D = 0.6$
 	
 1. As an alternative, calculate per Section 7.3.4
	Criteria: 1. Predominantly vertical cracks/spalled mortar across the full depth of each end of the spandrel. In over 1/3 of the courses, cracks go through the masonry units.
	 2. Possibly some deterioration of units at bottom ends of spandrel, but no vertical slip of the spandrel.
	
	 Typical Appearance:
	
		Replacement or enhancement is required for full restoration of seismic performance.
		 For partial restoration of performance:
		 • Stitch across crack with pins and drilled dowels.
		 • Repoint spalled mortar.
		 • Inject cracks.
		    $\lambda_K^* = 0.8$
		    $\lambda_Q^* = 0.8$
		    $\lambda_D^* = 1.0$

Extreme	*Criteria:* • Vertical load-carrying ability is threatened. *Typical Indications* One or more of the following: • Lintel support has separated from the pier. • Out-of-plane movement of the spandrel. • Spandrel has slipped vertically.	• Replacement or enhancement required.

URM1M	COMPONENT DAMAGE CLASSIFICATION GUIDE	System:	**URM**
		Component Type:	**Solid Wall**
		Behavior Mode:	**Out-of-Plane Flexural Response**

How to distinguish behavior mode:

By observation:

Out-of-plane failures are common in URM buildings. Usually they occur due to the lack of adequate wall ties, as discussed in Table 7-1. When ties are adequate, the wall may fail due to out-of-plane bending between floor levels. One mode of failure observed in experiments is rigid-body rocking motion occurring on three cracks: one at the top of the wall, one at the bottom, and one at midheight. As rocking increases, the mortar and masonry units at the crack locations can be degraded, and residual offsets can occur at the crack planes. The ultimate limit state is that the walls rock too far and overturn. Important variables are the vertical stress on the wall and the height-to-thickness ratio of the wall. Thus, walls at the top of buildings and slender walls are more likely to suffer damage.

By analysis:

None required.

Caution:

If horizontal cracks are located directly below wall-diaphragm ties, damage may be due to bed joint sliding associated with tie damage. For piers, if horizontal cracks are observed at the top and bottom of the pier but not at midheight, see URM2A. Confirm whether the face brick is unbonded to the backing brick. If so, the thickness in the h/t requirement is reduced to the thickness of the backing wythes.

Refer to Evaluation Procedures for: Out-of-plane wall behavior: See Section 7.3.5.

Level	Description of Damage	Typical Performance Restoration Measures
Insignificant For out-of-plane loads: $\lambda_{h/t} = 1.0$ For in-plane modes given previously, assume out-of-plane damage leads to Moderate damage for URM2B and Insignificant damage for all other modes.	*Criteria:* 1. Hairline cracks at floor/roof lines and midheight of stories. 2. No out-of-plane offset or spalling of mortar along cracks. *Typical Appearance:* 	Not necessary for restoration of structural performance. (Measures may be necessary for restoration of nonstructural characteristics.)

COMPONENT DAMAGE
CLASSIFICATION GUIDE **continued**

Level	Description of Damage		Typical Performance Restoration Measures
Slight		Not used.	
Moderate For <u>out-of-plane loads</u>: $\lambda_{h/t}= 0.9$ For <u>in-plane modes</u>, see Insignificant damage	*Criteria:*	1. Cracks at floor/roof lines and midheight of stories may have mortar spalls up to full depth of joint and possibly: 2. Out-of-plane offsets along cracks of up to 1/8". *Typical Appearance:* See Insignificant damage above.	• Repoint spalled mortar: • For <u>out-of-plane loads</u>: $\lambda_{h/t}= 1.0$ • For <u>in-plane loads</u>: use Moderate for URM2B and Insignificant for all other modes.
Heavy For <u>out-of-plane loads</u>: $\lambda_{h/t}= 0.6$ For <u>in-plane modes</u> given previously, assume out-of-plane damage leads to *Heavy* for all other modes.	*Criteria:* *Typical Appearance:*	1 Cracks at floor/roof lines and midheight of stories may have mortar spalls up to full depth of joint. 2 Spalling and rounding at edges of units along crack plane. 3 Out-of-plane offsets along cracks of up to 1/2".	Replacement or enhancement is required for full restoration of seismic performance. For <u>partial</u> restoration of out-of-plane performance: • Replace/drypack damaged units • Repoint spalled mortar $\lambda_{h/t}= 0.8$
Extreme	*Criteria:* *Typical Indications*	• Vertical-load-carrying ability is threatened • Significant out-of-plane or in-plane movement at top and bottom of piers ("walking"). • Significant crushing/spalling of bricks at crack locations.	• Replacement or enhancement required.

8: Infilled Frames

8.1 Introduction and Background

This section provides material relating to infilled frame (INF) construction that supports and supplements the Damage Classification Guides (or Component Guides) in Chapters 5 through 7. Following this introductory material, infilled frame component types are defined and discussed in Section 8.2. Inelastic behavior modes are also summarized in Section 8.2. The overall damage evaluation procedure uses conventional material properties as a starting point. Section 8.3 provides information on strength and deformation properties of infilled frame components. The information on infilled frame components has been generated from a review of available empirical and theoretical data listed in the Tabular Bibliography (Section 5.2 of FEMA 307) and References section of this document, or Section 5.3 of FEMA 307). These provide the user with further detailed resources on infilled frame component behavior.

Infilled frame construction has been in use for more than 200 years. The infilling of frames, in contrast with URM structures, is associated primarily with the construction of high-rise buildings—the frames being a means of carrying gravity loads, the infills a means of providing a building envelope and/or internal partitioning. In high-rise structures, the frames have been generally well-engineered in accordance with the state-of-knowledge of the day, whereas the infill panels were invariably considered to be "nonstructural". It was not until the 1950s that investigations began on the interaction between infill panels and the frames of buildings (Polyakov, 1956). This pioneering work undertaken in the former Soviet Union was strictly a reflection of Russian building practice. However, many of the theoretical techniques and other findings are still of relevance. The first study in the United States that investigated the lateral-load behavior of infilled frames, using specimens typical of U.S. construction practice (steel frames with brick infills), was reported by Benjamin and Williams (1958). These and other early studies were mostly concerned with the monotonic lateral-strength capacity of infilled frame systems.

Immediately following the advent of experimental investigations, analytical research began on the performance of infilled-frame systems. Over the years, several different methods of analysis were proposed for determining the composite strength of an infilled-frame system. These methods included elasticity solutions based on the Airy stress function, the finite-difference method, the finite-element method, and plastic methods of analysis. For a summary, see Maghaddam and Dowling (1987).

Although these methods of analysis have been shown to be reasonably successful in predicting the strength capacity of infilled-frame systems, each method has its roots in elasticity or rigid plasticity, making it either difficult or impossible to extend the findings to inelastic (elasto-plastic) behavior, especially if cyclic loading is to be considered. Therefore, it is not surprising that the equivalent-strut method of analysis has become the most popular approach for analyzing infilled frame systems. Early equivalent-strut methods, starting with Stafford-Smith (1966), used an equivalent single strut to represent infill behavior. It was later realized that such a simplification did not accurately capture all facets of frame/panel interaction. Therefore, several multiple-strut methods of analysis have been proposed (see for example Chrysostomou et al., 1988; Thiruvengadam, 1985; Mander et al., 1994). In spite of these attempts to enhance infilled frame analysis using a multiple-strut approach, there are still drawbacks—principally the inability to model force transfer-slip at the frame-panel interfaces (Gergely et al., 1994). Nonlinear finite element analysis, however, can be used if such a refinement is required (Shing et al., 1994; Mosalan et al., 1994), but difficulties remain, mostly due to computational limitations, on analyzing more than one panel at a time.

The general consensus is that a single equivalent-strut approach (two struts per panel for reversed cyclic loading analysis, one across each diagonal) may be successfully used for design and evaluation studies of infilled frame systems. Such an approach has been recently adopted by FEMA 273.

In spite of the general success of modeling infilled frames with solid panels, major difficulties still remain unresolved regarding the modeling approach for infilled frames with openings. Such frames, in practice, are commonplace and are perhaps the norm rather than the exception. However, only a limited amount of research has been undertaken on infilled frames with openings (e.g., Benjamin and Williams, 1958; Durrani and Luo, 1994; Coul, 1966; Dawe et al., 1985a,b; Holmes, 1961; Liauw, 1977; Mallick and Garg, 1971). Other strength analysis recommendations have been made for infills

---

with wide openings, but these have not been substantiated by experimental studies (Hamburger and Chakradeo, 1993; Freeman, 1994). For this reason, it is suggested that infilled frames with openings exceeding 50 percent of the panel area be treated either using other sections of this document (namely URM for infills with brick piers or reinforced concrete for cases in which infills surround steel columns) or by nonlinear finite element procedures such as discussed by Kariotis et al. (1996).

It is important to recognize that many behavior problems with infilled frames arise from discontinuities of infill, resulting from soft stories or checkered patterns, leading to a high concentration of forces to be transferred among components.

Other impediments to reliable modeling generalizations of infilled-frame systems are the large variation in construction practice over different geographic regions and changes of materials over time. Early infilled-frame construction generally consisted of clay brick (or sometimes stone masonry) and iron/steel frames. With time, concrete frames became popular and concrete masonry units or solid (poured) concrete were used for the infill panels. Concrete masonry or concrete infill panels may be either unreinforced or reinforced (and grouted or not in the case of concrete blocks).

Early research that investigated the seismic performance of infilled-frame specimens using reversed cyclic loading mostly focused on developing improved seismically-resistant design, analysis, and construction techniques for new structures (e.g., Axely and Bertero, 1979; Bertero and Brokken, 1983; Klingner and Bertero, 1976, 1978; Zarnic and Tomazevic, 1984, 1985a,b). Little research was done to investigate the seismic performance of existing structures with nonductile detailing. Although some studies have been conducted on infilled frames with deficient detailing (e.g., Gergely et al., 1993, 1994; Flannagan and Bennett, 1994; Mander et al., 1993a,b; Reinhorn et al., 1995), much work, especially experimental investigations, remains to be done.

8.2 Infilled Frame Masonry Component Types and Behavior Modes

8.2.1 Component Types

Infilled-frame elements are made up of infilled-panel and frame components, as summarized in Table 8-1.

The general characteristics of these basic components are summarized as follows:

a. Infilled Panels

Infilled panels are primarily categorized according to material and geometric configuration.

Materials. Clay brick masonry is perhaps the most commonly encountered type of infill material. The use of this traditional building material for infill construction dates back to the 1800s, when steel or iron frames were first used for high-rise construction. Generally twin or multi-wythe bricks are used, but other forms exist, such as cavity walls for exterior facades. Most often, brick masonry is unreinforced (see Section 7.1). In more modern buildings, reinforced, grouted-cavity wall construction may be found.

Hollow clay tile (HCT) is a relatively modern form of unreinforced masonry infill construction (see Section 7.1). The infills are often offset with respect to the centerlines of columns. HCT is often found on building facades. With steel frames, the clay tiles can be placed around the frames for aesthetic and fire protection purposes. HCT is very commonly used for interior partitions in framed buildings. As a material, HCT is generally very brittle and prone to force-controlled behavior.

Concrete masonry unit (CMU) construction is a form of infill using hollow concrete blocks laid up with mortar. CMU may be left hollow or filled with grout, either partially or completely. If grouted, steel reinforcement may or may not be present. The strength and ductility of the infill is highly dependent on the degree of grouting and reinforcement. Ungrouted infills are comparatively weak. This is because when the in-plane forces become large, compressive splitting of the face shells occurs with a complete loss of strength in masonry. Moreover, sliding-shear resistance relies entirely on the mortar in the bed joints. Grouted concrete masonry infills can be quite strong for normal bay sizes. Although early spalling of face shells may occur due to high in-plane lateral compression stresses, the grouted core has considerable ability in resisting additional loads, particularly if reinforced. Chapter 6 has additional information for reinforced CMU and Chapter 7 is applicable to unreinforced CMU.

Concrete infills are typically reinforced, though often minimally. In older buildings, the reinforcement is generally only for temperature and shrinkage control and is rarely provided to resist structural loads. In

Table 8-1   *Component Types for Infilled Frames*

Component Type		Description/Examples	Materials/Details

INPS	Solid infill panel	Space within frame components completely filled	Concrete
			   Reinforced
			   Unreinforced
			Masonry (clay brick, hollow clay tile, concrete block)
			   Reinforced
			   Unreinforced

INPO	Infill panel with openings	Doors and windows Horizontal or vertical gaps Partial-height infill Partial-width infill	Same as solid infill panel
		Sub-components similar to:	
	INP1 Strong pier	RC1 RM1 URM1	Concrete Reinforced masonry URM
	INP2 Weak pier	RC2 RM2 URM2	Concrete Reinforced masonry URM
	INP3 Weak spandrel (lintel)	RC3 RM3 URM3	Concrete Reinforced masonry URM
	INP4 Strong spandrel (lintel)	RC4 RM4 URM4	Concrete Reinforced masonry URM
INF1	Frame column	Vertical, gravity-load-carrying	Concrete Steel
INF2	Frame beam	Horizontal, gravity-load-carrying	Concrete Steel
INF3	Frame joint	Connection between column and beam components Rigid moment-resisting Partially-rigid Simple shear	Monolithic concrete Precast concrete Bolted steel Riveted steel Welded steel

modern buildings, however, the reinforced concrete infill may be well reinforced and act compositely with the surrounding frame. See Chapter 5 for additional information applicable to concrete infill.

Geometry. Infill may have a wide variety of geometric configurations. Aspect ratios (length/height of the planar space defined by the surrounding frame components) for infilled panels varies from approximately 1:1 to 3:1 with most ranging from 1.5:1 to 2.5:1. Infills may be configured in many forms to suit partitioning and/or facade requirements. It is not uncommon to find infills placed eccentrically to the axis of the frame components. For wider multi-wythe infills, entire rows of bricks may not engage with the frame at all. This leads to differential behavior and movements. For the purposes of evaluating infilled frame performance, only those bricks/blocks bounded by frames should be considered as part of the load infilled panel component.

For the purposes of damage evaluation, Table 8-1 identifies two categories for infilled panel components based on geometric configuration. Solid infilled panel components (INPS) are those that completely fill the planar space tightly within the surrounding frame components. Those with openings (INPO) may exhibit fundamentally different behavior.

Initial gaps at the top or sides of an infill affect performance of the solid panel configurations. These gaps can arise from the construction process not providing a tight infill, or in the case of concrete masonry units, from shrinkage. Until gaps are closed, normal frame behavior can be expected. When the gaps suddenly close, impacting forces on the infill can dramatically change the behavior patterns of the frame. Seismic gaps can be built into infill wall panels, although this practice is not common in the United States.

Perforations within the infill panels are the most significant parameter affecting seismic behavior of infilled systems. Doors and windows are the two most prevalent opening types. Openings located in the center portion of the infill can lead to weak infill behavior. On the other hand, partial-height infills (with windows spanning the entire top half of the bay) can be relatively strong. The frames are often relatively weak in column shear and when partial-height infill is present, this potentially leads to a short-column soft-story collapse mechanism. Partial-width infills are also relatively common: in this case, window openings extend the full

height between floors. Partial-width infill often has been placed on each side of a column component.

b. Frames

The frame components of infilled-frame seismic elements are categorized primarily by material.

Steel. Steel frames are common, especially for older structures. Steel frames are also popular for modern high-rise buildings and low-rise, light-weight, commercial, building construction. Column and beam components are most often I-sections (older) or wide-flange sections (newer). Built-up columns and double-channel beams are much less common. In older steel frames, the beam and column components are typically joined by semi-rigid riveted connections. More modern steel frames often use bolted or welded, or both, semi-rigid connections. Many of the frame systems are enclosed by concrete, the beams enclosed as a part of the floor system and the columns encased for fire protection. In these circumstances semi-rigid (or partially restrained) riveted connections will behave as fully restrained until the confining concrete cracks. Because of the relatively high shear capacity of steel columns, the fully restrained mode of behavior may be dominant.

Concrete. Concrete frames are also a common form of construction. Reinforced concrete frames may be classified as either ductile or nonductile for seismic performance, based primarily on the details of reinforcement. Contemporary structural design requires ductile detailing of the members. Ductile detailing requires closely-spaced transverse hoops in the beams, columns, and connections. If such members surround weak infill panels, they will suffer relatively less serious damage under lateral loads. Fully-ductile frames are relatively rare and, in the United States, are found only in the west, and seldom in infilled frame buildings. Typically, beams do not have adequate confinement and rarely is the bottom reinforcement continued through the joint. Concrete columns can, however, be designed with ductility, particularly in the western United States, where it is not uncommon for the better-built buildings to have spiral reinforcing. Under these circumstances the columns possess a relatively high shear capacity and displacement ductility, and the beam or the infill will be the deformation-limiting component.

Non-ductile frames are very common, particularly in regions of low-to-medium seismic risk. These frames are not detailed for ductility and may have one or more deficiencies: columns weaker than beams, lap splices in column hinge zones, and insufficient transverse reinforcement for confinement, for shear strength, and for longitudinal reinforcement stability. The beam/column joints in concrete frames need to transmit high shear forces. When infills are present, shear force demands are considerably higher, leaving the beam or column vulnerable to shear failure.

Precast, prestressed, concrete frames are also commonly encountered with infilled panels. Although in many respects similar to reinforced concrete, the connections between columns and beams in precast construction are distinctly different. When non-engineered infills are placed between columns, premature failure may occur at the beam/column connections, leading to unseating of the beams.

8.2.2 Panel and Frame Modeling and Interaction

Because of the highly nonlinear nature of the infill/frame interaction, proper modeling of the behavioral characteristics is best accomplished by a thorough analysis, material testing, and nonlinear, finite-element, modeling. Lacking the resources for that approach, an estimate of the behavior may be made by using a procedure similar to the identification of the appropriate inelastic lateral mechanism, as discussed in Section 2.4 of this volume. The frame is modeled conventionally as an assembly of column (indicated by INF1) and beam (INF2) components, and connection components (INF3). The solid infilled components (INPS) can be modeled as equivalent struts in accordance with the recommendations of FEMA 273. Infilled components with openings (INPO) can sometimes also be modeled as struts depending on the size and location of the openings. Alternatively, sub-component "piers" (INP1, INP2) and "spandrels" (INP3, INP4) can be used to represent the infilled component with openings. Appropriate force-deformation characteristics for the sub-components can be generated using the information in Chapter 5 for concrete, Chapter 6 for reinforced masonry, and Chapter 7 for unreinforced masonry.

To establish the inelastic force-deformation behavior of the frame and infill using the component method, the engineer must manually determine the bifurcation points defining the mode of behavior. Broadly speaking, the behavior can be separated into two conditions which depend primarily on the degree of the infill interaction with the frame. In the case where the openings are extensive, the components can be assembled as frame elements and piers, with the frame performance

modified by a potential for beam shear failure for cases with the infill primarily around the column, or by short-column effects where the infill is primarily around the beam. Small piers within the frame contribute little to the overall stiffness, but must be checked to ensure displacement compatibility with the frame-limited deformations.

For conditions where the infill is the controlling element, the degree of interaction is more complex. Initially, the defining characteristic is an uncracked panel. As the loading increases, the panel will experience bed-joint sliding or diagonal tension failure and transform the infill into an equivalent strut. Beam and column shears need to be investigated at this loading to ensure they are not the load-limiting condition. Following strut formation, corner crushing is often the next and final limiting condition. When checking corner crushing, the beam and column need to be checked for shear, and the column needs to be checked to verify that it has sufficient tension capacity to support the corner crushing. The tensile capacity is usually adequate for steel columns, but may be the limiting factor for lightly reinforced concrete columns, or columns having lap-splice problems.

8.2.3 Behavior Modes

a. Solid Panels

In cases where the infill component controls the stiffness, the events that define the shape of the force-deformation curve are bed-joint sliding, diagonal tension, corner crushing, general shear failure, and out-of-plane failure. Under small deformations the stiffness and behavior are dominated by the panel stiffness characteristics. As the deformation increases the panel characteristics will be a function of its element properties. When the masonry units are strong relative to the mortar, diagonal tension will result in a stair-stepped pattern of cracks through head and bed joints. When the mortar is stronger than the units (rare), cracks will develop through the units as well as the mortar and follow a line normal to the direction of the principal stress. With the stair-stepped cracks, shear can continue to be resisted after cracking by the development of a compressive stress normal to the bed joints, characterized as a compression strut. If the mortar is weak relative to the units, an infill panel may crack along the bed joints instead of along the diagonal. In this case, horizontal cracks may occur across several bed joints as an assembly of units slides to accommodate the deflected shape of the frame. Although this cracking mode may occur at lower shear forces, the overall frame-infill will possess greater inelastic deformation capacity because frame action will dominate. When the infill panel is sufficiently strong in shear, the compressive stress at the corners will fail in crushing. This mode will be the strongest and stiffest, but has limited deformation capacity because the crushing will be abrupt. Furthermore, the large forces generated in this mode will be distributed to the beam and column members, and may result in either column or beam shear failures. Table 8-2 presents four principal behavior modes for solid-infill panel components. Further explanation on the expected damage characteristics and likelihood of occurrence are given below.

i. Bed-Joint Sliding: This behavior mode commonly occurs in conjunction with other modes of failure. Bed-joint sliding is likely to occur when the bounding frame is strong and flexible (such as steel frames). If the mortar beds are relatively weak compared to the adjacent masonry units (especially bricks), a plane of weakness forms, usually near the mid-height level of the infill panel. Damage takes the form of minor crushing. There is really no limit to the displacement capacity of this behavior mode. Therefore, energy is continuously dissipated via Coulomb friction.

ii. Diagonal Cracking: Under lateral in-plane loading of an infill frame system, high compression stresses form across the diagonal of an infill. Transverse to these principal compression stresses and strains are tension strains. When the tensile strains exceed the cracking strain of the infill panel material, diagonal cracking occurs. These cracks commence in the center of the infill and run parallel to the compression diagonal. As interstory drifts increase, the diagonal cracks tend to propagate until they extend from one corner to the diagonally opposite corner.

This common form of cracking is evident in most infill panels that have been subjected to high lateral loads and sometimes occur with bed-joint sliding. Diagonal cracking behavior usually signals the formation of a new diagonal strut behavior mode.

iii. Corner Compression: Under lateral loading of infilled frames, some form of corner compression inevitably occurs. This is because of the high stress concentrations at each corner of the compression diagonal. For strong/stiff columns and beams, corner crushing is located over a relatively small region; whereas for weaker frames, especially concrete frames, corner crushing is more extensive and the damage extends into the concrete frame itself. In spite of the crushing damage that occurs, this is a relatively ductile failure mode. As interstory drifts increase, corner crushing becomes more pronounced to the extent that masonry units in the corner may fall out entirely. When this happens, crushing propagates towards the center of the beam and/or column.

iv. Out-of-Plane Failure: Ground shaking transverse to the plane of a wall may lead to an out-of-plane behavior mode. Experiments using air bags (Abrams, 1994), as well as shaking-table studies (Mander et al., 1994), show that for normal, infill panel, height-to-thickness ratios, considerable shaking is necessary to cause failure of the infill. However, out-of-plane failure may occur in the upper stories of high-rise buildings, where the floor accelerations are basically resonance amplifications of prominent sinusoidal ground motion input. In lower stories, when combined with high in-plane story shears, infill panels tend to progressively "walk-out" of the frame enclosure on each cycle of loading. Although complete out-of-plane failure is not common, there is some evidence that this behavior mode has occurred.

Table 8-2   Behavior Modes For Solid Infilled Panel Components

Behavior Mode	Description/Likelihood of Occurrence	Ductility	Figure	Paragraph (Section 8.2.3a)
Bed-joint sliding	Occurs in brick masonry, particularly when length of panel is large relative to height aspect ratio is large and the mortar strength is low.	High	8-2	i
Diagonal cracking	Likely to occur in some form	Moderate	8-1, 8-4, 8-5	ii
Corner compression	Crushing generally occurs with stiff columns.	Moderate	8-1	iii
Out-of-plane failure	More likely to occur in upper stories of buildings. However, out-of-plane "walking" is likely to occur in the bottom stories, due to concurrent in-plane loading.	Low	8-5	iv

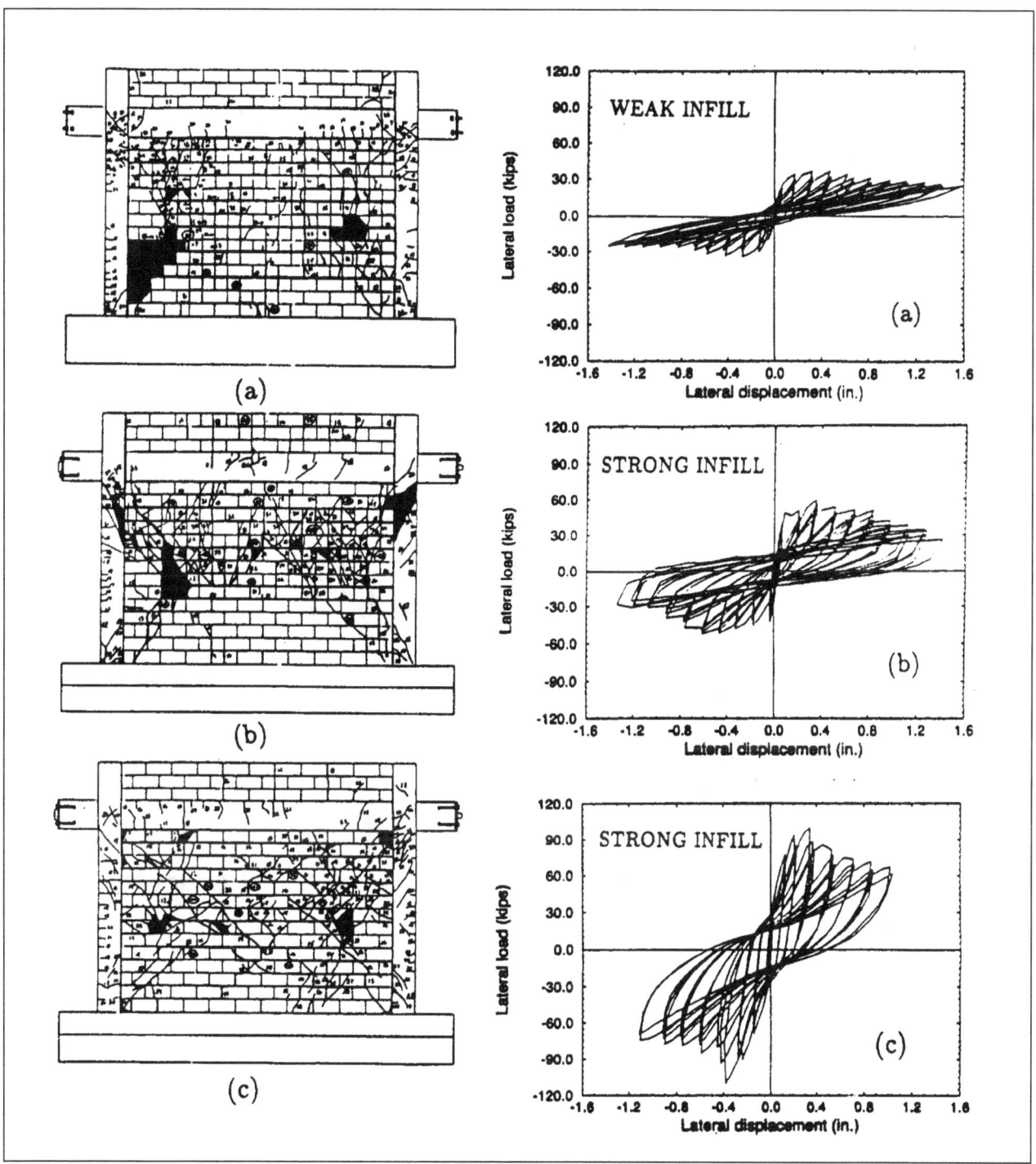

Figure 8-1   Ductile reinforced concrete frames with concrete masonry infills tested by Mehrabi et al. (1996). (The weak and strong infills were ungrouted and grouted, respectively). (a) Specimen 4, (b) Specimen 5, (c) Specimen 7; h/L aspect ratio = 0.67. Note 1 in. = 1.65% interstory drift

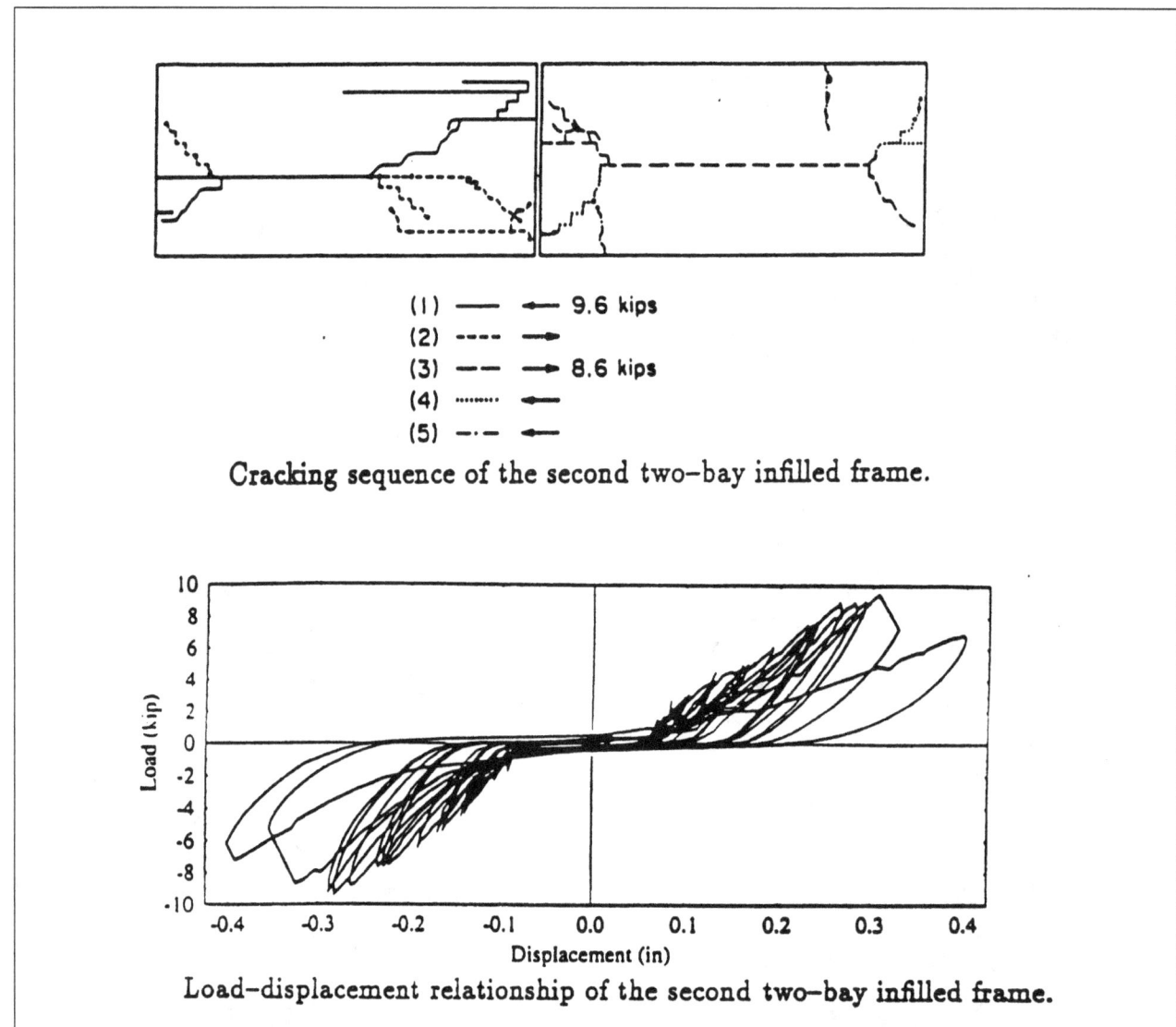

(1) ———— ⟵ 9.6 kips
(2) ----- ⟶
(3) —— ⟶ 8.6 kips
(4) ········· ⟵
(5) —·— ⟵

Cracking sequence of the second two-bay infilled frame.

Load-displacement relationship of the second two-bay infilled frame.

Figure 8-2   *Bed-joint sliding of a two-bay steel frame-block infill. Model study by Gergely et al. (1994).*

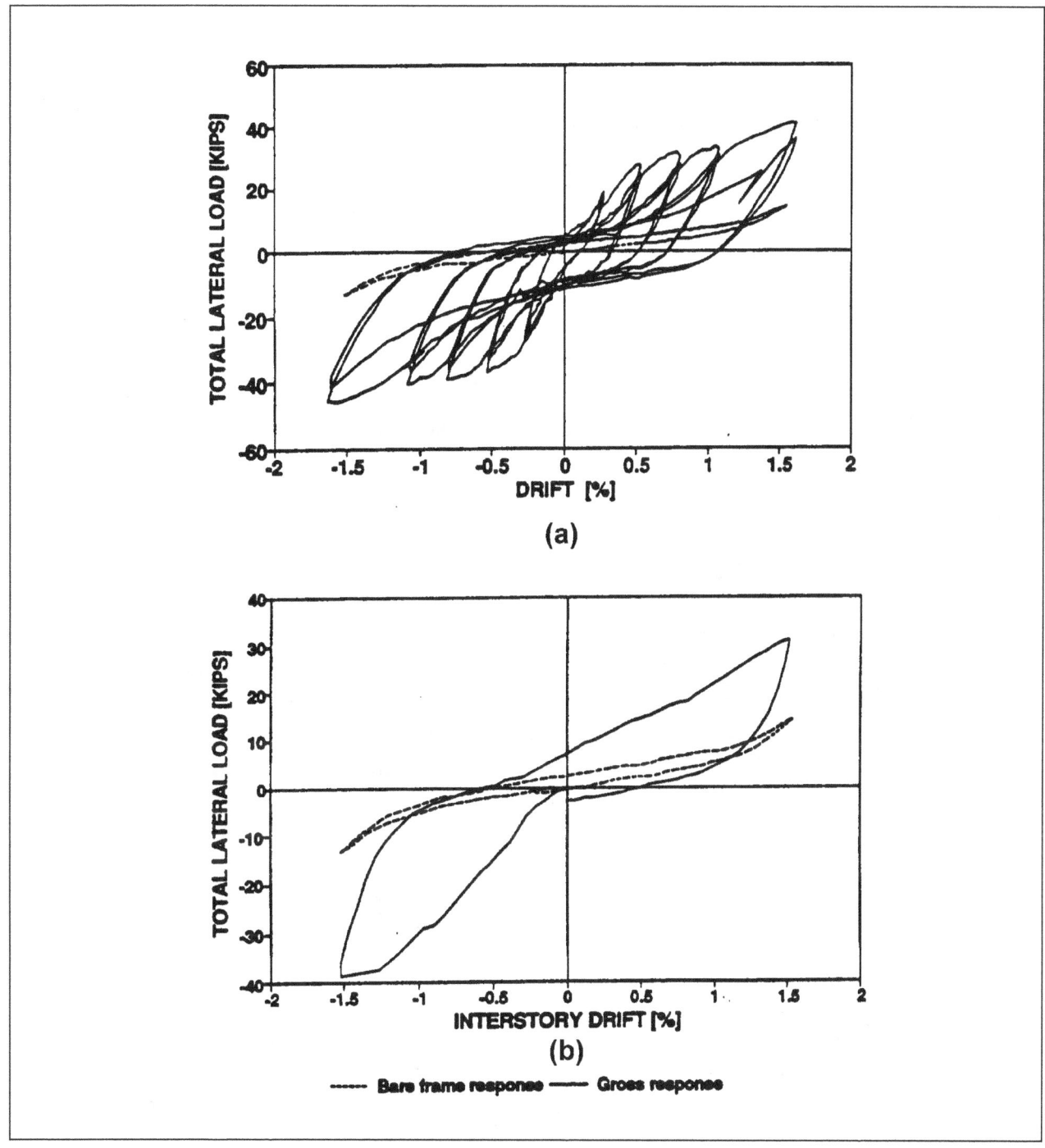

Figure 8-3   Specimen tested by Mander et al. (1993a). Steel frame-clay brick masonry infill. Top and seat angles
semi-rigid connections used to connect beams to columns.
(a) Original specimen.
(b) Specimen repaired with 1/2-inch ferrocement overlay and retested.

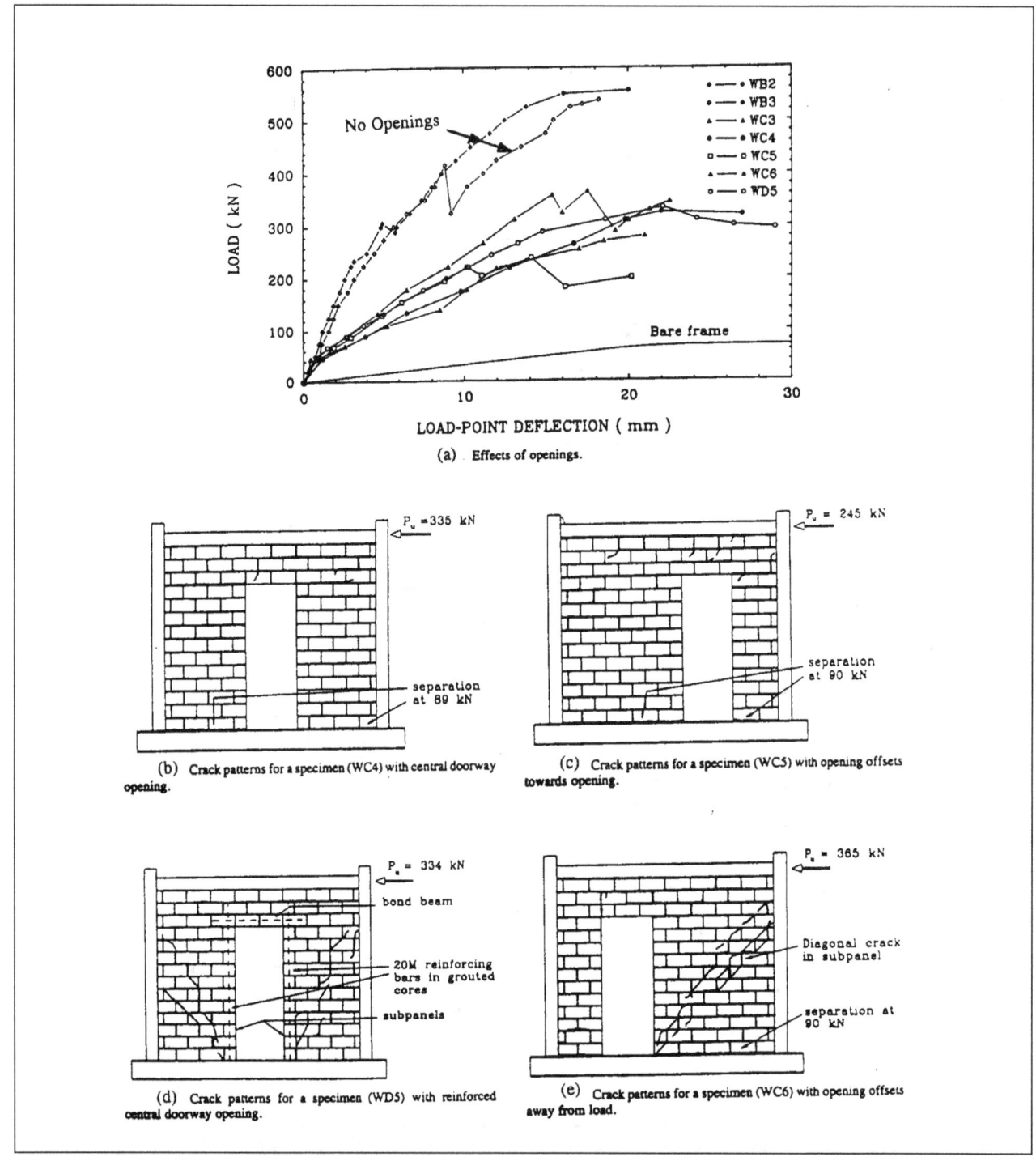

(a) Effects of openings.

(b) Crack patterns for a specimen (WC4) with central doorway opening.

(c) Crack patterns for a specimen (WC5) with opening offsets towards opening.

(d) Crack patterns for a specimen (WD5) with reinforced central doorway opening.

(e) Crack patterns for a specimen (WC6) with opening offsets away from load.

Figure 8-4   Effect of openings on the monotonic lateral-load performance of steel frame-masonry infill tested by Dawe and Seah (1988).

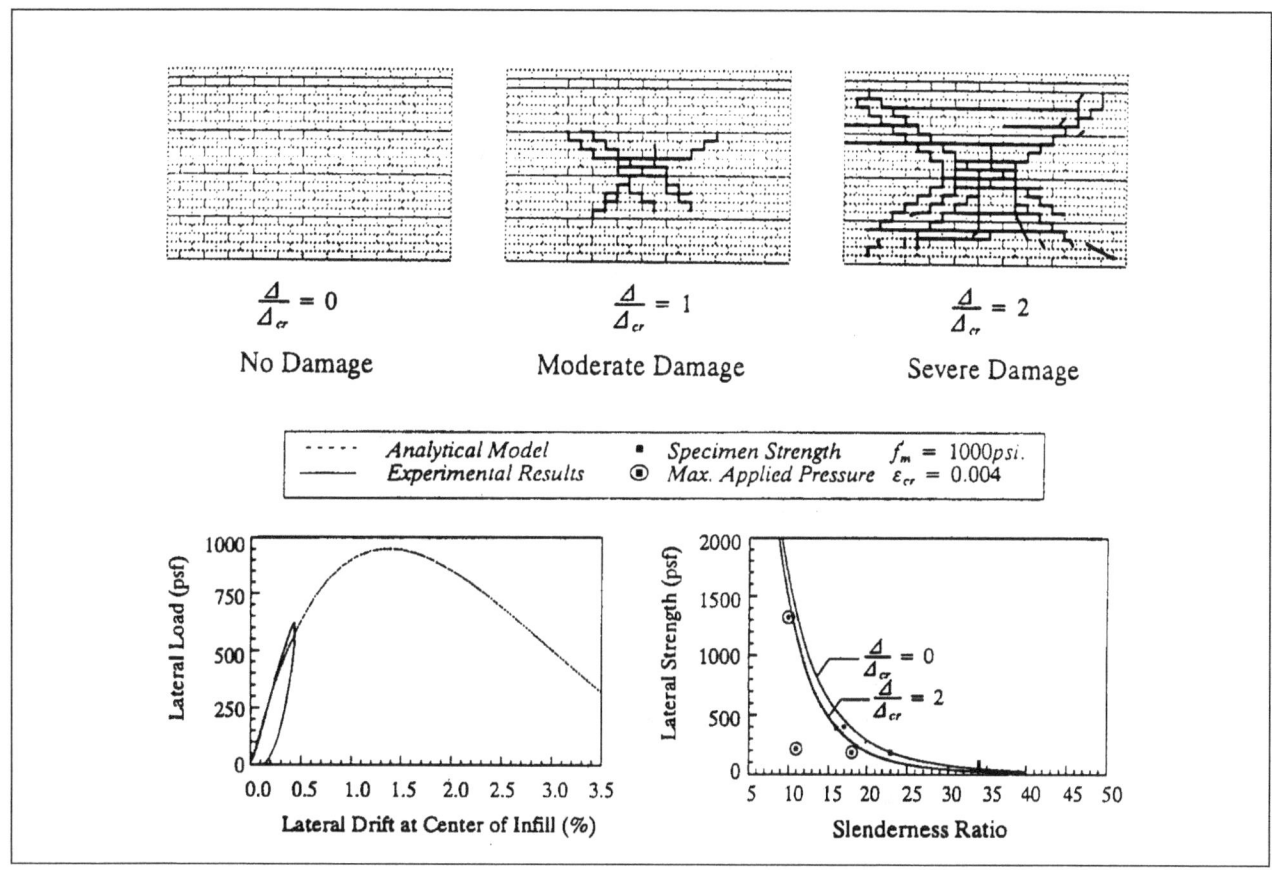

Figure 8-5   *Out-of-plane behavior of infilled masonry walls showing crack patterns and out-of-plane lateral load vs. lateral displacement of an air bag test. (From Angel and Abrams, 1994).*

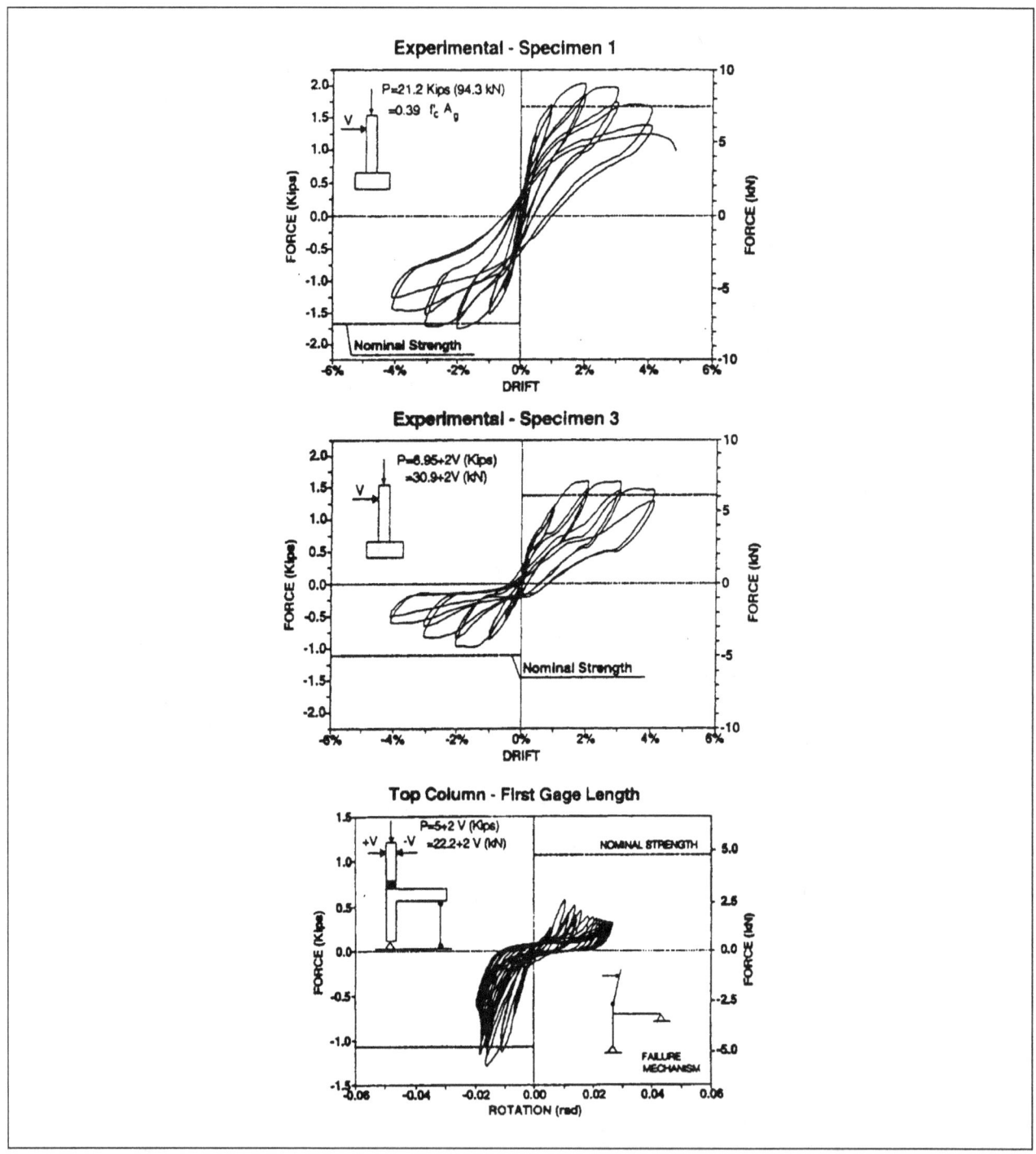

Figure 8-6   *Experiments conducted by Aycardi et al. (1994), showing the performance of nonductile frame members with lap splices at the base of the column.*
Specimen 1: Column with a moderately high level of axial load.
Specimen 3: Column with a lower level, variable, axial load.
Slab-beam-column subassemblage: Tested with variable axial load.
(Note: Deterioration was due to cyclic loading action, the weak zone being the beam rather than the column).

b. Panels With Openings

Infilled panels with openings are best viewed as assemblies of subcomponents of the appropriate material. The behavior modes for the subcomponents are discussed in the other material sections (Chapter 5 for concrete, Chapter 6 for reinforced masonry, and Chapter 7 for URM). These subcomponents interact with the surrounding frame and can alter the frame response. Principal types of interaction can that can occur are strong columns and strong piers inducing shear failure in the beams, strong spandrel components reducing the ductility by causing short-column effects, and by the infill inducing tension yielding or bar splice failures in the column. A discussion of the frame component behavior modes is given below in Section 8.2.3c.

c. Frame components

Table 8-3 and Table 8-4 present the principal behavior modes for steel and concrete frames, respectively, possessing infills. An explanation of these behavior modes follows.

i. Flexural Yielding in Steel: Flexural yielding of the frame is primarily associated with steel frames. When large lateral loads are imposed on moment frames, flexural yielding that leads to plastic hinging is to be expected adjacent to rigid connection. For infilled frames, this generally occurs at the base of fixed-base columns. This is generally evidenced by cracking of paint (if any) and buckling of compression flanges. As the rotational capacity of flexural plastic hinges in steel members is high, and unlikely to be attained in infilled-frame systems because the infills limit the interstory drifts, damage from this behavior mode is mostly cosmetic and generally not serious.

ii. Shear Yielding in Steel: When corner crushing occurs in a strong infill, the diagonal compression strut moves downward into the column providing a large shear force at the end-region of the column. For thin webbed steel members (non-compact sections) that are not confined by concrete, this may lead to web buckling and large localized shear deformations in the frame members. However, shear yielding in steel is ductile, and damage arising from this behavior mode is not serious.

iii. Bolted or Riveted Connection Failure: Steel-frame systems with infill panels generally have bolted or riveted semi-rigid, beam-to-column connections, and the connection is usually encased in concrete.

Under these conditions the joint will behave as a rigid joint until the concrete breaks. Under large lateral loads the concrete can fail, or, when there is no confinement, yielding of the connection may occur. Such behavior is not serious, because the ductility capacity of semi-rigid connections is considerable (Mander et al., 1994 and 1995). However, in the presence of infill panels, the large diagonal strut force puts the connection under considerable axial tension. This may cause prying in the connection angles, giving the appearance of serious damage. Nevertheless, the ductility capability of these connections is still considerable and they are capable of sustaining many cycles of loading before a low-cycle fatigue failure, by which time the infill panel itself will have inevitably failed (Mander et al., 1993a).

iv. Flexural Yielding in Reinforced Concrete Frame Components: This behavior mode is expected to occur in reinforced concrete frames with infill when the interstory drift ratios exceed about 0.005. Flexural yielding behavior occurs where moments are greatest—that is, at the ends of beam and column members. In order for a complete side-sway mechanism to form in a structural concrete frame, flexural yielding and plastic hinging must also occur at the base of the columns (ground-floor level). Flexural yielding behavior in reinforced concrete beam and column components is characterized by tensile cracking in the cover concrete (transverse to the axis of the member), coupled with some compression crushing in the cover concrete on the opposite face. High bending moments in frames are also generally associated with high shear forces. Shear demand, when coupled with bending moment, produces diagonal cracking. When flexural plastic-hinge rotations become substantial (note that this is unlikely for infilled frames as the infills substantially limit the amplitudes of interstory drifts), considerable crushing and loss of cover concrete is evident, often leaving the longitudinal and transverse reinforcement exposed. If such severe damage is evident in an infilled-frame system, it is likely to occur in the lowest story, where high story shear demand has caused the infill panel to fail, leading to subsequent, high, interstory drifts (see, for example, Klingner and Bertero, 1978).

v. Lap-Splice Slip in Concrete: Most older infilled-frame structures have not been specifically designed for earthquake resistance, and the frames possess nonductile details. This can lead to column lap splices occurring in potential plastic-hinge zones. For the lap-splice detail deficiency, the connection will undergo flexural yielding at large lateral drifts. Cracking associated with incipient hinging tends to show a combination of transverse (flexure) cracks along with longitudinal (splitting) cracks that run parallel to the longitudinal column reinforcement. These longitudinal cracks signal that the bars in the lap-splice zone have begun to slip. If the interstory drifts are substantial and the cyclic loading pronounced, then the bond within the lap-splice zone is destroyed. Moment transfer in the lap-splice zone becomes limited when spalling of the cover concrete is apparent. It should be noted, however, that this behavior mode does not necessarily lead to an unsafe situation, for the lack of moment capacity finally leads to a pinned connection, capable of transferring axial load and shear. For some experimental results on nonductile concrete columns with lap splices at the base of columns, see Aycardi et al. (1992, 1994).

vi. Column Tension Yielding: When the effect of the infill is substantial, the frame will behave more like a braced frame than moment frame, resulting in the columns resisting the lateral forces and overturning forces in tension and compression. For many older buildings the columns are lightly reinforced and may have more compression capacity than tension capacity, resulting in a tension yielding condition. This is a ductile mode, allowing larger displacements without causing an abrupt failure. The limiting deformation in this mode is the deformation capacity of the infill.

vii. Concrete Shear Failure: Infilled frames that possess strong infill panels generate large shear forces in the infill panels when under lateral side-sway. These shear forces must be transferred from the panel into the frame. If the infill panel is damaged in the corners as a result of corner crushing, then the diagonal

strut tends to move away from the panel corner. The high transverse forces from the diagonal strut then enter the frame some distance (typically about one member depth) away from the beam/column connection. This provides a very high shear demand over a short column (or sometimes beam) length. Damage to the frame members is indicated by large diagonal X-cracks and spalling of the cover concrete. Under very severe cases of damage (weak frame / strong infill), complete loss of cover concrete and bulging of the core may be expected. If this occurs in a column, it is a serious form of damage, because the ability of the column to transfer axial loads may be seriously impaired. Therefore, it is not surprising that the ductility capability of such shear-critical elements is low.

viii. Concrete Joint Failure: Beam/column joints are subjected to high shear forces when under lateral loading. These shear forces can be amplified when infills are present. For concrete frames with nonductile reinforcing steel details, there is generally a deficiency or complete absence of transverse reinforcement within the beam/column joint core. Therefore, the shear-strength capacity is inevitably less than the demand imposed, even at moderate interstory drifts. Consequently, this highly likely behavior mode leads to large X-cracks in the beam-column joint region. Under cyclic loading, the cover concrete spalls, the joint concrete bulges, and the longitudinal column reinforcement tends to buckle. Such behavior tends to keep the adjacent beam and column plastic-hinge regions from being severely damaged. However, the ability of the frame system to carry axial loads through the damaged joints is suspect, especially if the behavior mode is associated with an adjacent infill panel that is also near failure.

Table 8-3   Behavior Modes For Infilled Steel-Frame Components

Behavior Mode	Description, and Likelihood of Occurrence	Ductility	Figure	Paragraph (see Section 8.2.3c):
Flexural yielding	Hinges form at base of columns, and can occur adjacent to the beam/column joint if the members are weaker than the connection.	High	Similar to 8-1(a)	i
Shear yielding	Unlikely, except when infill causes short-column effect.			ii
Bolted or riveted connection failure	Likely.	High	8-3	iii

Table 8-4   Behavior Modes For Infilled Concrete-Frame Components

Behavior Mode	Description, and Likelihood of Occurrence	Ductility	Figure	Paragraph (see Section 8.2.3c):
Flexural yielding	Should always occur at ground-floor level. Probably will also occur adjacent to beam/column joint.	High	8-1c	iv
Lap-splice slip	Probable, will normally occur at ground-floor level.	Moderate-to-low	8-6	v
Column tension yielding	Lightly reinforced columns with strong infill	Moderate-to-high		vi
Shear failure	Probable in nonductile frames. Likely at partial-height infills.	Low	8-1b	vii
Joint failure	Probable with nonductile detailing.	Low	8-1a	viii

8.3   Infilled Frame Evaluation Procedures

8.3.1   Solid Infilled-Panel Components

This subsection gives equations for quantifying the stiffness, strength and deformation capacity of infilled panels. Note that Young's modulus and strength values for the infill panel are given in terms of masonry materials. For reinforced concrete infills, make the following substitutions: $E_c = 57000\sqrt{f'_{ce}}$ for E_m and f'_{ce} for f'_{me}.

a.   Stiffness

The effective width(s) of a diagonal compression strut that can be used to assess the stiffness and strength of an infill panel is initially calculated using the recommendations given in FEMA 273. The provisions are based on the early work of Mainstone (1971) and Mainstone and Weeks (1970) and are restated below for the convenience of the user.

The equivalent strut is represented by the actual infill thickness that is in contact with the frame (t_{inf}) and the diagonal length (r_{inf}) and an equivalent width, a, given by:

$$a = 0.175(\lambda_1 h_{col})^{-0.4} r_{inf} \qquad (8\text{-}1)$$

where

$$\lambda_1 = \left[\frac{E_{me} t_{inf} \sin 2\theta}{4 E_{fe} I_{col} h_{inf}}\right]^{\frac{1}{4}} \qquad (8\text{-}2)$$

and

h_{col} =   column height between centerlines of beams, in.

h_{inf} =   height of infill panel, in.

E_{fe} =   expected modulus of elasticity of frame material, psi.

E_{me} =   expected modulus of elasticity of infill material, psi.

I_{col} =   moment of inertial of column, in^4.

r_{inf} =   diagonal length of infill panel, in.

t_{inf} =   thickness of infill panel and equivalent strut, in.

θ =   angle whose tangent is the infill height-to-length aspect ratio, radians

$$\theta = \tan^{-1}\left(\frac{h_{inf}}{L_{inf}}\right) \qquad (8\text{-}3)$$

where

L_{inf} =   length of infill panel, in.

Only the masonry wythes in full contact with the frame elements need to be considered when computing in-plane stiffness, unless positive anchorage capable of transmitting in-plane forces from frame members to all masonry wythes is provided on all sides of the walls.

b. Strength

The strength capacity of an infill panel is a complex phenomenon. It is important to analyze several potential failure modes, as these will give an indication of potential crack and damage patterns. Four failure modes are possible, as described below.

i. <u>Sliding-Shear Failure.</u> The Mohr-Coulomb failure criteria can be used to assess the initial sliding-shear capacity of the infill:

$$V_{slide}^i = \left(\tau_0 + \sigma_y \tan\phi \right) L_{inf} t_{inf} = \mu N \qquad (8\text{-}4)$$

where τ_0 = cohesive capacity of the mortar beds, which, in the absence of data may be taken as

$$\tau_0 = \frac{f'_{me90}}{20} \qquad (8\text{-}5)$$

where ϕ = the angle of sliding friction of the masonry along a bed joint. Note that $\mu = \tan\phi$, where μ = coefficient of sliding friction along the bed joint. After the infill's cohesive bond strength is destroyed as a result of cyclic loading, the infill still has some ability to resist sliding through shear friction in the bed joints. As a result, the final Mohr-Coulomb failure criteria reduce to:

$$V_{slide}^i = \left(\sigma_y \tan\phi \right) L_{inf} t_{inf} = \mu N \qquad (8\text{-}6)$$

where N = vertical load in the panel. If deformations are small, then $V_{slide} \approx 0$ because σ_y may only result from the self-weight of the panel. However, if these interstory drifts become large, then the bounding columns impose a vertical load due to shortening of the height of the panel. The vertical shortening strain in the panel is given by

$$\varepsilon = \frac{\delta}{h} = \theta \frac{\Delta}{h} = \theta^2 \qquad (8\text{-}7)$$

where

δ = downward movement of the upper beam as a result of the panel drift angle, θ

h = interstory height (center-to-center of beams)

Δ = interstory drift (displacement)

θ = interstory drift angle

The axial load on the infill is

$$N = \varepsilon L_{inf} t_{inf} E_m \qquad (8\text{-}8)$$

where E_m = Young's modulus of the masonry, which in the absence of tests may be set at $550 f'_{me}$.

Substituting equations (8-7) and (8-8) into (8-6) gives

$$V_{slide}^i = \mu L_{inf} t_{inf} E_m \theta^2 \qquad (8\text{-}9)$$

ii. <u>Compression Failure.</u> For compression failure of the equivalent diagonal strut, a modified version of the method suggested by Stafford-Smith and Carter (1969) can be adopted. The shear force (horizontal component of the diagonal strut capacity) is calculated as

$$V_c = a t_{inf} f'_{m90} \cos\theta \qquad (8\text{-}10)$$

where

a = equivalent strut width, defined above

t_{inf} = infill thickness

f'_{me90} = expected strength of masonry in the *horizontal* direction, which may be set at 50% of the expected stacked prism strength f'_{me}.

iii. <u>Diagonal Tension Failure of Panel.</u> Using the recommendation of Saneinejad and Hobbs (1995), the cracking shear in the infill is given by

$$V_{cr} = \frac{2\sqrt{2}\, t_{inf} \sigma_{cr}}{\left(\dfrac{L_{inf}}{h_{inf}} + \dfrac{h_{inf}}{L_{inf}} \right)} \qquad (8\text{-}11)$$

The cracking capacity of masonry, σ_{cr}, is somewhat dependent on the orientation of the principal stresses with respect to the bed joints.

In the absence of tests results, the cracking strength may be taken as

$$\delta_{cr} = \frac{f'_{me90}}{20} \qquad (8\text{-}12)$$

or

$$\sigma_{cr} \approx v_{me} \qquad (8\text{-}13)$$

where v_{me} = cohesive strength of the masonry bed joint, which is given by

$$v_{me} = 20\sqrt{f'_{me}} \qquad (8\text{-}14)$$

where f'_{me} = expected comprehensive strength of a masonry prism.

iv. <u>General Shear Failure of Panel.</u> Based on the recommendations of FEMA 273, as well as Paulay and Priestley (1992), the initial and final contributions of shear carried by the infill panel may be defined as:

$$v_{mi} = A_{vh} 2\sqrt{f'_{me}} \qquad (8\text{-}15)$$

$$V_{mf} = 0.3 V_{mi} \qquad (8\text{-}16)$$

where

V_{mi} = available initial shear capacity that is consumed during the first half-cyclic (monotonic) loading

V_{mf} = final shear capacity as a result of cyclic-loading effects

A_{vh} = net horizontal shear area of the infill panel.

Note for a complete infill with no openings

$$A_{vh} = L_{inf} t_{inf} \qquad (8\text{-}17)$$

The above values give upper and lower bounds to the cyclic-loading resistance of the infill.

v. <u>The Effect of Infill Panel Reinforcement.</u> If either a masonry or concrete infill panel is reinforced, then the reinforcement should improve the shear strength of the panel. The shear demand carried by the reinforcement is given by the well known ACI 318-95 (ACI, 1995) provisions.

$$v_s = \rho_w f_{ye} A_{vh} \qquad (8\text{-}18)$$

where ρ_w = volumetric ratio of the reinforcement in the infill panel, f_{ye} = expected yield strength of the web reinforcement within the infill panel, and A_{vh} is defined above.

c. **Deformation Capacity of Solid Infilled-Panel Components**

There are no clear experimental results for the deformation capacities for each of the four behavior modes for infilled panel components, nor are there suitable analytical models available. Experiments show that diagonal cracking begins with the onset of nonlinear behavior at interstory drifts of 0.25% and is essentially complete (from corner to corner) in a panel by about 0.5%. Corner crushing begins at the same stage, but its extent depends on the amount of cyclic loading sustained. There is essentially no limit to the ability of an infill panel to deform in sliding shear—other behavior modes usually govern. Thus, limits imposed by the general shear behavior mode determine the displacement capacity of infill panels. Experimental evidence supports the following interstory drift limit states for different masonry infill panels:

Brick masonry	1.5%
Grouted concrete block masonry	2.0%
Ungrouted concrete block masonry	2.5%

8.3.2 Infilled-Panel Components with Openings

The strength of infill panels with openings is best assessed using rational models composed of subcomponents of the relevant materials. See Chapter 5 for concrete, Chapter 6 for reinforced masonry, and Chapter 7 for unreinforced masonry.

8.3.3 Out-of-Plane Behavior of Infilled-Panel Components

FEMA 273 as well as Angel and Abrams (1994) describe methods for assessing infill capacity to resist out-of-plane demands. Based on these recommendations, the following formulae can be used to assess the infill strength. In these expressions, w is the uniform pressure that causes out-of-plane failure of the wall.

$$w = \frac{2 f'_{me}}{(h/t)} \lambda R_1 R_2 \qquad (8\text{-}19)$$

where

$f'_{me} =$ expected masonry strength

$\lambda =$ slenderness parameter defined in Table 8-5

$R_1 =$ out-of-plane reduction factors, set at $R_1 = 1$ for no damage (See Table 8-5 for moderate and severe damage)

$R_2 =$ Stiffness-reduction factor for bending frame members, given by

$$R_2 = 0.35 + 71.4 \times (10)^{-9} EI \text{ not to exceed } 1 \qquad (8\text{-}20)$$

where

$EI =$ flexural rigidity of the weakest frame on the non-continuous side of the infill panel (units: k-in)

8.3.4 Steel-Frame Components

a. Flexure

The flexural strength of steel frames, based on conventional plastic concepts is given as

$$M_p = F_{ye} Z_x \qquad (8\text{-}21)$$

where

$Z_x =$ plastic-section modulus

$F_{ye} =$ expected yield strength

In the absence of specific data, F_{ye} is set initially at 48 ksi (330 MPa) and 55 ksi (380 MPa) for Grades A36 and 50, respectively.

b. Shear

Steel-frame shear-strength capacity of beams and columns is based on the relationship

$$V_p = 0.6 A_w F_{ye} \qquad (8\text{-}22)$$

where

$F_{ye} =$ expected yield strength, with default values defined above

$A_w =$ web area of the member

The 0.6 coefficient is based on the Von Mises yield criteria.

The critical section in a steel frame component with respect to shear is assumed to occur at an equivalent short beam or column formed between a frame joint and the compression strut associated with the infill panel. It may conservatively be assumed that the centroid of the diagonal strut force moves downward by an amount equal to

$$l_{ceff} = \frac{a}{\cos\theta_c} \qquad (8\text{-}23)$$

where

$a =$ effective width of a longitudinal compression struct. This is defined in Section Section 8.3.1a.

$l_{ceff} =$ effective length of a "short" fixed-fixed column.

$$\tan\theta_c = \frac{h_{inf} - \left(\frac{a}{\cos\theta_c}\right)}{L_{inf}} \qquad (8\text{-}24)$$

The shear demand is at a maximum when flexural plastic hinges form at each end of this so-called "short column", thus

$$V_{col} = \frac{2 M_p^{col}}{l_{ceff}} \qquad (8\text{-}25)$$

c. Joints

The strength and deformation capacity of riveted, bolted and welded connections along the panel zones are largely geometry-dependent. Due to wide variations in construction practice, the reader is referred to FEMA 273.

8.3.5 Concrete-Frame Components

a. Flexure

Flexural strength of reinforced concrete frames should be based on the nominal strength provisions of the ACI 318-95 code. However, expected strength values should be used for the material properties. Flexural deformation capacity depends on the amount of transverse reinforcement. If ductile detailing is used, then dependable plastic-hinge rotations of 0.035 radians can easily be attained. For nonductile detailing, experimental

Table 8-5   Out-of-plane infill strength parameters.

Height-to-thickness ratio $\dfrac{h}{t}$	Slenderness parameter λ	Strength-reduction factor R_1	
		Moderate Damage	*Severe Damage*
5	0.130	1.0	1.0
10	0.060	0.9	0.9
15	0.035	0.9	0.8
20	0.020	0.8	0.7
25	0.015	0.8	0.6
30	0.008	0.7	0.5
35	0.005	0.7	0.5
40	0.003	0.7	0.5

research by Aycardi et al. (1992, 1994) has shown that interstory drifts of 0.03 radians are possible. Because the drift demands on infilled-frame systems are generally not as great as for bare frames, it is likely that the rotational demands will be less than the rotational capacity.

b. Shear

The critical section for shear strength is similar to that for steel frames, as discussed in Section 8.3.4. If the "short column" member is shear-critical, then the corner-to-corner crack angle forms. This angle can be calculated from

$$\alpha_c = \tan^{-1}\frac{jd}{l_{ceff}} \qquad (8\text{-}26)$$

where

$jd =$ internal lever arm within the column member. In lieu of a precise analysis, this may be set at 80% of the overall member width.

Similarly, the beam should also be checked so that the effective length of the beam is given by

$$l_{ceff} = \frac{a}{\sin\theta_b} \qquad (8\text{-}27)$$

$$V_b = \frac{\left(M_p^+ + M_p^-\right)}{l_{ceff}} \qquad (8\text{-}28)$$

$$\tan\theta_b = \frac{L_{inf}}{L_{inf} - \dfrac{a}{\sin\theta_b}} \qquad (8\text{-}29)$$

The corner-to-corner crack angle forming in a reinforced concrete beam is:

$$\alpha_b = \tan^{-1}\frac{d-d'}{l_{ceff}} \qquad (8\text{-}30)$$

in which

$l_{ceff} =$ "short-beam" length

$d-d' =$ distance between centroids of top and bottom reinforcement

$M_p^+ =$ maximum positive moment generated by tensile yielding of the bottom reinforcement

$M_p^- =$ maximum negative moment generated by tensile yielding of the top beam steel, including the effects of slab steel, if any.

$$M_p^+ = A_s\left(1.25f_{ye}\right)(d-d') \qquad (8\text{-}31)$$

$$M_p^- = A_s'\left(1.25f_{ye}\right)(d-d') \qquad (8\text{-}32)$$

where

A_s = area of bottom steel. If this steel is not fully anchored and only extends a short distance into the joint, the value of A_s used in Equation 8-31 should be prorated by $\frac{l_{em}}{l_d}$ where l_{em} = embedment length and l_d = development length, as given by ACI 318.

A_s' = area of top steel, including slab steel

$1.25f_{ye}$ = expected overstrength of the tension reinforcement, the 1.25 factor allowing for strain-hardening effects, and f_y = probable/measured yield strength of the longitudinal beam reinforcement

Concrete-frame shear-strength capacity is initially based on the recommendations of ATC-40 and FEMA 273. This recommended design procedure generally provides a lower bound to the shear-strength capacity. The ultimate shear capacity is given by

$$V_u = V_s + V_c \qquad (8\text{-}33)$$

where V_s is the shear carried by the steel

$$V_s = A_{sh}f_{yh}\frac{d}{s} \qquad (8\text{-}34)$$

and V_c is the shear carried by concrete:

$$V_c = 3.5\lambda\left(k + \frac{N_u}{2000A_g}\right)\sqrt{f_c'}b_w d \qquad (8\text{-}35)$$

where

k = 1.0 in regions of low ductility demand and 0 in regions of moderate or high ductility demand,

λ = 0.75 for lightweight aggregate concrete and 1.0 for normal-weight aggregate concrete

N_u = axial compression force in pounds (equals 0 for tension force)

The approach recommended by Priestley et al. (1996) is less conservative and may provide an estimate of shear capacity that is more compatible with observations in the field, particularly in the presence of large diagonal cracks. In this approach, the shear capacity is given by:

$$V_n = V_s + V_p + V_c \qquad (8\text{-}36)$$

where V_s, V_p, and V_c are the shear demand carried by steel, compressive axial-strut force, and concrete mechanism, respectively. These are defined below.

V_s, the shear carried by the transverse reinforcement, is given by:

$$V_s = A_{sh}f_{yhe}\frac{jd}{s}\cot\theta \qquad (8\text{-}37)$$

where

A_{sh} = area of steel in one transverse hoop set

f_{yhe} = expected strength of the transverse reinforcement

jd = internal lever arm, which in lieu of a more precise analysis may be set at 0.8D

D = member depth

s = center to center spacing of the hoop sets

θ = corner-to-corner crack angle measured to the axis of the column

V_p, the shear demand carried by axial load (strut action) in a column, is given by

$$V_p = P\tan\theta \qquad (8\text{-}38)$$

where

P = axial load in the frame member

θ = as defined above.

V_c, the shear demand carried by the concrete is given by

$$V_c = k\sqrt{f_{ce}'}b_w d \qquad (8\text{-}39)$$

in which

b_w = web width

d = effective member depth, and

k = coefficient depending on the displacement ductility of the member and may be defined as follows:

$$k = 3.5 \text{ for } \mu \leq 2 \qquad (8\text{-}40)$$

$$k = 1.2 \text{ for } \mu = 4 \qquad (8\text{-}41)$$

$$k = 0.6 \text{ for } \mu \geq 8 \qquad (8\text{-}42)$$

To find k for $2 < \mu < 4$ and $4 < \mu < 8$, use linear interpolation between the above-specified ductility limits. For equations containing f'_{ce}, use psi units. Note that upper- and lower-bound values of V_c should be computed, representing the initial and the final (residual) shear capacity values using Equations 8-40 and 8-42, respectively.

c. Joints

FEMA 273 presents guidelines for assessing beam-column joint strength using the formula given below:

$$V_n = \lambda \gamma \sqrt{f'_{ce}} A_j \qquad (8\text{-}43)$$

where

$A_j =$	nominal gross-section area
$\lambda =$	0.75 or 1.0 for light weight and normal weight aggregate concrete, respectively
$\gamma =$	strength coefficient ranging from 4 to 12 and 8 to 20 for joints without and with appreciable transverse reinforcement, respectively.

Specific values of γ may be found in Table 6-8 of FEMA 273.

As an alternative to the FEMA 273 approach, the following procedure used in bridge-joint evaluation (Priestley, 1996) may be helpful for correlating behavior modes and observed damage patterns.

The nominal principal stresses on a joint are used to assess whether the joint will crack. A stress analysis that employs Mohr's circle is used to determine the major principal tension stress

$$\sigma_t = \frac{\sigma_x + \sigma_y}{2} + \sqrt{v_j^2 + \left(\frac{\sigma_x + \sigma_y}{2}\right)^2} \qquad (8\text{-}44)$$

where σ_x, σ_y and v_j are the average bounding actions on the joint, as defined below.

σ_y is the average normal stress on the column given by:

$$\sigma_y = \frac{P_{col}}{b_c h_c} \qquad (8\text{-}45)$$

where

$P_{col} =$	column axial load (tension positive)
$b_c =$	column depth
$h_c =$	column width

σ_x is the average normal stress on the beam, given by:

$$\sigma_x = \frac{P_b}{d_b b_b} \qquad (8\text{-}46)$$

in which

$d_b =$	overall beam depth
$b_b =$	beam width
$P_b =$	axial force in the beam (if any)

v_j, the average joint shear stress is given by:

$$v_j = \frac{V_{jh}}{b_j h_c} \qquad (8\text{-}47)$$

in which

$V_{jh} =$	horizontal joint shear force assuming column and beam
$b_j =$	smaller of b_c or b_b

$$\text{If: } \sigma_t < 3.5\sqrt{f'_{ce}} \qquad (8\text{-}48)$$

then the joint may be assumed to remain elastic and uncracked.

$$\text{If: } \sigma_t > 5\sqrt{f'_{ce}} \qquad (8\text{-}49)$$

$$\text{or } \sigma_t > 7\sqrt{f'_{ce}} \qquad (8\text{-}50)$$

then full diagonal cracking may be expected for exterior joints or corner joints, respectively, under biaxial response.

If σ_t is between the above limits, then some partial cracking may be expected. Moreover, if hinging occurs in the beam adjacent to the connection, then yield penetration of the longitudinal bars into the joint occurs. With sequential cycling, this eventually leads to joint failure.

Similarly, the principal compression stress, σ_c, should be checked such that

$$\sigma_c = \frac{\sigma_x + \sigma_y}{2} - \sqrt{v_j^2 + \left(\frac{\sigma_x + \sigma_y}{2}\right)^2} \quad (8\text{-}51)$$

If, for one-way frames

$$|\sigma_c| > 0.5 f'_{ce} \quad (8\text{-}52)$$

or, if for two-way joints where biaxial loading may occur

$$|\sigma_c| > 0.45 f'_{ce} \quad (8\text{-}53)$$

then joint failure may be expected due to compression crushing of the diagonal struts within the joints.

Degradation of joint strength after crack formation may be expected. For assessing the degraded strength, the following rules may be used:

$$\gamma_j < 0.005 \quad \text{no change in } \sigma_t \quad (8\text{-}54)$$

$$\gamma_j < 0.02 \quad \sigma_t = 1.2\sqrt{f'_{ce}} \quad (8\text{-}55)$$

$$\gamma_j < 0.04 \quad \sigma_t = 0 \quad (8\text{-}56)$$

where

$\gamma_j =$ joint rotation angle (in radians)

For intermediate values of γ_j linear interpolation may be used to determine σ_t. With this value for σ_t, the joint shear strength may be determined from Equations 8-44 and 8-47 as follows:

$$V_{jh} = b_j h_c \sqrt{\left(\sigma_t - \frac{\sigma_x + \sigma_y}{2}\right) - \left(\frac{\sigma_x + \sigma_y}{2}\right)^2} \quad (8\text{-}57)$$

d. Bond Slip of Lap-Splice Connections

Lap-splice connections often occur at the base of a column, particularly in older nonductile concrete frames. Provided that the lap length is sufficient to develop yield (i.e., $20d_b$), the nominal ultimate strength

capacity can be attained. However, postelastic deformations quickly degrade the bond-strength capacity, and within one inelastic cycle of loading, the lap splice may be assumed to have failed. This failure is evident if longitudinal (tensile) splitting cracks are noticed at the base of the column.

When the lap splice fails in bond, it does not generally lead to a catastrophic failure, as the column is still able to transfer moment due to the presence of the eccentric compression stress block that arises as a result of the axial load in the column. Thus, the following initial and final failure model may be assumed:

$$M_i = M_n \text{ for } \theta_p = 0 \quad (8\text{-}58)$$

and

$$M_f = P\left(\frac{d - d'}{2}\right) \text{ for } \theta_p > 0.025 \quad (8\text{-}59)$$

where

$\theta_p =$ plastic rotation of the connection in radians

$P =$ axial load that takes into account the variation in force due to the truss action of infilled frames. Note that axial load will increase for a compression side column, whereas for a tension side column the axial load will decrease to a point that tension could be induced. For the latter case assume $P = 0$.

$(d - d') =$ distance between the outer layers of reinforcement in the column.

For intermediate values of plastic hinge rotation ($\theta_p > 0.025$) the following interpolation may be used

$$M_l = M_n - \frac{\theta_p}{0.025}\left(M_n - M_f\right) \quad (8\text{-}60)$$

At large rotations the concrete crushes and may be severely damaged.

8.4   Infilled Frame Component Guides

The following Component Damage Classification Guides contain details of the behavior modes for infilled-frame components. Included are the distinguishing characteristics of the specific behavior mode, the description of damage at various levels of severity, and performance restoration measures. Information may not be included in the Component Damage Classification Guides for certain damage severity levels; in these instances, for the behavior mode under consideration, it is not possible to make refined distinctions with regard to severity of damage. See also Section 3.5 for general discussion of the use of the Component Guides and Section 4.4.3 for information on the modeling and acceptability criteria for components.

INPS1	COMPONENT DAMAGE CLASSIFICATION GUIDE	System:	**Infilled Frame**
		Component Type:	**Infill Panel**
		Behavior Mode:	**Corner crushing**
		Applicable Materials:	**Concrete Frame-Block Infill**

How to Distinguish Damage Type:

By Observation:
This type of damage occurs because ungrouted concrete block infills are inherently weak compared to adjacent concrete columns. Lateral movements create high corner strains leading to early failure of the corner concrete blocks. Some diagonal cracking and/or concrete bed joint sliding is also evident.

By Analysis (See Section 8.3):
The plastic limit methods of Liauw and Kwan (1983) or the simplified truss method of Stafford-Smith (1966) can be used to analyze the hierarchy of strength mechanisms.

Severity	Description of Damage	Performance Restoration Measures
Insignificant $\lambda_K = 0.9$ $\lambda_Q = 0.9$ $\lambda_D = 1.0$	*Criteria:* Separation of mortar around perimeter of panel and some crushing or mortar near corners of infill panel *Typical Appearance:*.	• Repoint spalled mortar. • Inject cracks.
Moderate $\lambda_K = 0.6$ $\lambda_Q = 0.8$ $\lambda_D = 0.8$	*Criteria*: Crushing of mortar, cracking of blocks including lateral movement of face shells. *Typical Appearance:*	• Remove and replace damaged units. • Inject cracks around perimeter of infill. • Apply composite overlay at damaged corners.
Heavy $\lambda_K = 0.5$ $\lambda_Q = 0.7$ $\lambda_D = 0.7$	*Criteria*: Loss of corner blocks through complete spalling of face shells. Diagonal (stairstep) cracking and/or bed joint sliding may also be evident. *Typical Appearance:*	• Remove and replace infill or apply composite overlay on infill.

INPS2	COMPONENT DAMAGE CLASSIFICATION GUIDE	System:	**Infilled Frame**
		Component Type:	**Infill Panel**
		Behavior Mode:	**Diagonal Tension**
		Applicable Materials:	**Masonry Units**

How to distinguish damage type:

By Observation:
In this type of damage, cracking occurs across the diagonals of the infill panel. If drifts are large, secondary cracking may also be expected at an angle of about 45 degrees to 65 degrees to the horizontal. For large drifts the corner strains are intense and crushing may also be observed.

By Analysis (See Section 8.3):
It is possible to determine the diagonal cracking strength by rational mechanics, or by simplified strut methods such as that proposed by Stafford-Smith et al. (1969).

Severity	Description of Damage	Performance Restoration Measures
Insignificant $\lambda_K = 0.7$ $\lambda_Q = 0.9$ $\lambda_D = 1.0$	*Criteria:* Initial hairline cracking occur on diagonals in masonry. This is mostly associated with breaking of the bond between mortar and bricks. Cracking mostly concentrated within center region of panel *Typical appearance:.* 	Measures not necessary for structural performance restoration. (Certain measures may be necessary for the restoration of nonstructural characteristics).
Moderate $\lambda_K = 0.4$ $\lambda_Q = 0.8$ $\lambda_D = 0.9$	*Criteria:* Hairline cracks fully extend along diagonals following the mortar courses in a stairstep fashion, but sometimes propagate through bricks. Some crushing and/or "walking-out" of the mortar may be observed. Cracks mostly closed due to confinement provided by frames. *Typical appearance* 	Repoint spalled mortar. Remove and replace damaged masonry units.
Heavy $\lambda_K = 0.2$ $\lambda_Q = 0.5$ $\lambda_D = 0.8$	*Criteria:* Cracks widen to about 1/8", and are usually associated with corner crushing. Much loss of mortar is evident. More than one diagonal crack is generally evident. Crushing/cracking of the bricks is also evident. Portions of the entire infill may "walk" out-of-plane under cyclic loading. *Typical appearance:* 	Remove and replace damaged infill, or patch spalls. Apply shotcrete, ferrocement or composite overlay.

INPS3	COMPONENT DAMAGE CLASSIFICATION GUIDE	System:	**Infilled Frame**
		Component Type:	**Infill Panel**
		Behavior Mode:	**Bed joint sliding.**
		Applicable Materials:	**Steel-Frame Brick Infill**

How to Distinguish Damage Type:

By Observation:
In this type of damage, crushing may initially occur at the corner of the infill. As the displacements become large to accommodate racking movements in the panel, movements occur along bed joints in the form of diagonal (stairstep) cracking or horizontal bed joint sliding. Such movements are a secondary outcome to corner crushing.

By Analysis (See Section 8.3):
The effective strut method of analysis suggested by Stafford-Smith (1966) and/or plastic limit methods suggested by Liauw and Kwan (1983) should be used to check the hierarchy of failure mechanisms.

Severity	Description of Damage	Performance Restoration Measures
Insignificant $\lambda_K = 0.8$ $\lambda_Q = 0.9$ $\lambda_D = 0.9$	*Criteria:* Crushing of mortar around perimeter of frame. This is particularly noticeable adjacent to the columns near the corners of the infill panels. *Typical appearance:* 	Repoint spalled mortar. Inject cracks.
Moderate $\lambda_K = 0.5$ $\lambda_Q = 0.8$ $\lambda_D = 0.8$	*Criteria:* Crushing of mortar and cracking of bricks extend over larger zones adjacent to beam and column *Typical appearance:.* 	Remove damaged bricks and replace.
Heavy $\lambda_K = 0.4$ $\lambda_Q = 0.7$ $\lambda_D = 0.7$	*Criteria:* Significant crushing of mortar and bricks extends around most of the perimeter frame, particularly along the height of the column. *Typical appearance:* 	Remove and replace infill, or patch spalls. Apply shotcrete, ferrocement or composite overlay on infill.

INPS4	COMPONENT DAMAGE CLASSIFICATION GUIDE	System:	**Infilled Frame**
		Component Type:	**Infill Panel**
		Behavior Mode:	**Corner crushing and diagonal cracking**
		Applicable Materials:	**Concrete Frame-Block Infill**

How to Distinguish Damage Type:

By Observation:
Damage is equally distributed between both the frame and the infill. Crushing of the blocks and severe flexural cracking of infill occurs. Distributed diagonal cracks also occur.

By Analysis (See Section 8.3):
Limit-strength analysis methods are necessary to determine strength distribution well into the inelastic range.

Severity	Description of Damage	Performance Restoration Measures
Insignificant $\lambda_K = 0.9$ $\lambda_Q = 1.0$ $\lambda_D = 1.0$	*Criteria:* Separation of mortar around frame occurs first in beam-to-infill interface. Some hairline cracks may be evident along mortar courses.	Repoint spalled mortar. Inject cracks.
Moderate $\lambda_K = 0.6$ $\lambda_Q = 0.8$ $\lambda_D = 0.8$	*Criteria:* For a ductile (strong column-weak beam) frame design, yielding of longitudinal reinforcement occurs first in beam, with minor cracking in columns. Compression splitting occurs in corner blocks. Some hairline X cracks may be expected in beam-column joint *Typical appearance:.* 	Remove and replace damaged masonry units. Inject cracks.
Heavy $\lambda_K = 0.5$ $\lambda_Q = 0.6$ $\lambda_D = 0.6$	*Criteria:* Extensive cracking in beam and column hinge zones, leading to spalling of cover concrete in frame. Diagonal cracking passes through blocks. Faceshells spall off in corners, and also across a critical shear plane at mid-height of infill *Typical appearance:.* 	Remove and replace infill. Remove and patch spalled and loose concrete in frame. Inject cracks.

INPS5	COMPONENT DAMAGE CLASSIFICATION GUIDE	System: **Infilled Frame**
		Component Type: **Infill Panel**
		Behavior Mode: **Out-of-Plane**
		Applicable Materials: **Masonry Infill**

How to Distinguish Damage Type:

By Inspection:
This type of damage occurs with strong out-of-plane shaking. When coupled with in-plane shaking, the panel could potentially fall out. This behavior makes it difficult to distinguish which of the two types of shaking caused the damage.

By Analysis (See Section 8.3):
Arching action analysis is necessary.

Severity	Description of Damage	Performance Restoration Measures
Insignificant $\lambda_K = 0.9$ $\lambda_Q = 1.0$ $\lambda_D = 1.0$	*Criteria:* Flexural cracking in the mortar beds around the perimeter, with hairline cracking in mortar bed at mid-height of panel.	Repoint spalled mortar.
Moderate $\lambda_K = 0.9$ $\lambda_Q = 0.8$ $\lambda_D = 1.0$	*Criteria:* Crushing and loss of mortar along top, mid-height, bottom and side mortar beds. Possibly some in-plane damage, as evidenced by hair-line X-cracks in the central panel area. *Typical appearance:* 	Apply shotcrete, ferrocement, or composite overlay to the infill.
Heavy $\lambda_K = 0.5$ $\lambda_Q = 0.6$ $\lambda_D = 0.9$	*Criteria:* Severe corner-to-corner cracking with some out-of-plane dislodgment of masonry. Top, bottom and mid-height mortar bed is completely crushed and/or missing. There is some out-of-plane dislodgment of masonry. Concurrent in-plane damage should also be expected, as evidenced by extensive X-cracking. *Typical appearance:* 	Remove and replace infill.

INF1C1	COMPONENT DAMAGE CLASSIFICATION GUIDE	System:	Infilled Frame
		Component Type:	Concrete Column
		Behavior Mode:	Column Snap through Shear Failure
		Applicable Materials:	Concrete Frame Masonry Infill

How to Distinguish Damage Type:

By Observation:
If infills are stiff and/or strong, then the frame is the weaker component. Cracking is not across a corner-to-corner diagonal, but on a flatter angle. Column cracking over a length equal to two member widths is severe and a sign of low frame shear capacity.

By Analysis (See Section 8.3):
Column shear capacity should be checked. Shear failure is generally associated with inadequate transverse (shear/confinement) reinforcement.

Severity	Description of Damage	Performance Restoration Measures
Insignificant $\lambda_K = 0.9$ $\lambda_Q = 0.9$ $\lambda_D = 1.0$	*Criteria:* Several flexural cracks form in columns near top corner of infill. *Typical appearance:* 	Remove and patch spalled and loose concrete. Inject cracks.
Moderate $\lambda_K = 0.7$ $\lambda_Q = 0.7$ $\lambda_D = 0.4$	*Criteria:* Flexure cracks change into shear X-cracks over a short length near column end. (Generally over about two column widths). Column cover in this vicinity will be loose. Some associated crushing may appear in the infill. *Typical appearance:* 	Remove and patch spalled and loose concrete. Apply composite overlay to damaged region of column.
Heavy $\lambda_K = 0.4$ $\lambda_Q = 0.2$ $\lambda_D = 0.4$	Criteria: Cracking in column may be so severe that transverse hoops have fractured about one member width away from column end (at middle of X-cracks). Cover concrete in this vicinity will be mostly spalled away. *Typical appearance:* 	Remove spalled and loose concrete. Remove and replace buckled or fractured reinforcing. Provide additional ties over length of replaced bars. Patch concrete. Inject cracks. Apply composite overlay to damaged region of column.

INF1C2	COMPONENT DAMAGE CLASSIFICATION GUIDE	System:	**Infilled Frame**
		Component Type:	**Concrete Column**
		Behavior Mode:	**Lap Splice Failure**
		Applicable Materials:	**Reinforced Concrete**

How to Distinguish Damage Type:

By Inspection:
Lack of sufficient lap length in hinge zone leads to eventual slippage. The cover spalls off due to high compression stresses, exposing the core concrete and damaged lap splice zone.

By Analysis (See Section 8.3):
Refer to FEMA 307.

Severity	Description of Damage	Performance Restoration Measures
Insignificant $\lambda_K = 0.9$ $\lambda_Q = 1.0$ $\lambda_D = 1.0$	*Criteria:* Flexural cracks at floor level. Slight hairline vertical cracks. *Typical appearance* FLEXURAL CRACK	Inject cracks in frame.
Moderate $\lambda_K = 0.8$ $\lambda_Q = 0.5$ $\lambda_D = 1.0$	*Criteria:* Tensile flexural cracks at floor slab level with some evidence of toe crushing over the bottom 1/2". Longitudinal splitting cracks loosen the cover concrete. *Typical appearance:* LONGITUDINAL SPLITTING TOE CRUSHING	Inject cracks in frame.
Heavy $\lambda_K = 0.5$ $\lambda_Q = 0.5$ $\lambda_D = 1.0$	*Criteria:* Significant spalling of the cover concrete over the length of the lap splice, exposing the core and reinforcing steel *Typical appearance:.* COVER SPALLS OFF	Remove spalled and loose concrete. Provide additional ties over the length of the exposed bars. Patch concrete. Apply composite overlay to damaged region of column.

INF3C	COMPONENT DAMAGE CLASSIFICATION GUIDE	System:	**Infilled Frame**
		Component Type:	**Concrete Frame**
		Behavior Mode:	**Connection Damage**
		Applicable Materials:	**Reinforced Concrete**

How to Distinguish Damage Type:

By Inspection:
Distress is caused by overstrength of members framing into the connection, leading to very high principal tension stresses.

By Analysis (See Section 8.3):
Refer to FEMA 307.

Severity	Description of Damage	Performance Restoration Measures
Insignificant $\lambda_K = 0.9$ $\lambda_Q = 1.0$ $\lambda_D = 1.0$	*Criteria:* Slight X hairline cracks in joint *Typical appearance:.*	Inject cracks.
Moderate $\lambda_K = 0.8$ $\lambda_Q = 0.5$ $\lambda_D = 0.9$	*Criteria:* X-cracks in joint become more extensive and widen to about 1/8". *Typical appearance:*	Inject cracks.
Heavy $\lambda_K = 0.5$ $\lambda_Q = 0.5$ $\lambda_D = 0.5$	*Criteria:* Extensive X-cracks in joint widen to about 1/4". Exterior joints show cover concrete spalling off from back of joint. Some side cover may also spall off. *Typical appearance:* SPALLED COVER	Remove spalled and loose concrete. Remove and replace buckled or fractured reinforcing. Provide additional ties over the length of the replaced bars. Patch concrete. Inject cracks.

INF3S	COMPONENT DAMAGE CLASSIFICATION GUIDE	System:	**Infilled Frame**
		Component Type:	**Framed Connection**
		Behavior Type:	**Simple Connection Damage**
		Applicable Materials:	**Steel Frame-Masonry Infill**

How to Distinguish Damage Type:

By Observation:
Damage to simple (semi-rigid) steel connections occur due to the high shears that must be transferred in the inelastic range.

By Analysis (See Section 8.3):
Plastic limit analysis of connections (see Mander et al. (1994)) is necessary. Fatigue failure is unlikely, but should be checked.

Severity	Description of Damage	Performance Restoration Measures
Insignificant $\lambda_K = 0.9$ $\lambda_Q = 1.0$ $\lambda_D = 1.0$	*Criteria:* As the frame racks, the connection yields and paint may flake off.	Unnecessary for restoration of structural performance. (Certain measures may be necessary for restoration of nonstructural characteristics).
Moderate $\lambda_K = 0.9$ $\lambda_Q = 1.0$ $\lambda_D = 1.0$	*Criteria:* As drifts increase, both prying and slip may be evident. Angles pull away from column face, leaving infill frame bay with a larger overall opening. Gaps may be apparent around the perimeter of the infill.	Repaint and retorque bolts if loosened.
Heavy $\lambda_K = 0.5$ $\lambda_Q = 1.0$ $\lambda_D = 0.9$	*Criteria:* Angles may show fatigue cracks or failure. If this is the case, the infill will also be showing signs of significant distress. STEEL COLUMN GAP STEEL BEAM POTENTIAL CRACKS SPLITTING	Remove angles and replace both bolts and angles. Remove infill and replace.

Glossary

Bearing wall A concrete or masonry wall that supports a portion of the building weight, in addition to its own weight, without a surrounding frame.

Behavior mode The predominant type of damage observed for a particular component. This is dependent on the relative magnitudes of the ratios of applied loads to component strength for axial, flexural and shearing actions.

Collapse Prevention A performance level whereby a building is extensively damaged, has little residual stiffness and strength, but remains standing; any other damage is acceptable.

Component A structural member such as a beam, column, or wall, that is an individual part of a structural element.

Coupled wall A wall element in which vertical pier components are joined at one or more levels by horizontal spandrel components.

Damage Physical evidence of inelastic deformation of a structural component caused by a damaging earthquake.

Damaging ground motion The ground motion that shook the building under consideration and caused resulting damage. This ground motion may or may not have been recorded at the site of the building. In some cases, it may be an estimate of the actual ground motion that occurred. It might consist of estimated time-history records or corresponding response spectra.

Direct method The determination of performance restoration measures from the observed damage without relative performance analysis.

Element An assembly of structural components (e.g., coupled shear walls, frames).

Global displacement capacity The maximum global displacement tolerable for a specific performance level. This global displacement limit is normally controlled by the acceptability of distortion of individual components or a group of components within the structure.

Global performance displacement demand The overall displacement of a representative point on a building subject to a performance ground motion. The representative point is normally at the roof level or at the effective center of mass for a given mode of vibration.

Global structure The assembly representing all of the structural elements of a building.

Immediate Occupancy A performance level whereby a building sustains minimal or no damage to its structural elements and only minor damage to its nonstructural components.

Inelastic lateral mechanism The plastic mechanism formed in an element, or assembly of elements, under the combined action of vertical and lateral loads. This is a unique mechanism for a specified pattern of lateral loads.

Infilled frame A concrete or steel frame with concrete or masonry panels installed between the beams and columns.

Life Safety A performance level whereby a building may experience extensive damage to structural and nonstructural components, but remains stable and has significant reserve capacity.

Nonlinear static procedure — A structural analysis technique in which the structure is modeled as an assembly of components capable of nonlinear force-deformation behavior, then subjected to a monotonically increasing lateral load in a specific pattern to generate a global force-displacement capacity curve. The displacement demand is determined with a spectral representation of ground motion using one of several alternative methods.

Performance ground motion — Hypothetical ground motion consistent with the specified seismic hazard level associated with a specific performance objective. This is characterized by time history record(s) or corresponding response spectra.

Performance objective — A goal consisting of a specific performance level for a building subject to a specific seismic hazard.

Performance level — A hypothetical damage state for a building used to establish design seismic performance objectives. The most common performance levels, in order of decreasing amounts of damage, are Collapse Prevention, Life Safety, and Immediate Occupancy.

Performance restoration measures — Actions that might be implemented for a damaged building that result in future performance equivalent to that of the building in its pre-event state for a specific performance objective. These hypothetical repairs would result in a restored performance index equal to the performance index of the pre-event building.

Pier — A vertical wall component.

Pre-existing condition — Physical evidence of inelastic deformation or deterioration of a structural component that existed before the damaging earthquake

Relative performance analysis — An analysis of a building in its damaged and pre-event condition to determine the effects of the damage on the capability of the building to meet specific seismic performance objectives.

Repair — An action taken to address a damaged component of a building.

Severity of damage — The relative intensity of damage to a particular component classified as insignificant, slight, moderate, heavy, or extreme.

Shear wall — A concrete or masonry panel, connected to the adjacent floor system, that resists in-plane lateral loads.

Spandrel — A wall component that spans horizontally.

Structural repairs — Repairs that address damage to components to restore structural properties.

List of General Symbols

d_c Global displacement capacity for pre-event structure for specified performance level.

d'_c Global displacement capacity for damaged structure for specified performance level.

d^*_c Global displacement capacity for repaired structure for specified performance level.

d_d Global displacement demand for pre-event structure for specified seismic hazard.

d'_d Global displacement demand for damaged structure for specified seismic hazard.

d^*_d Global displacement demand for repaired structure for specified seismic hazard.

d_e Maximum global displacement caused by the damaging ground motion.

λ_D Modification factor applied to component deformation acceptability limits to account for earthquake damage.

λ_K Modification factor for idealized component force-deformation curve to account for change in effective initial stiffness resulting from earthquake damage.

λ_Q Modification factor for idealized component force-deformation curve to account for change in expected strength resulting from earthquake damage.

RD Absolute value of the residual deformation in a structural component resulting from earthquake damage.

Q_{CE} Expected strength of a component or element at the deformation level under consideration in a deformation-controlled element.

References

ABK, 1981a, *Methodology for Mitigation of Seismic Hazards in Existing Unreinforced Masonry Buildings: Categorization of Buildings,* A Joint Venture of Agbabian Associates, S.B. Barnes and Associates, and Kariotis and Associates (ABK), Topical Report 01, c/o Agbabian Associates, El Segundo, California.

ABK, 1981b, *Methodology for Mitigation of Seismic Hazards in Existing Unreinforced Masonry Buildings: Diaphragm Testing,* A Joint Venture of Agbabian Associates, S.B. Barnes and Associates, and Kariotis and Associates (ABK), Topical Report 03, c/o Agbabian Associates, El Segundo, California.

ABK, 1981c, *Methodology for Mitigation of Seismic Hazards in Existing Unreinforced Masonry Buildings: Wall Testing, Out-of-Plane,* A Joint Venture of Agbabian Associates, S.B. Barnes and Associates, and Kariotis and Associates (ABK), Topical Report 04, c/o Agbabian Associates, El Segundo, California

ABK, 1984, *Methodology for Mitigation of Seismic Hazards in Existing Unreinforced Masonry Buildings: The Methodology,* A Joint Venture of Agbabian Associates, S.B. Barnes and Associates, and Kariotis and Associates (ABK), Topical Report 08, c/o Agbabian Associates, El Segundo, California.

Abrams, D.P., 1992, "Strength and Behavior of Unreinforced Masonry Elements," *Proceedings of Tenth World Conference on Earthquake Engineering,* Balkema Publishers, Rotterdam, The Netherlands, pp. 3475-3480.

Abrams, D.P. (Ed.), 1994, *Proceedings of the NCEER Workshop on Seismic Response of Masonry Infills,* National Center for Earthquake Engineering Research, Technical Report NCEER-94-0004.

Abrams, D.P., 1997, personal communication, May.

Abrams, D.P., and Epperson, G.S, 1989, "Evaluation of Shear Strength of Unreinforced Brick Walls Based on Nondestructive Measurements," *5th Canadian Masonry Symposium,* Department of Civil Engineering, University of British Columbia, Vancouver, British Columbia., Volume 2.

Abrams, D.P., and Epperson, G.S, 1989, "Testing of Brick Masonry Piers at Seventy Years", *The Life of Structures: Physical Testing,* Butterworths, London.

Abrams, D.P., and Paulson, T.J., 1989, "Measured Nonlinear Dynamic Response of Reinforced Concrete Masonry Building Systems," *Proceedings of the Fifth Canadian Masonry Symposium,* University of British Columbia, Vancouver, B.C., Canada.

Abrams, D.P., and Paulson, T.J., 1990, "Perceptions and Observations of Seismic Response for Reinforced Masonry Building Structures," *Proceedings of the Fifth North American Masonry Conference,* University of Illinois at Urbana-Champaign.

Abrams, D.P., and Shah, N., 1992, *Cyclic Load Testing of Unreinforced Masonry Walls,* College of Engineering, University of Illinois at Urbana, Advanced Construction Technology Center Report #92-26-10.

ACI, 1995, *Building Code Requirements for Reinforced Concrete,* American Concrete Institute, Report ACI 318-95, Detroit, Michigan.

ACI Committee 224, 1994, "Causes, Evaluation and Repair of Cracks in Concrete Structures", *Manual of Concrete Practice,* American Concrete Institute, Detroit, Michigan.

Adham, S.A., 1985, "Static and Dynamic Out-of-Plane Response of Brick Masonry Walls," *Proceedings of the 7th International Brick Masonry Conference,* Melbourne, Australia.

Agbabian, M., Adham, S, Masri, S.,and Avanessian, V., 1989, *Out-of-Plane Dynamic Testing of Concrete Masonry Walls,* U.S. Coordinated Program for Masonry Building Research, Report Nos. 3.2b-1 and 3.2b-2.

Al-Chaar, G., Angel, R. and Abrams, D., 1994, "Dynamic testing of unreinforced brick masonry infills," *Proc. of the Structures Congress '94,* Atlanta, Georgia., ASCE, 1: 791-796.

Alexander, C.M., Heidebrecht, A.C., and Tso, W.K., 1973, *Cyclic Load Tests on Shear Wall Panels,* Proceedings, Fifth World Conference on Earthquake Engineering, Rome, pp. 1116–1119.

Ali, Aejaz and Wight, J.K., 1991, *R/C Structural Walls with Staggered Door Openings,* Journal of Structural Engineering, ASCE, Vol. 117, No. 5, pp. 1514-1531.

Anderson, D.L., and Priestley, M.J.N., 1992, "In Plane Shear Strength of Masonry Walls," *Proceedings of the 6th Canadian Masonry Symposium*, Saskatoon, Saskatchewan.

Angel R., and Abrams, D.P., 1994, "Out-Of-Plane Strength evaluation of URM infill Panels," *Proceedings of the NCEER Workshop on Seismic Response of Masonry Infills,* D.P. Abrams editor, NCEER Technical Report NCEER-94-0004.

Antebi, J., Utku, S., and Hansen, R.J., 1960, *The Response of Shear Walls to Dynamic Loads*, MIT Department of Civil and Sanitary Engineering, Report DASA-1160, Cambridge, Massachusetts.

Anthoine, A., Magonette, G., and Magenes, G., 1995, "Shear-Compression Testing and Analysis of Brick Masonry Walls," *Proceedings of the 10th European Conference on Earthquake Engineering*, Duma, editor, Balkema: Rotterdam, The Netherlands.

Aristizabal-Ochoa, D., J., and Sozen, M.A., 1976, *Behavior of Ten-Story Reinforced Concrete Walls Subjected to Earthquake Motions*, Civil Engineering Studies Structural Research Series No. 431, Report UILU-ENG-76-2017, University of Illinois, Urbana, Illinois.

ASCE/ACI Task Committee 426, 1973, *The Shear Strength of Reinforced Concrete Members*, ASCE Journal of Structural Engineering, Vol. 99, No. ST6, pp 1091-1187.

ASTM, latest editions, Standard having the number C-67, American Society for Testing Materials Philadelphia, Pennsylvania.

ATC, 1983, *Seismic Resistance of Reinforced Concrete Shear Walls and Frame Joints: Implication of Recent Research for Design Engineers*, Applied Technology Council, ATC-11 Report, Redwood City, California.

ATC, 1989, *Procedures for the Post Earthquake Safety Evaluation of Buildings*, Applied Technology Council, ATC-20 Report, Redwood City, California.

ATC, 1992, *Guidelines for Cyclic Seismic Testing of Components of Steel Structures*, Applied Technology Council, ATC-24 Report, Redwood City, California.

ATC, 1995, *Addendum to the ATC-20 Post Earthquake Building Safety Evaluation Procedures*, Applied Technology Council, ATC-20-2 Report, Redwood City, California.

ATC, 1996, *The Seismic Evaluation and Retrofit of Concrete Buildings*, Applied Technology Council, ATC-40 Report, Redwood City, California.

ATC, 1997a, *NEHRP Guidelines for the Seismic Rehabilitation of Buildings*, prepared by the Applied Technology Council (ATC-33 project) for the Building Seismic Safety Council, published by the Federal Emergency Management Agency, Report No. FEMA 273, Washington, D.C.

ATC, 1997b, NEHRP *Commentary on the Guidelines for the Seismic Rehabilitation of Buildings*, prepared by the Applied Technology Council (ATC-33 project) for the Building Seismic Safety Council, published by the Federal Emergency Management Agency, Report No. FEMA 274, Washington, D.C.

ATC, 1998a, *Evaluation of Earthquake Damaged Concrete and Masonry Wall Buildings, Technical Resources*, prepared by the Applied Technology Council (ATC-43 project) for the Partnership for Response and Recovery, published by the Federal Emergency Management Agency, Report No. FEMA 307, Washington D.C.

ATC, 1998b, *Repair of Earthquake Damaged Concrete and Masonry Wall Buildings*, prepared by the Applied Technology Council (ATC-43 project) for the Partnership for Response and Recovery, published by the Federal Emergency Management Agency, Report No. FEMA 308, Washington D.C.

Atkinson, R.H.,Amadei, B.P.,Saeb, S., and Sture, S., 1989, "Response of Masonry Bed Joints in Direct Shear," *American Society of Civil Engineers Journal of the Structural Division*, Vol. 115, No. 9.

Atkinson, R.H., and Kingsley, G.R., 1985, *A Comparison of the Behavior of Clay and Concrete Masonry in Compression*, U.S. Coordinated Program for Masonry Building Research, Report No. 1.1-1.

Atkinson, R.H., Kingsley, G.R., Saeb, S., B. Amadei, B., and Sture, S., 1988, "A Laboratory and In-situ Study of the Shear Strength of Masonry Bed Joints," *Proceedings of the 8th International Brick/Block Masonry Conference*, Dublin.

Axely, J.W. and Bertero, V.V., 1979, "Infill panels: their influence on seismic response of buildings," Earthquake Eng. Research Center, University of California at Berkeley, Report No. EERC 79-28.

Aycardi, L.E., Mander, J.B., and Reinhorn, A.M., 1992, *Seismic Resistance of RC Frame Structures*

Designed only for Gravity Loads, Part II: Experimental Performance of Subassemblages, National Center for Earthquake Engineering Research, Technical Report NCEER-92-0028.

Aycardi, L.E., Mander, J.B., and Reinhorn, A.M., 1994, "Seismic resistance of reinforced concrete frame structures designed only for gravity loads: Experimental performance of subassemblages", *ACI Structural Journal*, (91)5: 552-563.

Azizinamini, Atorod, Glikin, J.D., and Oesterle, R.G., 1994, *Tilt-up Wall Test Results*, Portland Cement Association, RP322D, Skokie, Illinois.

Barda, Felix, 1972, *Shear Strength of Low-Rise Walls with Boundary Elements*, Ph.D. University, Lehigh University, Bethlehem, Pennsylvania.

Barda, Felix, Hanson, J.W., and Corley, W.G., 1976, *Shear Strength of Low-Rise Walls with Boundary Elements*, Research and Development Bulletin RD043.01D, preprinted with permission from ACI Symposium Reinforced Concrete Structures in Seismic Zones, American Concrete Institute.

Benjamin, J.R., and Williams, H.A., 1958, "The behavior of one-story shear walls," *Proc. ASCE*, ST. 4, Paper 1723: 30.

Benjamin, J.R., and Williams, H.A., 1958a, "The Behavior of One-Story Reinforced Concrete Shear Walls," *Journal of the Structural Division*, ASCE, Vol. 83, No. ST3.

Benjamin, J.R., and Williams, H.A., 1958b, Behavior of One-Story Reinforced Concrete Shear Walls Containing Openings, *ACI Journal*, Vol. 55, pp. 605-618.

Bertero, V.V. and Brokken, S.T., 1983, "Infills in seismic resistant building," *Proc. ASCE*, 109(6).

BIA, 1988, *Technical Notes on Brick Construction, No. 17*, Brick Institute of America, Reston, Virginia.

Blakeley, R.W.G., Cooney, R.C., and Megget, L.M., 1975, "Seismic Shear Loading at Flexural Capacity in Cantilever Wall Structures," *Bulletin of the New Zealand National Society for Earthquake Engineering*, Vol. 8, No. 4.

Blondet, M., and Mays, R.L., 1991, *The Transverse Response of Clay Masonry Walls Subjected to Strong Motion Earthquakes*, Vols. 1 through 4, U.S. Coordinated Program for Masonry Building Research, Report No. 3.2(b)-2.

Bracci, J.M., Reinhorn, A.M., and Mander, J.B., 1995, "Seismic resistance of reinforced concrete frame structures designed for gravity loads: Performance of structural system", *ACI Structural Journal*, (92)5: 597-609.

Brokken, S.T., and Bertero, V.V., 1981, "Studies on effects of infills in seismic resistance R/C construction,"Earthquake Engineering Research Centre, University of California at Berkeley, Report No. EERC 81-12.

BSSC, 1992, *NEHRP Handbook for the Seismic Evaluation of Existing Buildings*, prepared by the Building Seismic Safety Council for the Federal Emergency Management Agency, Report No. FEMA 178, Washington, D.C.

BSSC, 1994, *NEHRP Recommended Provisions for Seismic Regulations for New Buildings,* prepared by the Building Seismic Safety Council for the Federal Emergency Management Agency, Report No. FEMA 222A, Washington, D.C.

BSSC, 1997, *NEHRP Recommended Provisions for Seismic Regulations for New Buildings and Other Structures*, prepared by the Building Seismic Safety Council for the Federal Emergency Management Agency, Report No. FEMA 302, Composite Draft Copy, Washington, D.C.

Calvi, G.M., Kingsley, G.R., and Magenes, G., 1996, "Testing of Masonry Structures for Seismic Assessment," *Earthquake Spectra*, Earthquake Engineering Research Institute, Oakland, California, Vol. 12, No. 1, pp. 145-162.

Calvi, G.M., Macchi, G., and Zanon, P., 1985, "Random Cyclic Behavior of Reinforced Masonry Under Shear Action," *Proceedings of the Seventh International Brick Masonry Conference*, Melbourne, Australia.

Calvi, G.M, and Magenes, G., 1994, "Experimental Results on Unreinforced Masonry Shear Walls Damaged and Repaired," *Proceedings of the 10th International Brick Masonry Conference*, Vol. 2, pp. 509-518.

Calvi, G.M., Magenes, G., Pavese, A., and Abrams, D.P., 1994, "Large Scale Seismic Testing of an Unreinforced Brick Masonry Building," *Proceedings of Fifth U.S. National Conference on Earthquake Engineering*, Chicago, Illinois, July 1994, pp.127-136.

Cardenas, Alex E., 1973, "Shear Walls -- Research and Design Practice," *Proceedings, Fifth World Conference on Earthquake Engineering*, Rome, Italy.

Chang, G. A., and Mander, J.B.,1994, *Seismic Energy Based Fatigue Damage Analysis of Bridge Columns: Part I - Evaluation of Seismic Capacity*, Technical Report NCEER-94-0006, and *Part II - Evaluation of Seismic Demand*, Technical Report NCEER-94-0013, National Center for Earthquake Engineering Research, State University of New York at Buffalo.

Chen, S.J., Hidalgo, P.A., Mayes, R.L., and Clough, R.W., 1978, *Cyclic Loading Tests of Masonry Single Piers, Volume 2 – Height to Width Ratio of 1*, Earthquake Engineering Research Center Report No. UCB/EERC-78/28, University of California, Berkeley, California.

Chrysostomou, C.Z., Gergely, P, and Abel, J.F., 1988, *Preliminary Studies of the Effect of Degrading Infill Walls on the Nonlinear Seismic Response of Steel Frames*, National Center for Earthquake Engineering Research, Technical Report NCEER-88-0046.

City of Los Angeles, 1985, "Division 88: Earthquake Hazard Reduction in Unreinforced Masonry Buildings," *City of Los Angeles Building Code*, Los Angeles, California.

City of Los Angeles, 1991, *Seismic Reinforcement Seminar Notes*, Los Angeles Department of Building and Safety.

Corley, W.G., Fiorato, A.E., and Oesterle, R.G., 1981, *Structural Walls*, American Concrete Institute, Publication SP-72, Detroit, Michigan, pp. 77-131.

Costley, A.C., Abrams, D.P., and Calvi, G.M., 1994, "Shaking Table Testing of an Unreinforced Brick Masonry Building," *Proceedings of Fifth U.S. National Conference on Earthquake Engineering*, Chicago, Illinois, July 1994, pp.127-136.

Costley, A.C., and Abrams, D.P, 1996a, "Response of Building Systems with Rocking Piers and Flexible Diaphragms," *Worldwide Advances in Structural Concrete and Masonry, Proceedings of the Committee on Concrete and Masonry Symposium*, ASCE Structures Congress XIV, Chicago, Illinois.

Costley, A.C., and Abrams, D.P., 1996b, *Dynamic Response of Unreinforced Masonry Buildings with Flexible Diaphragms*, National Center for Earthquake Engineering Research, Technical Report NCEER-96-0001, Buffalo, New York.

Coul, A., 1966, "The influence of concrete infilling on the strength and stiffness of steel frames," *Indian Concrete Journal*.

Crisafulli, F.J., Carr, A.J., and Park, R., 1995, "Shear Strength of Unreinforced Masonry Panels," *Proceedings of the Pacific Conference on Earthquake Engineering*; Melbourne, Australia, Parkville, Victoria, 3: 77-86.

CRSI, No publication date given, *Evaluation of Reinforcing Steel in Old Reinforced Concrete Structures*, Concrete Reinforcing Steel Institute, Engineering Data Report No. 11, Chicago, Illinois.

Dawe, J.L. and McBride, R.T., 1985a, "Experimental investigation of the shear resistance of masonry panels in steel frames," *Proceedings of the 7th Brick Masonry Conf.*, Melbourne, Australia.

Dawe, J.L. and Young, T.C., 1985b, "An investigation of factors influencing the behavior of masonry infill in steel frames subjected to in-plane shear," *Proceedings of the 7th International. Brick Masonry Conference*, Melbourne, Australia.

Dawe, J.L., and Seah, C.K., 1988, "Lateral Load Resistance of Masonry Panels in Flexible Steel Frames," *Proceedings of the Eighth International Brick and Block Masonry Conference*, Trinity College, Dublin, Republic of Ireland.

Dhanasekar, K., Page, A.W., and Kleeman, P.W., 1985, "The behavior of brick masonry under biaxial stress with particular reference to infilled frames," *Proceedings of the 7th International. Brick Masonry Conference*, Melbourne, Australia.

Drysdale, R.G., Hamid, A.A., and Baker, L.R., 1994, *Masonry Structures, Behavior and Design*, Prentice Hall, New Jersey.

Durrani, A.J., and Luo, Y.H., 1994, "Seismic Retrofit of Flat-Slab Buildings with Masonry Infills," in *Proceedings of the NCEER Workshop on Seismic Response of Masonry Infills*, D.P. Abrams editor, National Center for Earthquake Engineering Research, Technical Report NCEER-94-0004.

El-Bahy, A., Kunath, S.K., Taylor, A.W., and Stone, W.C., 1997, *Cumulative Seismic Damage of Reinforced Concrete Bridge Piers*, Building and Fire Research Laboratory, National Institute of Standards and Technology (NIST), Draft Report, Gaithersburg, Maryland.

Epperson, G.S., and Abrams, D.P., 1989, *Nondestructive Evaluation of Masonry Buildings,* College of

Engineering, University of Illinois at Urbana, Advanced Construction Technology Center Report No. 89-26-03, Illinois.

Epperson, G.S., and Abrams, D.P., 1992, "Evaluating Lateral Strength of Existing Unreinforced Brick Masonry Piers in the Laboratory," *Journal of The Masonry Society*, Boulder, Colorado, Vol. 10, No. 2, pp. 86-93.

Fattal, S.G., 1993, *Strength of Partially-Grouted Masonry Shear Walls Under Lateral Loads*, National Institute of Standards and Technology, NISTIR 5147, Gaithersburg, Maryland.

Feng, Jianguo, 1986, "The Seismic Shear Strength of Masonry Wall," *Proceedings of the US-PRC Joint Workshop on Seismic Resistance of Masonry Structures*, State Seismological Bureau, PRC and National Science Foundation, USA, Harbin, China.

Flanagan, R.D. and Bennett, R.M., 1994, "Uniform Lateral Load Capacity of Infilled Frames," *Proceedings of the Structures Congress '94*, Atlanta, Georgia, ASCE, 1: 785-790.

Focardi, F. and Manzini, E., 1984, "Diagonal tension tests on reinforced and non-reinforced brick panels," *Proceedings of the 8th World Conference on Earthquake Engineering*, San Francisco, California, VI: 839-846.

Foltz, S., and Yancy, C.W.C., 1993, "The Influence of Horizontal Reinforcement on the Shear Performance of Concrete Masonry Walls", *Masonry: Design and Construction, Problems and Repair*, ASTM STP 1180, American Society for Testing and Materials, Philadelphia, Pennsylvania.

Freeman S. A., 1994 "The Oakland Experience During Loma Prieta - Case Histories," *Proceedings of the NCEER Workshop on Seismic Response of Masonry Infills*, D.P. Abrams editor, National Center for Earthquake Engineering Research, Technical Report NCEER-94-0004.

Gambarotta, L. and Lagomarsino, S., 1996, "A Finite Element Model for the Evaluation and Rehabilitation of Brick Masonry Shear Walls," *Worldwide Advances in Structural Concrete and Masonry, Proceedings of the Committee on Concrete and Masonry Symposium*, ASCE Structures Congress XIV, Chicago, Illinois.

Gergely, P., White, R.N., and Mosalam, K.M., 1994, "Evaluation and Modeling of Infilled Frames," *Proceedings of the NCEER Workshop on Seismic Response of Masonry Infills*, D.P. Abrams editor, National Center for Earthquake Engineering Research, Technical Report NCEER-94-0004, pp. 1-51 to 1-56.

Gergely, P., White, R.N., Zawilinski, D., and Mosalam, K., 1993, "The Interaction of Masonry Infill and Steel or Concrete Frames," *Proceedings of the 1993 National Earthquake Conference, Earthquake Hazard Reduction in the Central and Eastern United States: A Time for Examination and Action*; Memphis, Tennessee, II: 183-191.

Ghanem, G.M., Elmagd, S.A., Salama, A.E., and Hamid, A.A., 1993, "Effect of Axial Compression on the Behavior of Partially Reinforced Masonry Shear Walls," *Proceedings of the Sixth North American Masonry Conference*, Philadelphia, Pennsylvania.

Gulkan, P., and Sozen, M.A., 1974, "Inelastic Response of Reinforced Concrete Structures to Earthquake Motions," *Journal of the American Concrete Institute*, Detroit, Michigan, pp. 604-610.

Hamburger, R.O. and Chakradeo, A.S., 1993, "Methodology for seismic capacity evaluation of steel-frame buildings with infill unreinforced masonry," *Proceedings of the 1993 National Earthquake Conference, Earthquake Hazard Reduction in the Central and Eastern United States: A Time for Examination and Action*; Memphis, Tennessee, II: 173-182.

Hamid, A., Assis, G., and Harris, H., 1988, *Material Models for Grouted Block Masonry*, U.S. Coordinated Program for Masonry Building Research, Report No. 1.2a-1.

Hamid, A., Abboud, B., Farah, M., Hatem, K., and Harris, H., 1989, *Response of Reinforced Block Masonry Walls to Out-of-Plane Static Loads*, U.S. Coordinated Program for Masonry Building Research, Report No. 3.2a-1.

Hammons, M.I., Atkinson, R.H., Schuller, M.P., and Tikalsky, P.J., 1994, *Masonry Research for Limit-States Design, Construction Productivity Advancement Research (CPAR)* U.S. Army Corps of Engineers Waterways Experiment Station, Program Report CPAR-SL-94-1, Vicksburg Mississippi.

Hanson, R.D., 1996, "The Evaluation of Reinforced Concrete Members Damaged by Earthquakes", *Spectra*, Vol. 12, No. 3, Earthquake Engineering Research Institute.

Hart, G.C., and Priestley, M.J.N., 1989, *Design Recommendations for Masonry Moment-Resisting Wall Frames*, UC San Diego Structural Systems Research Project Report No. 89/02.

Hart, G.C., Priestley, M.J.N., and Seible, F., 1992, "Masonry Wall Frame Design and Performance," *The Structural Design of Tall Buildings*, John Wile & Sons, New York.

He, L., and Priestley, M.J.N., 1992, *Seismic Behavior of Flanged Masonry Walls*, University of California, San Diego, Department of Applied Mechanics and Engineering Sciences, Report No. SSRP-92/09.

Hegemier, G.A., Arya, S.K., Nunn, R.O., Miller, M.E., Anvar, A., and Krishnamoorthy, G., 1978, *A Major Study of Concrete Masonry Under Seismic-Type Loading*, University of California, San Diego Report No. AMES-NSF TR-77-002.

Hidalgo, P.A., Mayes, R.L., McNiven, H.D., and Clough, R.W., 1978, *Cyclic Loading Tests of Masonry Single Piers, Volume 1 – Height to Width Ratio of 2*, University of California, Berkeley, Earthquake Engineering Research Center Report No. UCB/EERC-78/27.

Hidalgo, P.A., Mayes, R.L., McNiven, H.D., and Clough, R.W., 1979, *Cyclic Loading Tests of Masonry Single Piers, Volume 3 – Height to Width Ratio of 0.5*, University of California, Berkeley, Earthquake Engineering Research Center Report No. UCB/EERC-79/12.

Hill, James A., 1994, "Lateral Response of Unreinforced Masonry Infill Frame," *Proceedings of the Eleventh Conference* (formerly Electronic Computation Conference) held in conjunction with ASCE Structures Congress '94 and International Symposium '94, Atlanta, Georgia, pp. 77-83.

Hill, James A, 1997, oral communicaton.

Holmes, M., 1961, "Steel frames with brickwork and concrete infilling," *Proceedings of the Institute of Civil Engineers*, 19: 473.

Hon, C.Y., and Priestley, M.J.N., 1984, *Masonry Walls and Wall Frames Under Seismic Loading*, Department of Civil Engineering, University of Canterbury, Research Report 84-15, New Zealand.

ICBO, 1994, *Uniform Code for Building Conservation*, International Conference of Building Officials, Whittier, California.

ICBO, 1994, *UBC Standard 21-6*, International Conference of Building Officials, Whittier, California.

ICBO, 1994, 1997, *Uniform Building Code,* International Conference of Building Officials, Whittier, California.

Igarashi, A., F. Seible, and Hegemier, G.A., 1993, *Development of the Generated Sequential Displacement Procedure and the Simulated Seismic Testing of the TCCMAR 3-Story In-plane Walls*, U.S. Coordinated Program for Masonry Building Research Report No. 3.1(b)-2.

Iliya, R., and Bertero, V.V., 1980, *Effects of Amount and Arrangement of Wall-Panel Reinforcement on Behavior of Reinforced Concrete Walls*, Earthquake Engineering Research Center, University of California at Berkeley, Report UCB-EERC-82-04.

Innamorato, D. , 1994, *The Repair of Reinforced Structural Masonry Walls using a Carbon Fiber, Polymeric Matrix Composite Overlay*, M.Sc. Thesis in Structural Engineering, University of California, San Diego.

Kadir, M.R.A., 1974, *The structural behavior of masonry infill panels in framed structures*, Ph.D. thesis, University of Edinburgh.

Kahn, L.F. and Hanson, R.D., 1977, "Reinforced concrete shear walls for a seismic strengthening," *Proceedings of the 6th World Conf. on Earthquake Engineering*, New Delhi, India, III: 2499-2504.

Kariotis, J.C., 1986, "Rule of General Application - Basic Theory," *Earthquake Hazard Mitigation of Unreinforced Pre-1933 Masonry Buildings*, Structural Engineers Association of Southern California, Los Angeles, California.

Kariotis, J.C., Ewing, R.D., and Hill, J.,1996, Lecture Notes, Retrofit of Existing Tilt-Up Buildings and In-fill Frame Seminar, Structural Engineers Association of Southern California, Los Angeles, California.

Kariotis, J.C., Ewing, R.D., and Johnson, A.W., 1985a, "Predictions of Stability for Unreinforced Brick Masonry Walls Shaken by Earthquakes," *Proceedings of the 7th International Brick Masonry Conference*, Melbourne, Australia.

Kariotis, J.C., Ewing, R.D., and Johnson, A.W., 1985b, "Strength Determination and Shear Failure Modes of Unreinforced Brick Masonry with Low Strength

Mortar," *Proceedings of the 7th International Brick Masonry Conference*, Melbourne, Australia.

Kariotis, J.C., Ewing, R.D., and Johnson, A.W., 1985c, "Methodology for Mitigation of Earthquake Hazards in Unreinforced Brick Masonry Buildings," *Proceedings of the 7th International Brick Masonry Conference*, Melbourne, Australia.

Kariotis, J.C., Rahman, M.A., and El-mustapha, A.M., 1993, "Investigation of Current Seismic Design Provisions for Reinforced Masonry Shear Walls," *The Structural Design of Tall Buildings*, Vol. 2, 163-191.

Kariotis, J.C., Hart, G., Youssef, N., Goo, J., Hill, J., and Ngelm, D., 1994, *Simulation of Recorded Response of Unreinforced Masonry (URM) Infill Frame Building*, California Division of Mines and Geology, SMIP 94-05, Sacramento, California.

Kawashima, K., and Koyama, T., 1988, "Effect of Number of Loading Cycles on Dynamic Characteristics of Reinforced Concrete Bridge Pier Columns," *Proceedings of the Japan Society of Civil Engineers, Structural Eng./Earthquake Eng.*, Vol. 5, No. 1.

Kingsley, G.R., 1994, *Seismic Design and Response of a Full-scale Five-Story Reinforced Masonry Research Building*, Doctoral Dissertation, University of California, San Diego, School of Structural Engineering.

Kingsley, G.R., 1995, "Evaluation and Retrofit of Unreinforced Masonry Buildings," *Proceedings of the Third National Concrete and Masonry Engineering Conference*, San Francisco, California, pp. 709-728.

Kingsley, G.R., Magenes, G., and Calvi, G.M., 1996, "Measured Seismic Behavior of a Two-Story Masonry Building," *Worldwide Advances in Structural Concrete and Masonry, Proceedings of the Committee on Concrete and Masonry Symposium*, ASCE Structures Congress XIV, Chicago, Illinois.

Kingsley, G.R., Noland, J.L., and Atkinson, R.H., 1987, "Nondestructive Evaluation of Masonry Structures Using Sonic and Ultrasonic Pulse Velocity Techniques," *Proceedings of the Fourth North American Masonry Conference*, Univ. of Calif. Los Angeles.

Kingsley, G.R., Seible, F., Priestley, M.J.N., and Hegemier, G.A., 1994, *The U.S.-TCCMAR Full-scale Five-Story Masonry Research Building Test, Part II: Design, Construction, and Test Results*, Struc-

tural Systems Research Project, Report No. SSRP-94/02, University of California at San Diego.

Klingner, R.E. and Bertero, V.V., 1976, *Infilled frames in earthquake resistant construction*, Earthquake Engineering Research Centre, University of California at Berkeley, Report No. EERC 76-32.

Klingner, R.E. and Bertero, V.V., 1978, "Earthquake resistance of infilled frames," *Journal of the Structural Division, Proc. ASCE*, (104)6.

Kodur, V.K.R., Erki, M.A., and Quenneville, J.H.P., 1995, "Seismic design and analysis of masonry-infilled frames," *Journal of Civil Engineering*, 22: 576-587.

Krawinkler, H., 1996, "Pushover Analysis: Why, How, When, and When not to Use It", *Proceedings of the 65th Annual Convention of the Structural Engineers Association of California*, Maui.

Kubota, T., and Murakami, M., 1988, "Flexural Failure Test of Reinforced Concrete Masonry Walls -- Effect of Lap Joint of Reinforcement," *Proceedings of the Fourth Meeting of the U.S.-Japan Joint Technical Coordinating Committee on Masonry Research*, San Diego, California.

Kubota, T., Okamoto, S., Nishitani, Y., and Kato, S., 1985, "A Study on Structural Behavior of Brick Panel Walls," *Proceedings of the Seventh International Brick Masonry Conference*, Melbourne, Australia.

Kürkchübasche, A.G., Seible, F., and Kingsley, G.R., 1994, *The U.S.-TCCMAR Five-story Full-scale Reinforced Masonry Research Building Test: Part IV, Analytical Models*, Structural Systems Research Project, Report No. TR - 94/04, University of California at San Diego.

Laursen, P. T., Seible, F., and Hegemier, G. A. , 1995, *Seismic Retrofit and Repair of Reinforced Concrete with Carbon Overlays,* Structural Systems Research Project, Report No. TR - 95/01, University of California at San Diego.

Leiva, G., and Klingner, R.E., 1991, *In-Plane Seismic Resistance of Two-story Concrete Masonry Shear Walls with Openings,* U.S. Coordinated Program for Masonry Building Research, Report No. 3.1(c)-2.

Leiva, G., and Klingner, R.E., 1994, "Behavior and Design of Multi-Story Masonry Walls under In-plane Seismic Loading," *The Masonry Society Journal*, Vol. 13, No. 1.

Liauw, T.C. and Lee, S.W., 1977, "On the behavior and the analysis of multi-story infilled frames subjected to lateral loading," *Proceedings of the Institute of Civil Engineers*, 63: 641-656.

Liauw, T.C., 1979, "Tests on multi-storey infilled frames subject to dynamic lateral loading," *ACI Journal*, (76)4: 551-563.

Liauw, T.C. and Kwan, K.H., 1983a, "Plastic theory of infilled frames with finite interface shear strength," *Proceedings of the Institute of Civil Engineers*, (2)75: 707-723.

Liauw, T.C. and Kwan, K.H., 1983b, "Plastic theory of non-integral infilled frames," *Proceedings of the Institute of Civil Engineers*, (2)75: 379-396.

Lybas, J.M. and Sozen, M.A., 1977, *Effect of Beam Strength and Stiffness on Dynamic Behavior of Reinforce Concrete Coupled Walls*, University of Illinois Civil Engineering Studies, Structural Research Series No. 444, Report UILU-ENG-77-2016 (two volumes), Urbana, Illinois.

Magenes, G., 1997, oral communication, May.

Magenes, G., and Calvi, G.M., 1992, "Cyclic Behavior of Brick Masonry Walls," *Proceedings of the Tenth World Conference*, Balkema: Rotterdam, The Netherlands.

Magenes, G. and Calvi, G.M., 1995, "Shaking Table Tests on Brick Masonry Walls," *10th European Conference on Earthquake Engineering*, Duma, editor, Balkema: Rotterdam, The Netherlands.

Maghaddam, H.A. and Dowling, P.J., 1987, *The State of the Art in Infilled Frames*, Civil Engineering Department, Imperial College, ESEE Research Report No. 87-2, London.

Maier, Johannes, 1991, "Shear Wall Tests," *Preliminary Proceedings, International Workshop on Concrete Shear in Earthquake*, Houston Texas.

Mainstone, R.J. and Weeks, G.A., 1970, "The influence of bounding frame on the racking stiffness and strength of brick walls," *2nd International Brick Masonry Conference*.

Mainstone, R.J., 1971, "On the stiffness and strength of infilled frames," *Proceedings of the Institute of Civil Engineers* Sup., pp. 57-90.

Mallick, D.V. and Garg, R.P., 1971, "Effect of openings on the lateral stiffness of infilled frames," *Proceedings of the Institute of Civil Engineers*, 49: 193-209.

Mander J. B., Aycardi, L.E., and Kim, D-K, 1994, "Physical and Analytical Modeling of Brick Infilled Steel Frames," *Proceedings of the NCEER Workshop on Seismic Response of Masonry Infills*, D.P. Abrams editor, National Center for Earthquake Engineering Research, Technical Report NCEER-94-0004.

Mander, J.B., Chen, S.S., and Pekcan, G., 1994, "Low-cycle fatigue behavior of semi-rigid top-and-seat angle connections", *AISC Engineering Journal*, (31)3: 111-122.

Mander, J.B., and Dutta, A., 1997, "How Can Energy Based Seismic Design Be Accommodated In Seismic Mapping," *Proceedings of the FHWA/NCEER Workshop on the National Representation of Seismic Ground Motion for New and Existing Highway Facilities, Technical Report,* National Center for Earthquake Engineering Research, NCEER-97-0010, pp 95-114.

Mander, J.B., Mahmoodzadegan, B., Bhadra, S., and Chen, S.S., 1996, *Seismic Evaluation of a 30-Year-Old Highway Bridge Pier and Its Retrofit*, Department of Civil Engineering, State University of New York, Technical Report NCEER-96-0008, Buffalo NY.

Mander, J.B., and Nair, B., 1993a, "Seismic Resistance of Brick-Infilled Steel Frames With and Without Retrofit," *The Masonry Society Journal*, 12(2): 24-37.

Mander, J.B., Nair, B., Wojtkowski, K., and Ma, J., 1993b, *An Experimental Study on the Seismic Performance of Brick Infilled Steel Frames*, National Center for Earthquake Engineering Research, Technical Report NCEER-93-0001.

Mander, J.B., Pekcan, G., and Chen, S.S., 1995, "Low-cycle variable amplitude fatigue modeling of top-and-seat angle connections," *AISC Engineering Journal*, American Institute of Steel Construction, (32)2: 54-62.

Mansur, M.A., Balendra, T., and H'Ng, S.C., 1991, "Tests on Reinforced Concrete Low-Rise Shear Walls Under Cyclic Loading," *Preliminary Proceedings, International Workshop on Concrete Shear in Earthquake*, Houston Texas.

Manzouri, T., Shing, P.B., Amadei, B., Schuller, M., and Atkinson, R., 1995, *Repair and Retrofit of Unreinforced Masonry Walls: Experimental Evaluation and Finite Element Analysis*, Department of Civil, Environmental and Architectural Engineering, University of Colorado: Boulder, Colorado, Report CU/SR-95/2.

Manzouri, T., Shing, P.B., Amadei, B., 1996, "Analysis of Masonry Structures with Elastic/Viscoplastic Models, 1996," *Worldwide Advances in Structural Concrete and Masonry, Proceedings of the Committee on Concrete and Masonry Symposium*, ASCE Structures Congress XIV, Chicago, Illinois.

Matsumura, A., 1988, "Effectiveness of Shear Reinforcement in Fully Grouted Hollow Clay Masonry Walls," *Proceedings of the Fourth Meeting of the U.S.-Japan Joint Technical Coordinating Committee on Masonry Research*, San Diego, California.

Matsuno, M., Yamazaki, Y., Kaminosono, T., Teshigawara, M., 1987, "Experimental and Analytical Study of the Three Story Reinforced Clay Block Specimen," *Proceedings of the Third Meeting of the U.S.-Japan Joint Technical Coordinating Committee on Masonry Research*, Tomamu, Hokkaido, Japan.

Mayes, R.L., 1993, Unpublished study in support of the U.S. Coordinated Program for Masonry Building Research Design and Criteria Recommendations for Reinforced Masonry.

Mehrabi, A.B., Shing, P.B., Schuller, M.P., and Noland, J.L., 1996, "Experimental evaluation of masonry-infilled RC frames," *Journal of Structural Engineering*.

Merryman, K.M., Leiva, G., Antrobus, N., and Klingner, 1990, *In-plane Seismic Resistance of Two-story Concrete Masonry Coupled Shear Walls*, U.S. Coordinated Program for Masonry Building Research, Report No. 3.1(c)-1.

Modena, C., 1989, "Italian Practice in Evaluating, Strengthening, and Retrofitting Masonry Buildings," *Proceedings of the International Seminar on Evaluating, Strengthening, and Retrofitting Masonry Buildings*, University of Texas, Arlington, Texas.

Mosalam, K.M., Gergely, P., and White, R., 1994 "Performance and Analysis of Frames with "URM" Infills," *Proceedings of the Eleventh Conference* (formerly Electronic Computation Conference) held in conjunction with ASCE Structures Congress '94 and International Symposium '94, Atlanta, Georgia, pp. 57-66.

NIST, 1997, *Development of Procedures to Enhance the Performance of Rehabilitated URM Buildings*,

National Institute of Standards and Technology, GCR 97-724-1, Gaithersburg, Maryland.

Noakowski, Piotr, 1985, *Continuous Theory for the Determination of Crack Width under the Consideration of Bond*, Beton-Und Stahlbetonbau, 7 u..

Noland, J.L., 1990, "1990 Status Report: U.S. Coordinated Program for Masonry Building Research," *Proceedings of the Fifth North American Masonry Conference*, Urbana-Champaign, Illinois.

Oesterle, R.G., Fiorato, A.E., Johal, L.S., Carpenter, J.E., Russell, H.G., and Corley, W.G., 1976, *Earthquake Resistant Structural Walls -- Tests of Isolated Walls*, Report to National Science Foundation, Portland Cement Association, Skokie, Illinois, 315 pp.

Oesterle, R.G., Aristizabal-Ochoa, J.D., Fiorato, A.E., Russsell, H.G., and Corley, W.G., 1979, *Earthquake Resistant Structural Walls -- Tests of Isolated Walls -- Phase II*, Report to National Science Foundation, Portland Cement Association, Skokie, Illinois.

Oesterle, R.G., Aristizabal, J.D., Shiu, K.N., and Corley, W.G., 1983, *Web Crushing of Reinforced Concrete Structural Walls*, Portland Cement Association, Project HR3250, PCA R&D Ser. No. 1714, Draft Copy, Skokie, Illinois.

Ogata, K., and Kabeyasawa, T., 1984, "Experimental Study on the Hysteretic Behavior of Reinforced Concrete Shear Walls Under the Loading of Different Moment-to-Shear Ratios," *Transactions, Japan Concrete Institute*, Vol. 6, pp. 274-283.

Ohkubo, M., 1991, *Current Japanese System on Seismic Capacity and Retrofit Techniques for Existing Reinforced Concrete Buildings and Post-Earthquake Damage Inspection and Restoration Techniques*, University of California, San Diego Structural Systems Research Project Report No. SSRP-91/02.

Okada, T., and Kumazawa, F., 1987, "Flexural Behavior of Reinforced Concrete Block Beams with Spirally-Reinforced Lap Splices," *Proceedings of the Third Meeting of the U.S.-Japan Joint Technical Coordinating Committee on Masonry Research*, Tomamu, Hokkaido, Japan.

Parducci, A. and Mezzi, M., 1980, "Repeated horizontal displacement of infilled frames having different stiffness and connection systems--Experimental analysis", *Proceedings of the 7th World Conference on Earthquake Engineering*, Istanbul, Turkey, 7: 193-196.

Paret, T.F., Sasaki, K.K., Eilbeck, D.H., and Freeman, S.A., 1996, "Approximate Inelastic Procedures to Identify Failure Mechanisms from Higher Mode Effects," *Proceedings of the Eleventh World Conference on Earthquake Engineering*, Paper 966.

Park, R., 1996, "A Static Force-Based Procedure for the Seismic Assessment of Existing R/C Moment Resisting Frames," *Proceedings of the Annual Technical Conference*, New Zealand National Society for Earthquake Engineering.

Park, R. and Paulay, T., 1975, *Reinforced Concrete Structures*, John Wiley & Sons, New York.

Paulay, T., 1971a, "Coupling Beams of Reinforced-Concrete Shear Walls," *Journal of the Structural Division*, ASCE, Vol. 97, No. ST3, pp. 843–862.

Paulay, T., 1971b, "Simulated Seismic Loading of Spandrel Beams," *Journal of the Structural Division*, ASCE, Vol. 97, No. ST9, pp. 2407-2419.

Paulay, T., 1977, "Ductility of Reinforced Cocnrete Shearwalls for Seismic Areas," *Reinforced Concrete Structures in Seismic Zones*, American Concrete Institute, ACI Publication SP-52, Detroit, Michigan, pp. 127-147.

Paulay, T., 1980, "Earthquake-Resisting Shearwalls -- New Zealand Design Trends," *ACI Journal*, pp. 144-152.

Paulay, T., 1986, "A Critique of the Special Provisions for Seismic Design of the Bulding Code Requirements for Reinforced Concrete," *ACI Journal*, pp. 274-283.

Paulay, T., and Binney, J.R., 1974, "Diagonally Reinforced Coupling Beams of Shear Walls," *Shear in Reinforced Concrete*, ACI Publication SP-42, American Concrete Institute, Detroit, Michigan, pp. 579-598.

Paulay, T. and Priestley, M.J.N., 1992, *Seismic Design of Reinforced Concrete and Masonry Buildings*, John Wiley & Sons, New York.

Paulay, T., and Priestley, M.J.N., 1993, "Stability of Ductile Structural Walls," *ACI Structural Journal*, Vol. 90, No. 4.

Paulay, T., Priestley, M.J.N., and Synge, A.J., 1982, "Ductility in Earthquake Resisting Squat Shearwalls," *ACI Journal*, Vol. 79, No. 4, pp.257-269.

Paulay, T., and Santhakumar, A.R., 1976, "Ductile Behavior of Coupled Shear Walls," *Journal of the Structural Division, ASCE*, Vol. 102, No. ST1, pp. 93-108.

Polyakov, S.V., 1956, *Masonry in framed buildings*, pub. by Gosudarstvennoe Izdatelstvo po Stroitelstvu I Arkhitekture (Translation into English by G.L.Cairns).

Prawel, S.P. and Lee, H.H., 1994, "Research on the Seismic Performance of Repaired URM Walls," *Proc. of the US-Italy Workshop on Guidelines for Seismic Evaluation and Rehabilitation of Unreinforced Masonry Buildings*, Department of Structural Mechanics, University of Pavia, Italy, June 22-24, Abrams, D.P. et al. (eds.) National Center for Earthquake Engineering Research, SUNY at Buffalo, pp. 3-17 - 3-25.

Priestley, M.J.N., 1977, "Seismic Resistance of Reinforced Concrete Masonry Shear Walls with High Steel Percentages," *Bulletin of the New Zealand National Society for Earthquake Engineering*, Vol. 10, No. 1, pp.1-16.

Priestley, M.J.N, 1985, "Seismic Behavior of Unreinforced Masonry Walls," *Bulletin of the New Zealand National Society for Earthquake Engineering*, Vol. 18, No. 2.

Priestley, M.J.N., 1986, "Flexural Strength of Rectangular Unconfined Shear Walls with Distributed Reinforcement," *The Masonry Society Journal*, Vol. 5, No. 2.

Priestley, M.J.N., 1990, *Masonry Wall-Frame Joint Test*, Report No. 90-5, Sequad Consulting Engineers, Solana Beach, California.

Priestley, M.J.N., 1996, "Displacement-based seismic assessment of existing reinforced concrete buildings," *Bulletin of the New Zealand National Soc. for Earthquake Eng.*, (29)4: 256-271.

Priestley, M.J.N. and Elder, D.M., 1982, *Seismic Behavior of Slender Concrete Masonry Shear Walls*, Dept. of Civil Engineering, University of Canterbury, Christchurch, Research Report ISSN 0110-3326, New Zealand.

Priestley M.J.N., Evison, R.J., and Carr, A.J., 1978, "Seismic Response of Structures Free to Rock on their Foundations," *Bulletin of the New Zealand National Society for Earthquake Engineering*, Vol. 11, No. 3, pp. 141–150.

Priestley, M.J.N., and Hart, G.C., 1989, *Design Recommendations for the Period of Vibration of Masonry Wall Buildings*, Structural Systems Research

Project, Report No. SSRP 89/05, University of California at San Diego.

Priestley, M.J.N., and Hon, C.Y., 1985, "Seismic Design of Reinforced Concrete Masonry Moment - Resisting Frames," *The Masonry Society Journal*, Vol. 4., No. 1.

Priestley, M.J.N., and Limin, He, 1990, "Seismic Response of T-Section Masonry Shear Walls," *The Masonry Society Journal*, Vol. 9, No. 1.

Priestley, M.J.N., Seible, F., and Calvi, G.M., 1996, *Seismic Design and Retrofit of Bridges*, John Wiley & Sons, New York, 686 pp.

Priestley, M.J.N., Verma, R., and Xiao, Y., 1994, "Seismic Shear Strength of Reinforced Concrete Columns," *ASCE Journal of Structural Engineering*, Vol. 120, No. 8.

Reinhorn, A.M., Madan, A., Valles, R.E., Reichmann, Y., and Mander, J.B., 1995, *Modeling of Masonry Infill Panels for Structural Analysis*, National Center for Earthquake Engineering Research, Technical Report NCEER-95-0018.

Riddington, J. and Stafford-Smith, B., 1977, "Analysis of infilled frames subject to racking with design recommendations," *Structural Engineers*, (52)6: 263-268.

Riddington, J.R., 1984, "The influence of initial gaps on infilled frame behavior," *Proceedings of the Institute of Civil Engineers*, (2)77: 259-310.

Rutherford & Chekene, 1990, *Seismic Retrofitting Alternatives for San Francisco's Unreinforced Masonry Buildings: Estimates of Construction Cost and Seismic Damage for the San Francisco Department of City Planning*, Rutherford and Chekene Consulting Engineers: San Francisco, California.

Rutherford & Chekene, 1997, *Development of Procedures to Enhance the Performance of Rehabilitated Buildings*, prepared by Rutherford & Chekene Consulting Engineers, published by the National Institute of Standards and Technology as Reports NIST GCR 97-724-1 and 97-724-2.

Saatcioglu, Murat, 1991, "Hysteretic Shear Response of Low-rise Walls," *Preliminary Proceedings, International Workshop on Concrete Shear in Earthquake*, Houston Texas.

Sachanski, S., 1960, "Analysis of earthquake resistance of frame buildings taking into consideration the carring capacity of the filling masonry," *Proceedings of the 2nd World Conference on Earthquake Engineering*, Japan, 138: 1-15.

Saneinejad, A and Hobbs, B., 1995, "Inelastic design of infilled frames," *Journal of Structural Engineering*, (121)4: 634-650.

SANZ, Standards Association of New Zealand, 1995, *Code of Practice for the Design of Concrete Structures*, NZS3101.

Sassi, H., and Ranous, R., 1996, "Shear Failure in Reinforced Concrete Walls," *From Experience*, Structural Engineers Association of Southern California, Whittier, California.

Schuller, M.P., Atkinson, R.H., and Borgsmiller, J.T., 1994, "Injection Grouting for Repair and Retrofit of Unreinforced Masonry," *Proceedings of the 10th International Brick and Block Masonry Conference*, Calgary, Alberta, Canada.

Schultz, A.E., 1996 "Seismic Resistance of Partially-Grouted Masonry Shear Walls," *Worldwide Advances in Structural Concrete and Masonry*, A.E. Schultz and S.L. McCabe, Eds., American Society of Civil Engineers, New York.

SEAOC, 1996, *Recommended Lateral Force Requirements and Commentary* ("Blue Book"), 6th Edition, Seismology Committee, Structural Engineers Association of California, Sacramento, California.

SEAOC/CALBO, 1990, "Commentary on the SEAOC-CALBO Unreinforced Masonry Building Seismic Strengthening Provisions," *Evaluation and Strengthening of Unreinforced Masonry Buildings*, 1990 Fall Seminar, Structural Engineers Association of Northern California, San Francisco, California.

SEAOSC, 1986, "RGA (Rule of General Application) Unreinforced Masonry Bearing Wall Buildings (Alternate Design to Division 88)," *Earthquake Hazard Mitigation of Unreinforced Pre-1933 Masonry Buildings*, Structural Engineers Association of Southern California: Los Angeles, California.

Seible, F., Hegemier, G.A., Igarashi, A., and Kingsley, G.R., 1994a, "Simulated Seismic-Load Tests on Full- Scale Five-Story Masonry Building," *ASCE Journal of Structural Engineering*, Vol. 120, No. 3.

Seible, F., Kingsley, G.R., and Kürkchübasche, A.G., 1995, "Deformation and Force Capacity Assessment Issues in Structural Wall Buildings," *Recent*

Developments in Lateral Force Transfer in Buildings – Thomas Paulay Symposium, ACI SP-157.

Seible, F., Okada, T., Yamazaki, Y., and Teshigawara, M., 1987, "The Japanese 5-story Full Scale Reinforced Concrete Masonry Test — Design and Construction of the Test Specimen," *The Masonry Society Journal*, Vol. 6, No. 2.

Seible, F., Priestley, M.J.N., Kingsley, G.R., and Kürkchübasche, A.G., 1991, *Flexural coupling of Topped Hollow Core Plank Floor Systems in Shear Wall Structures*, Structural Systems Research Project, Report No. SSRP-91/10, University of California at San Diego.

Seible, F., Priestley, M.J.N., Kingsley, G.R., and Kürkchübasche, A.G., 1994b, "Seismic Response of a Five-story Full-scale RM Research Building", *ASCE Journal of Structural Engineering*, Vol. 120, No. 3.

Seible, F., Yamazaki, Y., Kaminosono, T., and Teshigawara, M., 1987, "The Japanese 5-story Full Scale Reinforced Concrete Masonry Test -- Loading and Instrumentation of the Test Building," *The Masonry Society Journal*, Vol. 6, No. 2.

Shapiro, D., Uzarski, J., and Webster, M., 1994, *Estimating Out-of-Plane Strength of Cracked Masonry Infills*, University of Illinois at Urbana-Champaign, Report SRS-588, 16 pp.

Shen, J. and Zhu, R., 1994, "Earthquake response simulation of reinforced concrete frames with infilled brick walls by pseudo-dynamic test," *Proc. of the Second International Conference on Earthquake Resistant Construction and Design*, Berlin, A. Balkema, Rotterdam, pp. 955-962.

Sheppard, P.F. and Tercelj, S., 1985, "Determination of the Seismic Resistance of an Historical Brick Masonry Building by Laboratory Tests of Cut-Out Wall Elements," *Proceedings of the 7th International Brick Masonry Conference*, Melbourne, Australia.

Shibata, A., and Sozen, M.A., 1976, "Substitute-Structure Method for Seismic Design in R/C," *Journal of the Structural Division*, ASCE, vol. 102, no. 1, pp. 1-18.

Shiga, Shibata, and Takahashi, 1973, "Experimental Study on Dynamic Properties of Reinforced Concrete Shear Walls," *Proceedings, Fifth World Conference on Earthquake Engineering*, Rome, pp. 107–117

Shiga, Shibata, and Takahashi, 1975, "Hysteretic Behavior of Reinforced Concrete Shear Walls," *Proceedings of the Review Meeting*, US-Japan Cooperative Research Program in Earthquake Engineering with Emphasis on the Safety of School Buildings, Honolulu Hawaii, pp. 1157–1166

Shing, P.B., Brunner, J., and Lofti, H.R., 1993, "Analysis of Shear Strength of Reinforced Masonry Walls," *Proceedings of the Sixth North American Masonry Conference*, Philadelphia, Pennsylvania.

Shing, P.B., Mehrabi, A.B., Schuller, M., and Noland, J.D., 1994, "Experimental evaluation and finite element analysis of masonry-infilled R/C frames," *Proc. of the Eleventh Conference* (formerly Electronic Computation Conference) held in conjunction with ASCE Structures Congress '94 and International Symposium '94, Atlanta, Georgia., pp. 84-93.

Shing, P.B., Noland, J.L., Spaeh, H.P., Klamerus, E.W., and Schuller, M.P., 1991, *Response of Single Story Reinforced Masonry Shear Walls to In-plane Lateral Loads*, TCCMAR Report No. 3.1(a)-2.

Shing, P., Schuller, M., Hoskere, V., and Carter, E., 1990a, "Flexural and Shear Response of Reinforced Masonry Walls," *ACI Structural Journal*, Vol. 87, No. 6.

Shing, P., Schuller, M., and Hoskere, V., 1990b, "In-plane Resistance of Reinforced Masonry Shear Walls," *Proceedings of ASCE Journal of the Structural Division*, Vol 116, No. 3.

Shiu, K.N., Daniel, J.I., Aristizabal-Ochoa, J.D., A.E. Fiorato, and W.G. Corley, 1981, *Earthquake Resistant Structural Walls -- Tests of Wall With and Without Openings*, Report to National Science Foundation, Portland Cement Association, Skokie, Illinois.

Soric, Z. and Tulin, L.E., 1987, "Bond in Reinforced Concrete Masonry," *Proceedings Fourth North American Masonry Conference*, Los Angeles, California.

Sozen, M.A., and Moehle, J.P., 1993, *Stiffness of Reinforced Concrete Walls Resisting In-Plane Shear*, Electric Power Research Institute, EPRI TR-102731, Palo Alto, California.

Stafford-Smith, B.S., 1966, "Behavior of square infilled frames," *ASCE* (92)1: 381-403.

Stafford-Smith, B. and Carter, C., 1969, "A method of analysis for infilled frames," *Proceedings of the Institute of Civil Engineers*, 44: 31-48.

Tena-Colunga, A., and Abrams, D.P., 1992, *Response of an Unreinforced Masonry Building During the Loma Prieta Earthquake*, Department of Civil Engineering, University of Illinois at Urbana-Champaign, Structural Research Series No. 576.

Thiruvengadam, V., 1985, "On the natural frequencies of infilled frames," *Earthquake Engineering and Structural Dynamics*, 13: 401-419.

Thomas, F.G., 1953, "The strength of brickwork," *J. Instnt. Struct. Engrs.*, (31)2: 35-46.

Thurston, S., and Hutchinson, D., 1982, "Reinforced Masonry Shear Walls Cyclic Load Tests in Contraflexure, " *Bulletin of the New Zealand Society for Earthquake Engineering*, Vol. 15, No. 1.

TMS, 1994, *Performance of Masonry Structures in the Northridge, California Earthquake of January 17, 1994*, Richard E. Klingner, ed., The Masonry Society.

Tomazevic, M., and Anicic, 1989, "Research, Technology and Practice in Evaluating, Strengthening, and Retrofitting Masonry Buildings: Some Yugoslavian Experiences," *Proceedings of the International Seminar on Evaluating, Strengthening and Retrofitting Masonry Buildings*, The Masonry Society: Boulder, Colorado.

Tomazevic, M., and Lutman, M., 1988, "Design of Reinforced Masonry Walls for Seismic Actions," *Brick and Block Masonry*, J.W. DeCourcy, Ed., Elsevier Applied Science, London.

Tomazevic, M., Lutman, M., and Velechovsky, T., 1993, "The Influence of Lateral Load Time-history on the Behavior of Reinforced-Masonry Walls Failing in Shear," *Proceedings of the Sixth North American Masonry Conference*, Philadelphia, Pennsylvania.

Tomazevic, M., Lutman, M., and Weiss, P., 1996, "Seismic Upgrading of Old Brick-Masonry Urban Houses: Tying of Walls with Steel Ties," *Earthquake Spectra,* EERI: Oakland, California, Volume 12, No. 3.

Tomazevic, M., and Modena, C., 1988, "Seismic Behavior of Mixed, Reinforced Concrete, Reinforced Masonry Structural Systems," *Brick and Block Masonry*, J.W. DeCourcy, Ed., Elsevier Applied Science, London.

Tomazevic, M., and Weiss, P., 1990, "A Rational, Experimentally Based Method for the Verification of Earthquake Resistance of Masonry Buildings," *Proceedings of the Fourth U.S. National Conference on Earthquake Engineering*, EERI: Oakland, California.

Tomazevic, M., and Zarnic, R., 1985, "The Effect of Horizontal Reinforcement on the Strength and Ductility of Masonry Walls at Shear Failure," *Proceedings of the Seventh International Brick Masonry Conference*, Melbourne, Australia.

Turnsek, V., and Sheppard, P., 1980, "The Shear and Flexural Resistance of Masonry Walls," *Proceedings of the International Research Conference on Earthquake Engineering*, Skopje, Yugoslavia.

Vallenas, J.M., Bertero, V.V., Popov, E.P., 1979, *Hysteretic Behavior of Reinforced Concrete Structural Walls*, Earthquake Engineering Research Center, University of California, Report No. UCB/EERC-79/20, Berkeley, California, 234 pp.

Vulcano A., and Bertero, V.V., 1987, *Analytical Models for Predicting the Lateral Response of RC Shear Walls*, University of California, UCB/EERC Report No. 87-19, Berkeley, California.

Wallace, J. W., 1994, "New Methodology for Seismic Design of RC Shear Walls," *Journal of the Structural Engineering Division,* American Society of Civil Engineers, New York, New York, Vol. 120, No. 3, pp. 863-884.

Wallace, J. W., 1995, "Seismic Design of RC Structural Walls, Part I: New Code Fromat," *Journal of the Structural Engineering Division,* American Society of Civil Engineers, New York, New York, Vol. 121, No. 1, pp. 75-87.

Wallace, J. W., 1996, "Evaluation of UBC-94 Provisions for Seismic Design of RC Structural Walls," *Earthquake Spectra*, EERI, Oakland, California.

Wallace, J. W. and Moehle, J.P., 1992, "Ductility and Detailing Requirements for Bearing Wall Buildings," *Journal of Structural Engineering, ASCE*, Vol. 118, No. 6, pp. 1625–1644.

Wallace, J. W. and Thomsen IV, J.H., 1995, "Seismic Design of RC Shear Walls (Parts I and II)," *Journal of Structural Engineering, -ASCE*, pp. 75-101.

Wang, T.Y., Bertero, V.V.,Popov, E.P., 1975, *Hysteretic Behavior of Reinforced Concrete Framed Walls*, Earthquake Engineering Research Center, University of California at Berkeley, Report No. UCB/EERC-75/23, Berkeley, California, 367 pp.

Weeks, J., Seible, F., Hegemier, G., and Priestley, M.J.N., 1994, *The U.S.-TCCMAR Full-scale Five-story Masonry Research Building Test, Part V - Repair and Retest*, University of California Structural Systems Research Project Report No. SSRP-94/05.

Wight, James K. (editor), 1985, *Earthquake Effects on Reinforced Concrete Structures*, US-Japan Research, ACI Special Publication SP-84, American Concrete Institute, Detroit Michigan.

Willsea, F.J., 1990, "SEAOC/CALBO Recommended Provisions", *Evaluation and Strengthening of Unreinforced Masonry Buildings*, 1990 Fall Seminar, Structural Engineers Association of Northern California, San Francisco, California.

Wiss, Janney, Elstner and Associates, 1970, *Final Report on Bar Tests*, for the Committee of Concrete Reinforcing Bar Producers, American Iron and Steel Institute (Job #70189), Northbrook, Illinois.

Wood, R.H., 1978, "Plasticity, composite action and collapse design of unreinforced shear wall panels in frames," *Proceedings of the Institute of Civil Engineers*, 2: 381-411.

Wood, S. L, 1990, "Shear Strength of Low-Rise Reinforced Concrete Walls," *ACI Structural Journal*, Vol. 87, No. 1, pp 99-107.

Wood, S. L., 1991, "Observed Behavior of Slender Reinforced Concrete Walls Subjected to Cyclic Loading," *Earthquake-Resistant Concrete Structures, Inelastic Response and Design*, ACI Special Publication SP-127, S.K. Ghosh, Editor, American Concrete Institute, Detroit Michigan, pp. 453–478

Xu, W., and Abrams, D.P., 1992, *Evaluation of Lateral Strength and Deflection for Cracked Unreinforced Masonry Walls*, U.S. Army Research Office: Triangle Park, North Carolina, Report ADA 264-160.

Yamamoto, M., and Kaminosono, T., 1986, "Behavior of Reinforced Masonry Walls with Boundary Beams," *Proceedings of the Second Meeting of the Joint Technical Coordinating Committee on Masonry Research*, Keystone, Colorado.

Yamazaki, Y., F. Seible, H., Mizuno, H., Kaminosono, T., Teshigawara, M., 1988b, "The Japanese 5-story Full Scale Reinforced Concrete Masonry Test -- Forced Vibration and Cyclic Load Test Results," *The Masonry Society Journal*, Vol. 7, No. 1.

Yamazaki, Y., Kaminosono, T., Teshigawara, M., and Seible, F., 1988a, "The Japanese 5-story Full Scale Reinforced Concrete Masonry Test -- Pseudo Dynamic and Ultimate Load Test Results," *The Masonry Society Journal*, Vol. 7, No. 2.

Yoshimura, K. and Kikuchi, K., 1995, "Experimental study on seismic behavior of masonry walls confined by R/C frames," *Proceedings of the Pacific Conference on Earthquake Engineering*, Melbourne, Australia, Parkville, Victoria, 3: 97-106.

Young, J.M., and Brown, R.H., 1988, *Compressive Stress Distribution of Grouted Hollow Clay Masonry Under Stain Gradient*, U.S. Coordinated Program for Masonry Building Research, Report No. 1.2b-1.

Zarnic, R. and Tomazevic, M., 1984, "The behavior of masonry infilled reinforced concrete frames subjected to seismic loading," *Proceedings of the 8th World Conference on Earthquake Engineering*, California, VI: 863-870.

Zarnic, R. and Tomazevic, M., 1985a, "Study of the behavior of masonry infilled reinforced concrete frames subjected to seismic loading," *Proc. 7th Intl. Brick Masonry Conf.*, Melbourne, Australia.

Zarnic, R. and Tomazevic, M., 1985b, *Study of the behavior of masonry infilled reinforced concrete frames subjected to seismic loading, Part 2*. A report to the research community of Slovenia, ZRMK/IKPI - 8502, Ljubljana.

Zsutty, T.C., 1990, "Applying Special Procedures", *Evaluation and Strengthening of Unreinforced Masonry Buildings*, 1990 Fall Seminar, Structural Engineers Association of Northern California, San Francisco, California.

ATC-43 Project Participants

Atc Management

Mr. Christopher Rojahn,
Principal Investigator
Applied Technology Council
555 Twin Dolphin Drive, Suite 550
Redwood City, CA 94065

Mr. Craig Comartin,
Co-PI and Project Director
7683 Andrea Avenue
Stockton, CA 95207

Technical Management Committee

Prof. Dan Abrams
University of Illinois
1245 Newmark Civil Eng'g. Lab., MC 250
205 North Mathews Avenue
Urbana, IL 61801-2397

Mr. James Hill
James A. Hill & Associates, Inc.
1349 East 28th Street
Signal Hill, CA 90806

Mr. Andrew T. Merovich
A.T. Merovich & Associates, Inc.
1163 Francisco Blvd., Second Floor
San Rafael, CA 94901

Prof. Jack Moehle
Earthquake Engineering Research Center
University of California at Berkeley
1301 South 46th Street
Richmond, CA 94804

FEMA/PARR Representatives

Mr. Timothy McCormick
PaRR Task Manager
Dewberry & Davis
8401 Arlington Boulevard
Fairfax, VA 22031-4666

Mr. Mark Doroudian
PaRR Representative
42 Silkwood
Aliso Viejo, CA 92656

Prof. Robert D. Hanson
FEMA Technical Monitor
74 North Pasadena Avenue, CA-1009-DR
Parsons Bldg., West Annex, Room 308
Pasadena, CA 91103

Materials Working Group

Dr. Joe Maffei, Group Leader
Consulting Structural Engineer
148 Hermosa Avenue
Oakland California

Mr. Brian Kehoe, Lead Consultant
Wiss, Janney, Elstner Associates, Inc.
2200 Powell Street, Suite 925
Emeryville, CA 94608

Dr. Greg Kingsley
KL&A of Colorado
805 14th Street
Golden, CO 80401

Mr. Bret Lizundia
Rutherford & Chekene
303 Second Street, Suite 800 North
San Francisco, CA 94107

Prof. John Mander
SUNY at Buffalo
Department of Civil Engineering
212 Ketter Hall
Buffalo, NY 14260

Analysis Working Group

Prof. Mark Aschheim, Group Leader
University of Illinois at Urbana
2118 Newmark CE Lab
205 North Mathews, MC 250
Urbana, IL 61801

Prof. Mete Sozen, Senior Consultant
Purdue University, School of Engineering
1284 Civil Engineering Building
West Lafayette, IN 47907-1284

Project Review Panel

Mr. Gregg J. Borchelt
Brick Institute of America
11490 Commerce Park Drive, #300
Reston, VA 20191

Dr. Gene Corley
Construction Technology Labs
5420 Old Orchard Road
Skokie, IL 60077-1030

Mr. Edwin Huston
Smith & Huston
Plaza 600 Building, 6th & Stewart, #620
Seattle, WA 98101

Prof. Richard E. Klingner
University of Texas
Civil Engineering Department
Cockbell Building, Room 4-2
Austin, TX 78705

Mr. Vilas Mujumdar
Office of Regulation Services
Division of State Architect
General Services
1300 I Street, Suite 800
Sacramento, CA 95814

Dr. Hassan A. Sassi
Governors Office of Emergency Services
74 North Pasadena Avenue
Pasadena, CA 91103

Mr. Carl Schulze
Libby Engineers
4452 Glacier Avenue
San Diego, CA 92120

Mr. Daniel Shapiro
SOH & Associates
550 Kearny Street, Suite 200
San Francisco, CA 94108

Prof. James K. Wight
University of Michigan
Department of Civil Engineering
2368 G G Brown
Ann Arbor, MI 48109-2125

Mr. Eugene Zeller
Long Beach Department of Building & Safety
333 W. Ocean Boulevard, Fourth Floor
Long Beach, CA 90802

Applied Technology Council Projects And Report Information

One of the primary purposes of Applied Technology Council is to develop resource documents that translate and summarize useful information to practicing engineers. This includes the development of guidelines and manuals, as well as the development of research recommendations for specific areas determined by the profession. ATC is not a code development organization, although several of the ATC project reports serve as resource documents for the development of codes, standards and specifications.

Applied Technology Council conducts projects that meet the following criteria:

1. The primary audience or benefactor is the design practitioner in structural engineering.

2. A cross section or consensus of engineering opinion is required to be obtained and presented by a neutral source.

3. The project fosters the advancement of structural engineering practice.

A brief description of several major completed projects and reports is given in the following section. Funding for projects is obtained from government agencies and tax-deductible contributions from the private sector.

ATC-1: This project resulted in five papers that were published as part of *Building Practices for Disaster Mitigation, Building Science Series 46*, proceedings of a workshop sponsored by the National Science Foundation (NSF) and the National Bureau of Standards (NBS). Available through the National Technical Information Service (NTIS), 5285 Port Royal Road, Springfield, VA 22151, as NTIS report No. COM-73-50188.

ATC-2: The report, *An Evaluation of a Response Spectrum Approach to Seismic Design of Buildings*, was funded by NSF and NBS and was conducted as part of the Cooperative Federal Program in Building Practices for Disaster Mitigation. Available through the ATC office. (Published 1974, 270 Pages)

ABSTRACT: This study evaluated the applicability and cost of the response spectrum approach to seis-

mic analysis and design that was proposed by various segments of the engineering profession. Specific building designs, design procedures and parameter values were evaluated for future application. Eleven existing buildings of varying dimensions were redesigned according to the procedures.

ATC-3: The report, *Tentative Provisions for the Development of Seismic Regulations for Buildings* (ATC-3-06), was funded by NSF and NBS. The second printing of this report, which includes proposed amendments, is available through the ATC office. (Published 1978, amended 1982, 505 pages plus proposed amendments)

ABSTRACT: The tentative provisions in this document represent the results of a concerted effort by a multi-disciplinary team of 85 nationally recognized experts in earthquake engineering. The provisions serve as the basis for the seismic provisions of the 1988 *Uniform Building Code* and the 1988 and subsequent issues of the *NEHRP Recommended Provisions for the Development of Seismic Regulation for New Buildings*. The second printing of this document contains proposed amendments prepared by a joint committee of the Building Seismic Safety Council (BSSC) and the NBS.

ATC-3-2: The project, Comparative Test Designs of Buildings Using ATC-3-06 Tentative Provisions, was funded by NSF. The project consisted of a study to develop and plan a program for making comparative test designs of the ATC-3-06 Tentative Provisions. The project report was written to be used by the Building Seismic Safety Council in its refinement of the ATC-3-06 Tentative Provisions.

ATC-3-4: The report, *Redesign of Three Multistory Buildings: A Comparison Using ATC-3-06 and 1982 Uniform Building Code Design Provisions*, was published under a grant from NSF. Available through the ATC office. (Published 1984, 112 pages)

ABSTRACT: This report evaluates the cost and technical impact of using the 1978 ATC-3-06 report, *Tentative Provisions for the Development of Seismic Regulations for Buildings*, as amended by a joint committee of the Building Seismic Safety Council and the National Bureau of Standards in 1982. The evaluations are based on studies of three existing California buildings redesigned in accordance with the ATC-3-06 Tentative Provisions and the 1982 *Uniform Building Code*. Included in the report are recommendations to code implementing bodies.

ATC-3-5: This project, Assistance for First Phase of ATC-3-06 Trial Design Program Being Conducted by the Building Seismic Safety Council, was funded by the Building Seismic Safety Council to provide the services of the ATC Senior Consultant and other ATC personnel to assist the BSSC in the conduct of the first phase of its Trial Design Program. The first phase provided for trial designs conducted for buildings in Los Angeles, Seattle, Phoenix, and Memphis.

ATC-3-6: This project, Assistance for Second Phase of ATC-3-06 Trial Design Program Being Conducted by the Building Seismic Safety Council, was funded by the Building Seismic Safety Council to provide the services of the ATC Senior Consultant and other ATC personnel to assist the BSSC in the conduct of the second phase of its Trial Design Program. The second phase provided for trial designs conducted for buildings in New York, Chicago, St. Louis, Charleston, and Fort Worth.

ATC-4: The report, *A Methodology for Seismic Design and Construction of Single-Family Dwelling*s, was published under a contract with the Department of Housing and Urban Development (HUD). Available through the ATC office. (Published 1976, 576 pages)

ABSTRACT: This report presents the results of an in-depth effort to develop design and construction details for single-family residences that minimize the potential economic loss and life-loss risk associated with earthquakes. The report: (1) discusses the ways structures behave when subjected to seismic forces, (2) sets forth suggested design criteria for conventional layouts of dwellings constructed with conventional materials, (3) presents construction details that do not require the designer to perform analytical calculations, (4) suggests procedures for efficient plan-checking, and (5) presents recommendations including details and schedules for use in the field by construction personnel and building inspectors.

ATC-4-1: The report, *The Home Builders Guide for Earthquake Design*, was published under a contract with HUD. Available through the ATC office. (Published 1980, 57 pages)

ABSTRACT: This report is an abridged version of the ATC-4 report. The concise, easily understood text of the Guide is supplemented with illustrations and 46 construction details. The details are provided to ensure that houses contain structural features that are properly positioned, dimensioned and constructed to resist earthquake forces. A brief description is included on how earthquake forces impact on houses and some precautionary constraints are given with respect to site selection and architectural designs.

ATC-5: The report, *Guidelines for Seismic Design and Construction of Single-Story Masonry Dwellings in Seismic Zone 2*, was developed under a contract with HUD. Available through the ATC office. (Published 1986, 38 pages)

ABSTRACT: The report offers a concise methodology for the earthquake design and construction of single-story masonry dwellings in Seismic Zone 2 of the United States, as defined by the 1973 *Uniform Building Code*. The Guidelines are based in part on shaking table tests of masonry construction conducted at the University of California at Berkeley Earthquake Engineering Research Center. The report is written in simple language and includes basic house plans, wall evaluations, detail drawings, and material specifications.

ATC-6: The report, *Seismic Design Guidelines for Highway Bridges*, was published under a contract with the Federal Highway Administration (FHWA). Available through the ATC office. (Published 1981, 210 pages)

ABSTRACT: The Guidelines are the recommendations of a team of sixteen nationally recognized experts that included consulting engineers, academics, state and federal agency representatives from throughout the United States. The Guidelines embody several new concepts that were significant departures from then existing design provisions. Included in the Guidelines are an extensive commentary, an example demonstrating the use of the

Guidelines, and summary reports on 21 bridges redesigned in accordance with the Guidelines. The guidelines have been adopted by the American Association of Highway and Transportation Officials as a guide specification.

ATC-6-1: The report, *Proceedings of a Workshop on Earthquake Resistance of Highway Bridges*, was published under a grant from NSF. Available through the ATC office. (Published 1979, 625 pages)

ABSTRACT: The report includes 23 state-of-the-art and state-of-practice papers on earthquake resistance of highway bridges. Seven of the twenty-three papers were authored by participants from Japan, New Zealand and Portugal. The Proceedings also contain recommendations for future research that were developed by the 45 workshop participants.

ATC-6-2: The report, *Seismic Retrofitting Guidelines for Highway Bridges*, was published under a contract with FHWA. Available through the ATC office. (Published 1983, 220 pages)

ABSTRACT: The Guidelines are the recommendations of a team of thirteen nationally recognized experts that included consulting engineers, academics, state highway engineers, and federal agency representatives. The Guidelines, applicable for use in all parts of the United States, include a preliminary screening procedure, methods for evaluating an existing bridge in detail, and potential retrofitting measures for the most common seismic deficiencies. Also included are special design requirements for various retrofitting measures.

ATC-7: The report, *Guidelines for the Design of Horizontal Wood Diaphragms*, was published under a grant from NSF. Available through the ATC office. (Published 1981, 190 pages)

ABSTRACT: Guidelines are presented for designing roof and floor systems so these can function as horizontal diaphragms in a lateral force resisting system. Analytical procedures, connection details and design examples are included in the Guidelines.

ATC-7-1: The report, *Proceedings of a Workshop of Design of Horizontal Wood Diaphragms*, was

published under a grant from NSF. Available through the ATC office. (Published 1980, 302 pages)

ABSTRACT: The report includes seven papers on state-of-the-practice and two papers on recent research. Also included are recommendations for future research that were developed by the 35 workshop participants.

ATC-8: This report, *Proceedings of a Workshop on the Design of Prefabricated Concrete Buildings for Earthquake Loads*, was funded by NSF. Available through the ATC office. (Published 1981, 400 pages)

ABSTRACT: The report includes eighteen state-of-the-art papers and six summary papers. Also included are recommendations for future research that were developed by the 43 workshop participants.

ATC-9: The report, *An Evaluation of the Imperial County Services Building Earthquake Response and Associated Damage*, was published under a grant from NSF. Available through the ATC office. (Published 1984, 231 pages)

ABSTRACT: The report presents the results of an in-depth evaluation of the Imperial County Services Building, a 6-story reinforced concrete frame and shear wall building severely damaged by the October 15, 1979 Imperial Valley, California, earthquake. The report contains a review and evaluation of earthquake damage to the building; a review and evaluation of the seismic design; a comparison of the requirements of various building codes as they relate to the building; and conclusions and recommendations pertaining to future building code provisions and future research needs.

ATC-10: This report, *An Investigation of the Correlation Between Earthquake Ground Motion and Building Performance*, was funded by the U.S. Geological Survey (USGS). Available through the ATC office. (Published 1982, 114 pages)

ABSTRACT: The report contains an in-depth analytical evaluation of the ultimate or limit capacity of selected representative building framing types, a discussion of the factors affecting the seismic performance of buildings, and a summary and comparison of seismic design and seismic risk parameters currently in widespread use.

ATC-10-1: This report, *Critical Aspects of Earthquake Ground Motion and Building Damage Potential*, was co-funded by the USGS and the NSF. Available through the ATC office. (Published 1984, 259 pages)

ABSTRACT: This document contains 19 state-of-the-art papers on ground motion, structural response, and structural design issues presented by prominent engineers and earth scientists in an ATC seminar. The main theme of the papers is to identify the critical aspects of ground motion and building performance that currently are not being considered in building design. The report also contains conclusions and recommendations of working groups convened after the Seminar.

ATC-11: The report, *Seismic Resistance of Reinforced Concrete Shear Walls and Frame Joints: Implications of Recent Research for Design Engineers*, was published under a grant from NSF. Available through the ATC office. (Published 1983, 184 pages)

ABSTRACT: This document presents the results of an in-depth review and synthesis of research reports pertaining to cyclic loading of reinforced concrete shear walls and cyclic loading of joint reinforced concrete frames. More than 125 research reports published since 1971 are reviewed and evaluated in this report. The preparation of the report included a consensus process involving numerous experienced design professionals from throughout the United States. The report contains reviews of current and past design practices, summaries of research developments, and in-depth discussions of design implications of recent research results.

ATC-12: This report, *Comparison of United States and New Zealand Seismic Design Practices for Highway Bridges*, was published under a grant from NSF. Available through the ATC office. (Published 1982, 270 pages)

ABSTRACT: The report contains summaries of all aspects and innovative design procedures used in New Zealand as well as comparison of United States and New Zealand design practice. Also included are research recommendations developed

at a 3-day workshop in New Zealand attended by 16 U.S. and 35 New Zealand bridge design engineers and researchers.

ATC-12-1: This report, *Proceedings of Second Joint U.S.-New Zealand Workshop on Seismic Resistance of Highway Bridges*, was published under a grant from NSF. Available through the ATC office. (Published 1986, 272 pages)

ABSTRACT: This report contains written versions of the papers presented at this 1985 Workshop as well as a list and prioritization of workshop recommendations. Included are summaries of research projects being conducted in both countries as well as state-of-the-practice papers on various aspects of design practice. Topics discussed include bridge design philosophy and loadings; design of columns, footings, piles, abutments and retaining structures; geotechnical aspects of foundation design; seismic analysis techniques; seismic retrofitting; case studies using base isolation; strong-motion data acquisition and interpretation; and testing of bridge components and bridge systems.

ATC-13: The report, *Earthquake Damage Evaluation Data for California*, was developed under a contract with the Federal Emergency Management Agency (FEMA). Available through the ATC office. (Published 1985, 492 pages)

ABSTRACT: This report presents expert-opinion earthquake damage and loss estimates for industrial, commercial, residential, utility and transportation facilities in California. Included are damage probability matrices for 78 classes of structures and estimates of time required to restore damaged facilities to pre-earthquake usability. The report also describes the inventory information essential for estimating economic losses and the methodology used to develop loss estimates on a regional basis.

ATC-14: The report, *Evaluating the Seismic Resistance of Existing Buildings*, was developed under a grant from the NSF. Available through the ATC office. (Published 1987, 370 pages)

ABSTRACT: This report, written for practicing structural engineers, describes a methodology for performing preliminary and detailed building seismic evaluations. The report contains a state-of-practice review; seismic loading criteria; data collection procedures; a detailed description of the building classification system; preliminary and detailed analysis procedures; and example case studies, including nonstructural considerations.

ATC-15: The report, *Comparison of Seismic Design Practices in the United States and Japan*, was published under a grant from NSF. Available through the ATC office. (Published 1984, 317 pages)

ABSTRACT: The report contains detailed technical papers describing design practices in the United States and Japan as well as recommendations emanating from a joint U.S.-Japan workshop held in Hawaii in March, 1984. Included are detailed descriptions of new seismic design methods for buildings in Japan and case studies of the design of specific buildings (in both countries). The report also contains an overview of the history and objectives of the Japan Structural Consultants Association.

ATC-15-1: The report, *Proceedings of Second U.S.-Japan Workshop on Improvement of Building Seismic Design and Construction Practices*, was published under a grant from NSF. Available through the ATC office. (Published 1987, 412 pages)

ABSTRACT: This report contains 23 technical papers presented at this San Francisco workshop in August, 1986, by practitioners and researchers from the U.S. and Japan. Included are state-of-the-practice papers and case studies of actual building designs and information on regulatory, contractual, and licensing issues.

ATC-15-2: The report, *Proceedings of Third U.S.-Japan Workshop on Improvement of Building Structural Design and Construction Practices*, was published jointly by ATC and the Japan Structural Consultants Association. Available through the ATC office. (Published 1989, 358 pages)

ABSTRACT: This report contains 21 technical papers presented at this Tokyo, Japan, workshop in July, 1988, by practitioners and researchers from the U.S., Japan, China, and New Zealand. Included are state-of-the-practice papers on various topics, including braced steel frame buildings, beam-column joints in reinforced concrete buildings, summaries of comparative U. S. and Japanese design, and base isolation and passive energy dissipation devices.

ATC-15-3: The report, *Proceedings of Fourth U.S.-Japan Workshop on Improvement of Building Structural Design and Construction Practices*, was published jointly by ATC and the Japan Structural Consultants Association. Available through the ATC office. (Published 1992, 484 pages)

ABSTRACT: This report contains 22 technical papers presented at this Kailua-Kona, Hawaii, workshop in August, 1990, by practitioners and researchers from the United States, Japan, and Peru. Included are papers on postearthquake building damage assessment; acceptable earth-quake damage; repair and retrofit of earthquake damaged buildings; base-isolated buildings, including Architectural Institute of Japan recommendations for design; active damping systems; wind-resistant design; and summaries of working group conclusions and recommendations.

ATC-15-4: The report, *Proceedings of Fifth U.S.-Japan Workshop on Improvement of Building Structural Design and Construction Practices*, was published jointly by ATC and the Japan Structural Consultants Association. Available through the ATC office. (Published 1994, 360 pages)

ABSTRACT: This report contains 20 technical papers presented at this San Diego, California workshop in September, 1992. Included are papers on performance goals/acceptable damage in seismic design; seismic design procedures and case studies; construction influences on design; seismic isolation and passive energy dissipation; design of irregular structures; seismic evaluation, repair and upgrading; quality control for design and construction; and summaries of working group discussions and recommendations.

ATC-16: This project, Development of a 5-Year Plan for Reducing the Earthquake Hazards Posed by Existing Nonfederal Buildings, was funded by FEMA and was conducted by a joint venture of ATC, the Building Seismic Safety Council and the Earthquake Engineering

Research Institute. The project involved a workshop in Phoenix, Arizona, where approximately 50 earthquake specialists met to identify the major tasks and goals for reducing the earthquake hazards posed by existing non-federal buildings nationwide. The plan was developed on the basis of nine issue papers presented at the workshop and workshop working group discussions. The Workshop Proceedings and Five-Year Plan are available through the Federal Emergency Management Agency, 500 "C" Street, S.W., Washington, DC 20472.

ATC-17: This report, *Proceedings of a Seminar and Workshop on Base Isolation and Passive Energy Dissipation*, was published under a grant from NSF. Available through the ATC office. (Published 1986, 478 pages)

ABSTRACT: The report contains 42 papers describing the state-of-the-art and state-of-the-practice in base-isolation and passive energy-dissipation technology. Included are papers describing case studies in the United States, applications and developments worldwide, recent innovations in technology development, and structural and ground motion issues. Also included is a proposed 5-year research agenda that addresses the following specific issues: (1) strong ground motion; (2) design criteria; (3) materials, quality control, and long-term reliability; (4) life cycle cost methodology; and (5) system response.

ATC-17-1: This report, *Proceedings of a Seminar on Seismic Isolation, Passive Energy Dissipation and Active Control,* was published under a grant from NSF. Available through the ATC office. (Published 1993, 841 pages)

ABSTRACT: The 2-volume report documents 70 technical papers presented during a two-day seminar in San Francisco in early 1993. Included are invited theme papers and competitively selected papers on issues related to seismic isolation systems, passive energy dissipation systems, active control systems and hybrid systems.

ATC-18: The report, *Seismic Design Criteria for Bridges and Other Highway Structures: Current and Future*, was published under a contract from the Multi-disciplinary Center for Earthquake Engineering Research (formerly NCEER), with funding from the

Federal Highway Administration. Available through the ATC office. (Published 1997, 152 pages)

ABSTRACT: This report documents the findings of a 4-year project to review and assess current seismic design criteria for new highway construction. The report addresses performance criteria, importance classification, definitions of seismic hazard for areas where damaging earthquakes have longer return periods, design ground motion, duration effects, site effects, structural response modification factors, ductility demand, design procedures, foundation and abutment modeling, soil-structure interaction, seat widths, joint details and detailing reinforced concrete for limited ductility in areas with low-to-moderate seismic activity. The report also provides lengthy discussion on future directions for code development and recommended research and development topics.

ATC-19: The report, *Structural Response Modification Factors* was funded by NSF and NCEER. Available through the ATC office. (Published 1995, 70 pages)

ABSTRACT: This report addresses structural response modification factors (R factors), which are used to reduce the seismic forces associated with elastic response to obtain design forces. The report documents the basis for current R values, how R factors are used for seismic design in other countries, a rational means for decomposing R into key components, a framework (and methods) for evaluating the key components of R, and the research necessary to improve the reliability of engineered construction designed using R factors.

ATC-20: The report, *Procedures for Postearthquake Safety Evaluation of Buildings*, was developed under a contract from the California Office of Emergency Services (OES), California Office of Statewide Health Planning and Development (OSHPD) and FEMA. Available through the ATC office (Published 1989, 152 pages)

ABSTRACT: This report provides procedures and guidelines for making on-the-spot evaluations and decisions regarding continued use and occupancy of earthquake damaged buildings. Written specifically for volunteer structural engineers and building inspectors, the report includes rapid and detailed evaluation procedures for inspecting buildings and posting them as "inspected" (apparently safe), "limited entry" or "unsafe". Also included are special procedures for evaluation of essential buildings (e.g., hospitals), and evaluation procedures for non-structural elements, and geotechnical hazards.

ATC-20-1: The report, *Field Manual: Postearthquake Safety Evaluation of Buildings*, was developed under a contract from OES and OSHPD. Available through the ATC office (Published 1989, 114 pages)

ABSTRACT: This report, a companion Field Manual for the ATC-20 report, summarizes the postearthquake safety evaluation procedures in brief concise format designed for ease of use in the field.

ATC-20-2: The report, *Addendum to the ATC-20 Postearthquake Building Safety Procedures* was published under a grant from the NSF and funded by the USGS. Available through the ATC office. (Published 1995, 94 pages)

ABSTRACT: This report provides updated assessment forms, placards, and procedures that are based on an in-depth review and evaluation of the widespread application of the ATC-20 procedures following five earthquakes occurring since the initial release of the ATC-20 report in 1989.

ATC-20-3: The report, *Case Studies in Rapid Postearthquake Safety Evaluation of Buildings,* was funded by ATC and R. P. Gallagher Associates. Available through the ATC office. (Published 1996, 295 pages)

ABSTRACT: This report contains 53 case studies using the ATC-20 Rapid Evaluation procedure. Each case study is illustrated with photos and describes how a building was inspected and evaluated for life safety, and includes a completed safety assessment form and placard. The report is intended to be used as a training and reference manual for building officials, building inspectors, civil and structural engineers, architects, disaster workers, and others who may be asked to perform safety evaluations after an earthquake.

ATC-20-T: The report, *Postearthquake Safety Evaluation of Buildings Training Manual* was developed under

a contract with FEMA. Available through the ATC office. (Published 1993, 177 pages; 160 slides)

ABSTRACT: This training manual is intended to facilitate the presentation of the contents of the ATC-20 and ATC-20-1. The training materials consist of 160 slides of photographs, schematic drawings and textual information and a companion training presentation narrative coordinated with the slides. Topics covered include: posting system; evaluation procedures; structural basics; wood frame, masonry, concrete, and steel frame structures; nonstructural elements; geotechnical hazards; hazardous materials; and field safety.

ATC-21: The report, *Rapid Visual Screening of Buildings for Potential Seismic Hazards: A Handbook*, was developed under a contract from FEMA. Available through the ATC office. (Published 1988, 185 pages)

ABSTRACT: This report describes a rapid visual screening procedure for identifying those buildings that might pose serious risk of loss of life and injury, or of severe curtailment of community services, in case of a damaging earthquake. The screening procedure utilizes a methodology based on a "sidewalk survey" approach that involves identification of the primary structural load resisting system and building materials, and assignment of a basic structural hazards score and performance modification factors based on observed building characteristics. Application of the methodology identifies those buildings that are potentially hazardous and should be analyzed in more detail by a professional engineer experienced in seismic design.

ATC-21-1: The report, *Rapid Visual Screening of Buildings for Potential Seismic Hazards: Supporting Documentation*, was developed under a contract from FEMA. Available through the ATC office. (Published 1988, 137 pages)

ABSTRACT: Included in this report are (1) a review and evaluation of existing procedures; (2) a listing of attributes considered ideal for a rapid visual screening procedure; and (3) a technical discussion of the recommended rapid visual screening procedure that is documented in the ATC-21 report.

ATC-21-2: The report, *Earthquake Damaged Buildings: An Overview of Heavy Debris and Victim Extrication*, was developed under a contract from FEMA. (Published 1988, 95 pages)

ABSTRACT: Included in this report, a companion volume to the ATC-21 and ATC-21-1 reports, is state-of-the-art information on (1) the identification of those buildings that might collapse and trap victims in debris or generate debris of such a size that its handling would require special or heavy lifting equipment; (2) guidance in identifying these types of buildings, on the basis of their major exterior features, and (3) the types and life capacities of equipment required to remove the heavy portion of the debris that might result from the collapse of such buildings.

ATC-21-T: The report, *Rapid Visual Screening of Buildings for Potential Seismic Hazards Training Manual* was developed under a contract with FEMA. Available through the ATC office. (Published 1996, 135 pages; 120 slides)

ABSTRACT: This training manual is intended to facilitate the presentation of the contents of the ATC-21 report. The training materials consist of 120 slides and a companion training presentation narrative coordinated with the slides. Topics covered include: description of procedure, building behavior, building types, building scores, occupancy and falling hazards, and implementation.

ATC-22: The report, *A Handbook for Seismic Evaluation of Existing Buildings (Preliminary)*, was developed under a contract from FEMA. Available through the ATC office. (Originally published in 1989; revised by BSSC and published as the *NEHRP Handbook for Seismic Evaluation of Existing Buildings* in 1992, 211 pages)

ABSTRACT: This handbook provides a methodology for seismic evaluation of existing buildings of different types and occupancies in areas of different seismicity throughout the United States. The methodology, which has been field tested in several programs nationwide, utilizes the information and procedures developed for and documented in the ATC-14 report. The handbook includes checklists, diagrams, and sketches designed to assist the user.

ATC-22-1: The report, *Seismic Evaluation of Existing Buildings: Supporting Documentation*, was developed under a contract from FEMA. (Published 1989, 160 pages)

ABSTRACT: Included in this report, a companion volume to the ATC-22 report, are (1) a review and evaluation of existing buildings seismic evaluation methodologies; (2) results from field tests of the ATC-14 methodology; and (3) summaries of evaluations of ATC-14 conducted by the National Center for Earthquake Engineering Research (State University of New York at Buffalo) and the City of San Francisco.

ATC-23A: The report, *General Acute Care Hospital Earthquake Survivability Inventory for California, Part A: Survey Description, Summary of Results, Data Analysis and Interpretation*, was developed under a contract from the Office of Statewide Health Planning and Development (OSHPD), State of California. Available through the ATC office. (Published 1991, 58 pages)

ABSTRACT: This report summarizes results from a seismic survey of 490 California acute care hospitals. Included are a description of the survey procedures and data collected, a summary of the data, and an illustrative discussion of data analysis and interpretation that has been provided to demonstrate potential applications of the ATC-23 database.

ATC-23B: The report, *General Acute Care Hospital Earthquake Survivability Inventory for California, Part B: Raw Data*, is a companion document to the ATC-23A Report and was developed under the above-mentioned contract from OSHPD. Available through the ATC office. (Published 1991, 377 pages)

ABSTRACT: Included in this report are tabulations of raw general site and building data for 490 acute care hospitals in California.

ATC-24: The report, *Guidelines for Seismic Testing of Components of Steel Structures*, was jointly funded by the American Iron and Steel Institute (AISI), American Institute of Steel Construction (AISC), National Center for Earthquake Engineering Research (NCEER), and NSF. Available through the ATC office. (Published 1992, 57 pages)

ABSTRACT: This report provides guidance for most cyclic experiments on components of steel structures for the purpose of consistency in experimental procedures. The report contains recommendations and companion commentary pertaining to loading histories, presentation of test results, and other aspects of experimentation. The recommendations are written specifically for experiments with slow cyclic load application.

ATC-25: The report, *Seismic Vulnerability and Impact of Disruption of Lifelines in the Conterminous United State*s, was developed under a contract from FEMA. Available through the ATC office. (Published 1991, 440 pages)

ABSTRACT: Documented in this report is a national overview of lifeline seismic vulnerability and impact of disruption. Lifelines considered include electric systems, water systems, transportation systems, gas and liquid fuel supply systems, and emergency service facilities (hospitals, fire and police stations). Vulnerability estimates and impacts developed are presented in terms of estimated first approximation direct damage losses and indirect economic losses.

ATC-25-1: The report, *A Model Methodology for Assessment of Seismic Vulnerability and Impact of Disruption of Water Supply Systems*, was developed under a contract from FEMA. Available through the ATC office. (Published 1992, 147 pages)

ABSTRACT: This report contains a practical methodology for the detailed assessment of seismic vulnerability and impact of disruption of water supply systems. The methodology has been designed for use by water system operators. Application of the methodology enables the user to develop estimates of direct damage to system components and the time required to restore damaged facilities to pre-earthquake usability. Suggested measures for mitigation of seismic hazards are also provided.

ATC-28: The report, *Development of Recommended Guidelines for Seismic Strengthening of Existing Buildings, Phase I: Issues Identification and Resolution*, was developed under a contract with FEMA. Available through the ATC office. (Published 1992, 150 pages)

ABSTRACT: This report identifies and provides resolutions for issues that will affect the development of guidelines for the seismic strengthening of existing buildings. Issues addressed include: implementation and format, coordination with other efforts, legal and political, social, economic, historic buildings, research and technology, seismicity and mapping, engineering philosophy and goals, issues related to the development of specific provisions, and nonstructural element issues.

ATC-29: The report, *Proceedings of a Seminar and Workshop on Seismic Design and Performance of Equipment and Nonstructural Elements in Buildings and Industrial Structures*, was developed under a grant from NCEER and NSF. Available through the ATC office. (Published 1992, 470 pages)

ABSTRACT: These Proceedings contain 35 papers describing state-of-the-art technical information pertaining to the seismic design and performance of equipment and nonstructural elements in buildings and industrial structures. The papers were presented at a seminar in Irvine, California in 1990. Included are papers describing current practice, codes and regulations; earthquake performance; analytical and experimental investigations; development of new seismic qualification methods; and research, practice, and code development needs for specific elements and systems. The report also includes a summary of a proposed 5-year research agenda for NCEER.

ATC-29-1: The report, *Proceedings Of Seminar On Seismic Design, Retrofit, And Performance Of Nonstructural Components*, was developed under a grant from NCEER and NSF. Available through the ATC office. (Published 1998, 518 pages)

ABSTRACT: These Proceedings contain 38 papers presenting current research, practice, and informed thinking pertinent to seismic design, retrofit, and performance of nonstructural components. The papers were presented at a seminar in San Francisco, California, in 1998. Included are papers describing observed performance in recent earthquakes; seismic design codes, standards, and procedures for commercial and institutional buildings; seismic design issues relating to industrial and hazardous material facilities; design, analysis, and testing; and seismic evaluation and rehabilitation of conventional and essential facilities, including hospitals.

ATC-30: The report, *Proceedings of Workshop for Utilization of Research on Engineering and Socioeconomic Aspects of 1985 Chile and Mexico Earthquakes*, was developed under a grant from the NSF. Available through the ATC office. (Published 1991, 113 pages)

ABSTRACT: This report documents the findings of a 1990 technology transfer workshop in San Diego, California, co-sponsored by ATC and the Earthquake Engineering Research Institute. Included in the report are invited papers and working group recommendations on geotechnical issues, structural response issues, architectural and urban design considerations, emergency response planning, search and rescue, and reconstruction policy issues.

ATC-31: The report, *Evaluation of the Performance of Seismically Retrofitted Buildings*, was developed under a contract from the National Institute of Standards and Technology (NIST, formerly NBS) and funded by the USGS. Available through the ATC office. (Published 1992, 75 pages)

ABSTRACT: This report summarizes the results from an investigation of the effectiveness of 229 seismically retrofitted buildings, primarily unreinforced masonry and concrete tilt-up buildings. All buildings were located in the areas affected by the 1987 Whittier Narrows, California, and 1989 Loma Prieta, California, earthquakes.

ATC-32: The report, *Improved Seismic Design Criteria for California Bridges: Provisional Recommendations,* was funded by the California Department of Transportation (Caltrans). Available through the ATC office. (Published 1996, 215 Pages)

ABSTRACT: This report provides recommended revisions to the current *Caltrans Bridge Design Specifications* (BDS) pertaining to seismic loading, structural response analysis, and component design. Special attention is given to design issues related to reinforced concrete components, steel components, foundations, and conventional bearings. The recommendations are based on recent research in the field of bridge seismic design and the performance

of Caltrans-designed bridges in the 1989 Loma Prieta and other recent California earthquakes.

ATC-34: The report, *A Critical Review of Current Approaches to Earthquake Resistant Design,* was developed under a grant from NCEER and NSF. Available through the ATC office. (Published, 1995, 94 pages)

ABSTRACT. This report documents the history of U. S. codes and standards of practice, focusing primarily on the strengths and deficiencies of current code approaches. Issues addressed include: seismic hazard analysis, earthquake collateral hazards, performance objectives, redundancy and configuration, response modification factors (R factors), simplified analysis procedures, modeling of structural components, foundation design, nonstructural component design, and risk and reliability. The report also identifies goals that a new seismic code should achieve.

ATC-35: This report, *Enhancing the Transfer of U.S. Geological Survey Research Results into Engineering Practice* was developed under a contract with the USGS. Available through the ATC office. (Published 1996, 120 pages)

ABSTRACT: The report provides a program of recommended "technology transfer" activities for the USGS; included are recommendations pertaining to management actions, communications with practicing engineers, and research activities to enhance development and transfer of information that is vital to engineering practice.

ATC-35-1: The report, *Proceedings of Seminar on New Developments in Earthquake Ground Motion Estimation and Implications for Engineering Design Practice*, was developed under a cooperative agreement with USGS. Available through the ATC office. (Published 1994, 478 pages)

ABSTRACT: These Proceedings contain 22 technical papers describing state-of-the-art information on regional earthquake risk (focused on five specific regions--California, Pacific Northwest, Central United States, and northeastern North America); new techniques for estimating strong ground motions as a function of earthquake source, travel path, and site parameters; and new developments specifically applicable to geotechnical engineering and the seismic design of buildings and bridges.

ATC-37: The report, *Review of Seismic Research Results on Existing Buildings,* was developed in conjunction with the Structural Engineers Association of California and California Universities for Research in Earthquake Engineering under a contract from the California Seismic Safety Commission (SSC). Available through the Seismic Safety Commission as Report SSC 94-03. (Published, 1994, 492 pages)

ABSTRACT. This report describes the state of knowledge of the earthquake performance of nonductile concrete frame, shear wall, and infilled buildings. Included are summaries of 90 recent research efforts with key results and conclusions in a simple, easy-to-access format written for practicing design professionals.

ATC-40: The report, *Seismic Evaluation and Retrofit of Concrete Buildings,* was developed under a contract from the California Seismic Safety Commission. Available through the ATC office. (Published, 1996, 612 pages)

ABSTRACT. This 2-volume report provides a state-of-the-art methodology for the seismic evaluation and retrofit of concrete buildings. Specific guidance is provided on the following topics: performance objectives; seismic hazard; determination of deficiencies; retrofit strategies; quality assurance procedures; nonlinear static analysis procedures; modeling rules; foundation effects; response limits; and nonstructural components. In 1997 this report received the Western States Seismic Policy Council "Overall Excellence and New Technology Award."

ATC-44: The report, *Hurricane Fran, South Carolina, September 5, 1996: Reconnaissance Report*, is available through the ATC office. (Published 1997, 36 pages.)

ABSTRACT: This report represents ATC's expanded mandate into structural engineering problems arising from wind storms and coastal flooding. It contains information on the causative hurricane; coastal impacts, including storm surge, waves, structural forces and erosion; building codes; observations and interpretations of damage; and lifeline performance. Conclusions address man-made beach nourishment, the effects of missile-like debris, breaches in the sandy barrier islands, and the timing and duration of such investigations.

ATC-R-1: The report, *Cyclic Testing of Narrow Plywood Shear Walls,* was developed with funding from the Henry J. Degenkolb Memorial Endowment Fund of the Applied Technology Council. Available through the ATC office (Published 1995, 64 pages)

ABSTRACT: This report documents ATC's first self-directed research program: a series of static and dynamic tests of narrow plywood wall panels having the standard 3.5-to-1 height-to-width ratio and anchored to the sill plate using typical bolted, 9-inch, 5000-lb. capacity hold-down devices. The report provides a description of the testing program and a summary of results, including comparisons of drift ratios found during testing with those specified in the seismic provisions of the 1991 *Uniform Building Code.*

ATC BOARD OF DIRECTORS (1973-Present)

Milton A. Abel	(1979-85)	John F. Meehan*	(1973-78)
James C. Anderson	(1978-81)	Andrew T. Merovich	(1996-99)
Thomas G. Atkinson*	(1988-94)	David L. Messinger	(1980-83)
Albert J. Blaylock	(1976-77)	Stephen McReavy	(1973)
Robert K. Burkett	(1984-88)	Bijan Mohraz	(1991-97)
James R. Cagley	(1998-2001)	William W. Moore*	(1973-76)
H. Patrick Campbell	(1989-90)	Gary Morrison	(1973)
Arthur N. L. Chiu	(1996-99)	Robert Morrison	(1981-84)
Anil Chopra	(1973-74)	Ronald F. Nelson	(1994-95)
Richard Christopherson*	(1976-80)	Joseph P. Nicoletti*	(1975-79)
Lee H. Cliff	(1973)	Bruce C. Olsen*	(1978-82)
John M. Coil*	(1986-87, 1991-97)	Gerard Pardoen	(1987-91)
Eugene E. Cole	(1985-86)	Stephen H. Pelham	(1998-2001)
Edwin T. Dean	(1996-99)	Norman D. Perkins	(1973-76)
Robert G. Dean	(1996-2001)	Richard J. Phillips	(1997-2000)
Edward F. Diekmann	(1978-81)	Maryann T. Phipps	(1995-96)
Burke A. Draheim	(1973-74)	Sherrill Pitkin	(1984-87)
John E. Droeger	(1973)	Edward V. Podlack	(1973)
Nicholas F. Forell*	(1989-96)	Chris D. Poland	(1984-87)
Douglas A. Foutch	(1993-97)	Egor P. Popov	(1976-79)
Paul Fratessa	(1991-92)	Robert F. Preece*	(1987-93)
Sigmund A. Freeman	(1986-89)	Lawrence D. Reaveley*	(1985-91)
Barry J. Goodno	(1986-89)	Philip J. Richter*	(1986-89)
Mark R. Gorman	(1984-87)	John M. Roberts	(1973)
Gerald H. Haines	(1981-82, 1984-85)	Charles W. Roeder	(1997-2000)
William J. Hall	(1985-86)	Arthur E. Ross*	(1985-91, 1993-94)
Gary C. Hart	(1975-78)	C. Mark Saunders*	(1993-2000)
Lyman Henry	(1973)	Walter D. Saunders*	(1975-79)
James A. Hill	(1992-95)	Lawrence G. Selna	(1981-84)
Ernest C. Hillman, Jr.	(1973-74)	Wilbur C. Schoeller	(1990-91)
Ephraim G. Hirsch	(1983-84)	Samuel Schultz*	(1980-84)
William T. Holmes*	(1983-87)	Daniel Shapiro*	(1977-81)
Warner Howe	(1977-80)	Jonathan G. Shipp	(1996-99)
Edwin T. Huston*	(1990-97)	Howard Simpson*	(1980-84)
Paul C. Jennings	(1973-75)	Mete Sozen	(1990-93)
Carl B. Johnson	(1974-76)	Donald R. Strand	(1982-83)
Edwin H. Johnson	(1988-89, 1998-2001)	James L. Stratta	(1975-79)
Stephen E. Johnston*	(1973-75, 1979-80)	Scott Stedman	(1996-97)
Joseph Kallaby*	(1973-75)	Edward J. Teal	(1976-79)
Donald R. Kay	(1989-92)	W. Martin Tellegen	(1973)
T. Robert Kealey*	(1984-88)	John C. Theiss*	(1991-98)
H. S. (Pete) Kellam	(1975-76)	Charles H. Thornton*	(1992-99)
Helmut Krawinkler	(1979-82)	James L. Tipton	(1973)
James S. Lai	(1982-85)	Ivan Viest	(1975-77)
Gerald D. Lehmer	(1973-74)	Ajit S. Virdee*	(1977-80, 1981-85)
James R. Libby	(1992-98)	J. John Walsh	(1987-90)
Charles Lindbergh	(1989-92)	Robert S. White	(1990-91)
R. Bruce Lindermann	(1983-86)	James A. Willis*	(1980-81, 1982-86)
L. W. Lu	(1987-90)	Thomas D. Wosser	(1974-77)
Walter B. Lum	(1975-78)	Loring A. Wyllie	(1987-88)
Kenneth A. Luttrell	(1991-98)	Edwin G. Zacher	(1981-84)
Newland J. Malmquist	(1997-2000)	Theodore C. Zsutty	(1982-85)
Melvyn H. Mark	(1979-82)		
John A. Martin	(1978-82)	* President	

ATC EXECUTIVE DIRECTORS (1973-Present)

Ronald Mayes	(1979-81)	Roland L. Sharpe	(1973-79)
Christopher Rojahn	(1981-present)		

Applied Technology Council
Sponsors, Supporters, and Contributors

Sponsors
Structural Engineers Association of California
James R. & Sharon K. Cagley
John M. Coil
Burkett & Wong

Supporters
Charles H. Thornton
Degenkolb Engineers
Japan Structural Consultants Association

Contributors
Lawrence D. Reaveley
Omar Dario Cardona Arboleda
Edwin T. Huston
John C. Theiss
Reaveley Engineers
Rutherford & Chekene
E. W. Blanch Co.